Corrosion Control

Samuel A. Bradford

VAN NOSTRAND REINHOLD
New York

Library of Congress Catalog Card Number 92-25295
ISBN 0-442-01088-5

Printed in the United States of America.

Van Nostrand Reinhold
115 Fifth Avenue
New York, New York 10003

Chapman and Hall
2–6 Boundary Row
London, SE 1 8HN, England

Thomas Nelson Australia
102 Dodds Street
South Melbourne 3205
Victoria, Australia

Nelson Canada
1120 Birchmount Road
Scarborough, Ontario M1K 5G4, Canada

16 15 14 13 12 11 10 9 8 7 6 5 4 3 2 1

Library of Congress Cataloging-in-Publication Data

Bradford, Samuel A., 1928–
      Corrosion control / Samuel A. Bradford.
            p.   cm.
      Includes bibliographical references and index.
      ISBN 0-442-01088-5
      1. Corrosion and anti-corrosives.   I. Title.
TA462.B648   1992
620.1′623- -dc20   92-25295
                                                                CIP

To my parents,
Phariss Cleino Bradford (1905–1986)
and Arthur Lenox Bradford (1904–1987)

Contents

v

Preface

Human beings undoubtedly became aware of corrosion just after they made their first metals. These people probably began to control corrosion very soon after that by trying to keep metal away from corrosive environments. "Bring your tools in out of the rain" and "Clean the blood off your sword right after battle" would have been early maxims. Now that the mechanisms of corrosion are better understood, more techniques have been developed to control it.

My corrosion experience extends over 10 years in industry and research and over 20 years teaching corrosion courses to university engineering students and industrial consulting. During that time I have developed an approach to corrosion that has successfully trained over 1500 engineers.

This book treats corrosion and high-temperature oxidation separately. Corrosion is divided into three groups: (1) chemical dissolution including uniform attack, (2) electrochemical corrosion from either metallurgical or environmental cells, and (3) corrosive–mechanical interactions. It seems more logical to group corrosion according to mechanisms than to arbitrarily separate them into 8 or 20 different types of corrosion as if they were unrelated.

University students and industry personnel alike generally are afraid of chemistry and consequently approach corrosion theory very hesitantly. In this text the electrochemical reactions responsible for corrosion are summed up in only five simple half-cell reactions. When these are combined on a polarization diagram, which is explained in detail, the electrochemical processes become obvious.

The purpose of this text is to train engineers and technologists not just to understand corrosion but to control it. Materials selection, coatings, chemical inhibitors, cathodic and anodic protection, and equipment design are covered in separate chapters. High-temperature oxidation is discussed in the final two chapters—one on oxidation theory and one on controlling oxidation by alloying and with coatings. Accompanying most of the chapters are questions and problems ($\sim$ 300 in total); some are simple calculations but others are real problems with more than one possible answer. This text uses the metric SI units (Systéme Internationale d'Unités), usually with English units in parentheses, except in the discussion of some real problems that were originally reported in English units where it seems silly to refer to a 6-in. pipe as 15.24-cm pipe. Units are not converted in the Memo questions because each industry works completely in one set of units.

For those who want a text stripped bare of any electrochemical theory at all, the starred (*) sections and starred chapter can be omitted without loss of continuity. However, the author strongly urges the reader to work through them. They are not beyond the abilities of any high school graduate who is interested in technology.

Acknowledgments

I wish to thank Dean F. D. Otto and the Faculty of Engineering, University of Alberta, for granting me a leave to write this text and I want to express my gratitude to my wife Evelin for her encouragement and understanding while I concentrated on it.

This text could not have been written without the help and cooperation of a great many people. Among them were Mrs. Tina Barker who took many of the photographs, Charles Bradford who drew the figures, and Benjamin Bradford who provided and maintained the word processor.

1

Introduction

Any time corrosion comes up in casual conversation, people talk about their old cars. Everyone who owns a car over 5-years old has first-hand experience with rusting, along with the bitter knowledge of what it costs in reliability and resale value (see Fig. 1-1). In recent years, automobile manufacturers have faced the problem and have begun to control corrosion by improving design, by sacrificial and inhibiting coatings, and by greater use of plastics.

Chemical plants, with their tremendous variety of aqueous, organic, and gaseous corrodents, come up with nearly every type of corrosion imaginable. It becomes quite a challenge to control corrosion of the equipment without interfering with chemical processes. Petroleum refineries have the best reputation for corrosion control, partly because the value of their product gives them the money to do it correctly and partly because the danger of fire is always present if anything goes wrong. The cost of corrosion-resistant materials and expensive chemical inhibitors is considered to be necessary insurance.

Ships, especially the huge supertankers, illustrate another type of corrosion problem. Seawater is very corrosive to steel and many other metals. Some metals that corrode only slightly, such as stainless steels, are likely to crack in seawater by the combination of corrosion and high stresses. Corrosion can cause the loss of a ship and its crew as well as damage to a fragile environment. Corrosion control commonly involves several coats of paint plus cathodic protection, as well as designing to minimize stress concentration.

1.1 WHAT IS CORROSION?

Corrosion is the damage to metal caused by reaction with its environment.

"Damage" is specified purposely to exclude processes, such as chemical milling, anodizing of aluminum, and bluing of steel, which modify the metal intentionally. All sorts of chemical and electrochemical processes are used

1

FIGURE 1-1.   Photograph of the author's mobile corrosion laboratory.

industrially to react with metals, but they are designed to improve the metal, not damage it. Thus these processes are not considered to be corrosion.

"Metal" is mentioned in the definition of corrosion, but any material can be damaged by its environment: plastics swell in solvents, concrete dissolves in sewage, wood rots, and so on. These results are all very serious problems that occur by various mechanisms, but they are not included in this definition. Metals, whether they are attacked uniformly or pit or crack in corrosion, are all corroded by the same basic mechanisms, which are quite different from those of other materials. This text concentrates on metals.

Rusting is a type of corrosion but it is the corrosion of ferrous metals (irons and steels) *only*, producing that familiar brownish-red corrosion product, rust.

The environment that corrodes a metal can be anything; air, water, and soil are common but everything from tomato juice to blood contacts metals, and most environments are corrosive.

Corrosion is a natural process for metals that causes them to react with their environment to form more stable compounds. In a perfect world the right material would always be selected, equipment designs would have no flaws, no mistakes would be made in operation, and corrosion would *still* occur—but at an acceptable rate.

1.2 THE COST OF CORROSION

Everybody is certain that *his/her* problems are bigger than anyone else's. This assumption applies to corrosion engineers also, who for years complained that corrosion is an immense problem. To see just how serious corrosion really is, the governments of several nations commissioned studies in the 1970s and 1980s, which basically arrived at numbers showing that corrosion is indeed a major problem. The study in the United States estimated the direct costs of corrosion to be approximately 4.9% of the gross national product for an industrialized nation. Of that 4.9%, roughly 1–2% is avoidable by properly applying technology already available—approximately $200 per person per year *wasted*.

This cost is greater than the financial cost of all the fires, floods, hurricanes, and earthquakes in the nation, even though these *other* natural disasters make headlines.

- How often have you seen a headline, "Corrosion ate up $300 million yesterday"?

Direct costs include parts and labor to replace automobile mufflers, metal roofing, condenser tubes, and all other corroded metal. Also, an entire machine may have to be scrapped because of the corrosion of one small part. Automobile corrosion alone costs $16 billion annually. Direct costs cover repainting of metals, although this expense is difficult to put precise numbers on, since much metal is painted for appearance as well as for corrosion protection. Also included is the cost of corrosion protection such as the capital costs of cathodic protection, its power and maintenance, the costs of chemical inhibitors, and the extra costs of corrosion-resistant materials.

Indirect costs are much more difficult to determine, although they are probably at least as great as the direct costs that were surveyed. Indirect costs include plant shutdowns, loss or contamination of products, loss of efficiency, and the overdesign necessary to allow for corrosion. Approximately 20% of electronic failures are caused by corrosion.

- An 8-in. oil pipeline 225 miles long with a 5/8-in.-wall thickness was installed several years ago with no corrosion protection. With protection it would have had a 1/4-in.-wall, which would save 3700 tons of steel (~ $1 million) and actually would increase internal capacity by 5%.

Corrosion leads to a depletion of our resources—a very real expense, but one that is not counted as a direct cost. It is estimated that 40% of our steel production goes to replace the steel lost to corrosion. Many metals, especially those essential in alloying, such as chromium and nickel, cannot be recycled by today's technology. Energy resources are also lost to corrosion because energy must be used to produce replacement metals.

Human resources are wasted. The time and ingenuity of a great many engineers and technicians are required in the daily battle against corrosion. Too often corrosion work is assigned to the new engineer or technologist because it is a quick way for him/her to get to know the people, the plant operation, and its problems. Then, if they are any good they get promoted, and the learning cycle has to begin again with another inexperienced trainee.

1.3 SAFETY AND ENVIRONMENTAL FACTORS

Not all corrosion is gradual and silent. Many serious accidents and explosions are initiated because of corrosion of critical components, causing personal injury and death. Environmental damage is another danger; oil pipeline leaks, for example, take years to heal.

- A few years ago the corrosion failure of an expansion joint in a chemical plant in England released poisonous vapors that killed 29 people.

Too often engineers take their cue from management whose motto is "Profit is the name of this game." For engineers, getting the job done well and safely must take precedence over cost. Certainly, cost is a consideration; any engineer who uses tantalum in a situation that could be handled by steel deserves to be fired. But where tantalum is needed, an engineer who takes a major risk by gambling with steel should be kicked out of the profession.

All decisions in engineering involve some risks, but the secret of successful engineering is to minimize the consequences of those risks. In simple terms, do not gamble with human life or irreparable environmental damage.

1.4 CORROSION ORGANIZATIONS AND JOURNALS

Everyone involved in corrosion, even on a part-time basis, can benefit by associating with other workers in the field and specialists in the various areas related to corrosion. The organization set up for this purpose is the National Association of Corrosion Engineers, more often called NACE, which has chapters in every major North American city. These local groups usually meet monthly for dinner and a technical presentation, and to talk over their corrosion problems, sometimes to brag about their successes. The chapters often sponsor special training courses and participate in district meetings where technical papers are presented. The NACE hosts a national conference annually for technical presentations, committee meetings, and a major trade show.

The cost of student membership in NACE is ridiculously low ($10/year as of this writing) but for that the student receives notice of chapter meetings, discounts on corrosion books and reports, and subscriptions to both Materials Performance (MP) and Corrosion journals. This opportunity to make contact with real life is one that should not be missed. Write to National Association of Corrosion Engineers, P.O. Box 218340, Houston, TX 77218.

Those involved in corrosion testing or in evaluating coatings and materials should be aware of the standards and test procedures published by the American Society for Testing and Materials (ASTM), 1916 Race Street, Philadelphia, PA 19103–1187. Specifications in industrial contracts commonly require ASTM standards to be met. The ASTM members annually receive a free volume of the latest test methods in their area of interest.

A lot of people around the world are working to solve corrosion problems. While each situation has some unique aspects, very similar corrosion problems have been faced successfully by others and the results published. Why try to reinvent the wheel? A list of the major corrosion journals published in English is given below.

MP (Materials Performance) (National Association of Corrosion Engineers, P.O. Box 218340, Houston, TX 77218) published monthly, approximately 90 pages per issue. Industrial reports on corrosion, corrosion-resistant materials, and coatings. Practical and nontheoretical. Includes industry news items, about 20% commercial advertising with employment ads.

Corrosion (National Association of Corrosion Engineers, P.O. Box 218340, Houston, TX 77218) published monthly, approximately 90 pages per issue. Research reports on corrosion, corrosion products, oxidation, testing, and protection methods. Most papers are theoretical but a Corrosion Engineering section contains reports on extensive research into industrial problems. Quality of papers and editing is very good.

Corrosion Science (Pergamon Press Inc., Maxwell House, Fairview Park, Elmsford, NY 10523) published approximately monthly, about 100 pages per issue with 10 papers. An international journal edited and printed in England with articles covering basic research in corrosion and oxidation, test methods, inhibitors, and so on. Short communications, book reviews, and a list of corrosion meetings throughout the world are included. Well edited, articles are top quality.

Corrosion Reviews (Freund Publishing House Ltd., P.O. Box 35010, Tel Aviv, Israel) published occasionally, approximately 150 pages. Lengthy review articles with numerous references on all aspects of corrosion, protective coatings, and treatments. Most reviews are invited articles by experts in their specialization. Extremely valuable for thorough surveys of a specific subject, including information more recent than that available in books.

Oxidation of Metals (Plenum Publishing Corporation, 233 Spring Street, New York, NY 10013) published bimonthly, approximately 200 pages, 10 articles per issue. Specialized research papers on oxidation of metals, structures and properties of oxide films, effects of alloying and surface treatments. Excellent quality.

British Corrosion Journal (Institute of Metals, 1 Carlton House Terrace, London SW1Y 5DB, England) published quarterly, approximately 70 pages, 8 articles per issue. Includes corrosion news, coming conferences, conference reports, and book reviews. Articles are mainly on new technology (European and worldwide), test methods and monitoring. Good quality, not theoretical.

Protection of Metals (Consultants Bureau, 233 Spring Street, New York, NY 10013) the English translation of *Zashchita Metallov*, published 6 times a year approximately 6 months after Russian publication. About 125 pages with 15 research articles, both theoretical and practical, plus about 15 short communications. A large percentage of the articles concern alloy development for corrosion and oxidation. Good quality, translated into good English.

Corrosion Prevention & Control (Scientific Surveys Ltd., P.O. Box 21, Beaconsfield, Bucks. HP9 1NS, England) published monthly, approximately 25 pages/issue, 5 papers. Publishes original research papers on all topics related to corrosion, articles on new technology, and review articles on corrosion. Included is a section on news in the corrosion and metal finishing industries, new publications, and events. Some commercial advertising. Papers and editing are of a reasonably good quality.

Anti-Corrosion Methods and Materials (Sawell Publications Ltd., 127 Stanstead Rd., London SE23 1JE, England) monthly, approximately 25 pages, about one-half commercial advertising. Short articles on protection equipment and corrosion-resistant materials. Details on industrial protection methods. Practical; any theory presented is at an elementary level.

Corrosion Abstracts (National Association of Corrosion Engineers, P.O. Box 218340, Houston, TX 77218) published bimonthly, approximately 80 pages per issue. Brief abstracts of the world's corrosion literature, indexed simply so that information on any subject is readily found. Some abstracts are too brief but the reference allows the reader to find the original source.

2
Basic Corrosion Theory

2.1 THERMODYNAMICS

Engineering metals are unstable on this planet. While humans thrive in the earth's environment of oxygen, water, and warm temperatures, their metal tools and equipment all corrode if given the opportunity. The metals try to lower their energy by spontaneously reacting to form solutions or compounds that have a greater thermodynamic stability.

The driving force for metallic corrosion is the Gibbs energy change, ΔG, which is the change in free energy of the metal and environment combination brought about by the corrosion. If a reaction is to be spontaneous, as corrosion reactions certainly are, ΔG for the process must be negative.

The term ΔG is only the difference between the Gibbs energies of the final and initial states of the reaction and, therefore, is independent of the various intermediate stages. Therefore, the corrosion reaction can be arbitrarily divided into either real or hypothetical steps, and the ΔG values are summed up for all the steps to find the true Gibbs energy change for the reaction. The units of ΔG are now commonly given in joules per mole (J/mol) of metal, or in the older units of calories per mole (cal/mol).

In corrosion measurements, the driving force is more often expressed in *volts* (V), which can be found from the equation

$$E = -\Delta G / nF \tag{2-1}$$

where E is the driving force (in volts, V) for the corrosion process, n is the number of moles of electrons per mole of metal involved in the process, and F is a constant called the "faraday," which is the electrical charge carried by a mole of electrons (or 96,490 C). Remember that joules = volts × coulombs. With ΔG being negative and with the minus sign in Eq. 2-1, spontaneous processes always have a positive voltage, E.

2.2 ELECTRODE REACTIONS

Aqueous corrosion is electrochemical. The principles of electrochemistry, established by Michael Faraday in the early nineteenth century, are basic to an understanding of corrosion and corrosion prevention.

The Corrosion Cell

Every electrochemical corrosion cell must have four components.

1. The anode, which is the metal that is corroding.
2. The cathode, which is a metal or other electronic conductor whose surface provides sites for the environment to react.
3. The electrolyte (the aqueous environment), in contact with both the anode and the cathode to provide a path for ionic conduction.
4. The electrical connection between the anode and the cathode to allow electrons to flow between them.

The components of an electrochemical cell are illustrated schematically in Figure 2-1. Anodes and cathodes are usually located quite close to one another and may even be on the same piece of metal. If any component were to be missing in the cell, electrochemical corrosion could not occur. Thus, analyzing the corrosion cell may provide the clue to stopping the corrosion.

Anode Reactions

Corrosion reactions can be separated into anode and cathode half-cell reactions to better understand the process. The anode reaction is quite simple—the anode metal M corrodes and goes into solution in the electrolyte as metal ions.

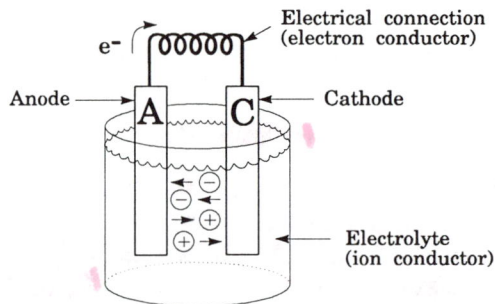

FIGURE 2-1.   The components of an electrochemical corrosion cell.

$$M \rightarrow M^{n+} + ne^- \qquad (2\text{-}2)$$

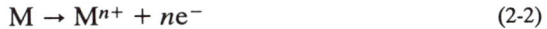

where n is the number of electrons (e^-) released by the metal. Chemists call this an "oxidation" (ox), which means a loss of electrons by the metal atoms. The electrons produced do not flow into the solution[1] but remain behind on the corroding metal, where they migrate through the electronic conductor to the cathode, as indicated in Figure 2-1.

For example, if steel is corroding, the anode reaction is

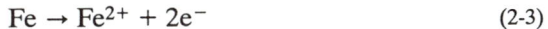

$$Fe \rightarrow Fe^{2+} + 2e^- \qquad (2\text{-}3)$$

or if aluminum is corroding the reaction is

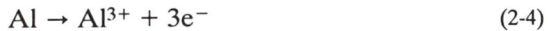

$$Al \rightarrow Al^{3+} + 3e^- \qquad (2\text{-}4)$$

Cathode Reactions

The cathode reaction consumes the electrons produced at the anode. If it did not, the anode would become so loaded with electrons that all reaction would cease immediately. At the cathode, some reducible species in the electrolyte adsorbs and picks up electrons, although the cathode itself does not react. Chemists call this a "reduction" (red) because the valence of the reactant is reduced.

Since it is the corrosive environment that reacts on the cathode, and many different corrosives can attack metals, several cathode reactions are possible.

1. The most common reaction is the one seen in nature and in neutral or basic solutions containing dissolved oxygen.

$$O_2 + 2H_2O + 4e^- \rightarrow 4OH^- \qquad (2\text{-}5)$$

- Workmen have collapsed after entering rusting storage tanks. The O_2 content of the air inside may be only 5%.

2. The next most important reaction is the one in acids.[2]

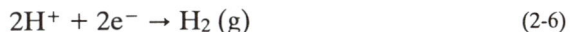

$$2H^+ + 2e^- \rightarrow H_2\,(g) \qquad (2\text{-}6)$$

where (g) indicates a gas.

---

[1]Bradford's law: Electrons can't swim.

[2]Bear in mind that the hydrogen ion (H^+) actually includes attached water molecules as H_3O^+ (the hydronium ion). For simplicity, the unreacting molecules of hydration are omitted in this text.

This reaction is not important in water. There are not enough hydrogen ions.

• Some metals react so violently with strong acid that they *float*, buoyed up by H_2 bubbles.

Remember that water with carbon dioxide (CO_2) gas dissolved in it makes H_2CO_3, (carbonic acid), which ionizes to H^+ ions; similarly SO_2 and SO_3 make sulfurous (H_2SO_3) and sulfuric (H_2SO_4) acids. Figure 2-2 shows the electrode processes that occur on a metal corroding in an acid.

One of the two cathode reactions given above will cover most corrosion processes, but a few less likely ones may show up.

3. In aerated acids and in oxidizing acids such as HNO_3 (nitric acid) and $HClO_4$ (perchloric acid), which can generate oxygen as they decompose, the cathode reaction is quite powerful.

$$O_2 + 4H^+ + 4e^- \rightarrow 2H_2O \qquad (2\text{-}7)$$

Because of this reaction, mixtures of O_2 and CO_2 gases in water can be 10–40 times as corrosive as a solution of either gas alone.

4. In some chemical processes the solutions may contain oxidizing agents that can be reduced in valence, as, for example,

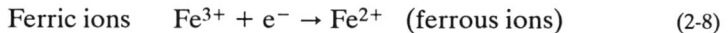

$$\text{Ferric ions} \quad Fe^{3+} + e^- \rightarrow Fe^{2+} \quad \text{(ferrous ions)} \qquad (2\text{-}8)$$

$$\text{Cupric ions} \quad Cu^{2+} + 2e^- \rightarrow Cu^0 \quad \text{(copper metal)} \qquad (2\text{-}9)$$

Reaction 2-5 (O_2 reduction) may produce rust or some other corrosion product that could cover the metal to slow down attack. Reactions 2-6 – 2-8 normally do not. Reaction 2-9 (Cu plating) even increases the corrosion rate.

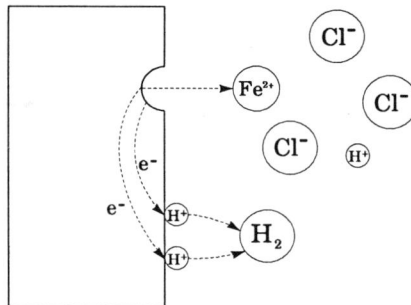

FIGURE 2-2.   Anode and cathode processes for corrosion of iron in unaerated hydrochloric acid.

- Many a contractor has learned to his/her sorrow that copper roofs or copper flashing do not mix with aluminum gutters. Rainwater containing Cu^{2+} ions flows into the aluminum gutters and plates out metallic copper while dissolving aluminum (Reactions 2-9 and 2-4). Areas plated with copper become excellent cathodes, concentrating all attack on the unplated aluminum.

Cell Reactions

The total cell reaction will be a coupling of an anode reaction with a cathode reaction. For example, the corrosion of steel by seawater is a combination of Reactions 2-3 and 2-5.

$$2Fe + 2H_2O + O_2 \rightarrow 2Fe^{2+} + 4OH^- \rightarrow 2Fe(OH)_2\downarrow \qquad (2\text{-}10)$$

The gelatinous ferrous hydroxide is then oxidized further.

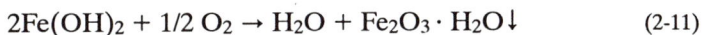

$$2Fe(OH)_2 + 1/2\,O_2 \rightarrow H_2O + Fe_2O_3 \cdot H_2O\downarrow \qquad (2\text{-}11)$$

producing the familiar brown rust.

When analyzing a corrosion problem, first identify all the components of the corrosion cell and determine the electrode reactions. Failure to do this at the beginning can lead to entirely erroneous conclusions and worthless attempted remedies.

2.3 ELECTRODE POTENTIALS

The standard electromotive force series as given in Table 2-1 lists the elements in order of their standard *reduction* potentials, $E°$. The most noble (unreactive) metals, such as gold, have the most positive reduction potentials and usually make the best cathodes. The metals at the negative end of the series corrode readily and thus tend to be anodes. The potentials listed are for half-cells with the ions all at 1 M concentration, and gases at 1 atm pressure, at *equilibrium*, measured against the standard hydrogen electrode (SHE) (Fig. 2-3).

While real corrosion processes are very unlikely to take place in 1 M solutions and almost never reach equilibrium, the standard series is useful in identifying anode and cathode reactions along with a rough estimate of how serious a driving force (voltage) the corrosion cell has.

- Chemistry teachers often point out that copper will not corrode in hydrochloric acid (HCl) because the copper reduction potential is above

TABLE 2-1   Standard Electrode Potentials

Reaction	E^{oa} (V)	
$Au^{3+} + 3e^- = Au$	1.498	Noble (cathodic)
$ClO_3^- + 6H^+ + 6e^- = Cl^- + 3H_2O$	1.451	
$Cl_2 + 2e^- = 2Cl^-$	1.358	
$Cr_2O_7^{2-} + 14H^+ + 6e^- = 2Cr^{3+} + 7H_2O$	1.232	
$O_2 + 4H^+ + 4e^- = 2H_2O$	1.229	
$Pt^{2+} + 2e^- = Pt$	1.118	
$NO_3^- + 4H^+ + 3e^- = NO + 2H_2O$	0.957	
$Pd^{2+} + 2e^- = Pd$	0.951	
$Ag^+ + e^- = Ag$	0.800	
$Hg_2^{2+} + 2e^- = 2Hg$	0.797	
$Fe^{3+} + e^- = Fe^{2+}$	0.771	
$I_2 + 2e^- = 2I^-$	0.536	
$O_2 + 2H_2O + 4e^- = 4OH^-$	0.401	
$Cu^{2+} + 2e^- = Cu$	0.342	
Cu^{2+} (sat) $+ 2e^- = Cu$ ($CuSO_4$)	0.316	Reference electrode
$AgCl + e^- = Ag + Cl^-$ (0.1 M KCl)	0.288	Reference electrode
$Hg_2Cl_2 + 2e^- = 2Hg + 2Cl^-$ (sat KCl)	0.241	Reference electrode
$2H^+ + 2e^- = H_2$	0.0000	
$Pb^{2+} + 2e^- = Pb$	−0.126	
$Sn^{2+} + 2e^- = Sn$	−0.138	
$Mo^{3+} + 3e^- = Mo$	−0.200	
$Ni^{2+} + 2e^- = Ni$	−0.257	
$Co^{2+} + 2e^- = Co$	−0.277	
$Cd^{2+} + 2e^- = Cd$	−0.403	
$Fe^{2+} + 2e^- = Fe$	−0.447	
$Cr^{3+} + 3e^- = Cr$	−0.744	
$Zn^{2+} + 2e^- = Zn$	−0.762	
$Nb^{3+} + 3e^- = Nb$	−1.099	(Columbium, Cb)
$Ti^{2+} + 2e^- = Ti$	−1.630	
$Al^{3+} + 3e^- = Al$	−1.662	
$Be^{2+} + 2e^- = Be$	−1.847	
$Mg^{2+} + 2e^- = Mg$	−2.372	
$Na^+ + e^- = Na$	−2.711	
$K^+ + e^- = K$	−2.931	
$Li^+ + e^- = Li$	−3.040	Reactive (anodic)

Source: Adapted from Weast, 1988.
*Notes: Ions at 1 M concn, 25°C (77°F), 1 atm.
Sat = saturated.

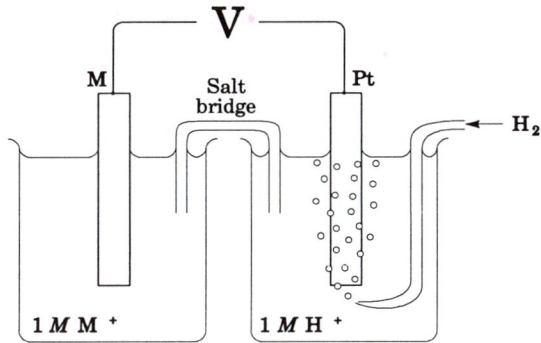

FIGURE 2-3.   Arrangement for measuring standard emf of a metal M against the standard hydrogen electrode.

hydrogen on the standard series. But a skeptical student who puts a penny in an open beaker of HCl finds that the copper *does* slowly corrode. Oxygen from the air dissolves in the acid, making the $O_2 + H^+$ cathode reaction (2-7) possible with $E° = 1.229$ V, well above the value of 0.342 V for copper.

*The Nernst Equation

Standard reduction potentials can be corrected to the concentrations of the real environment with the Nernst equation.

$$E = E° - \frac{RT}{nF} \ln \frac{[\text{red}]}{[\text{ox}]} \tag{2-12}$$

in which E is the real potential, R is the gas constant (8.314 J/K · mol), T is the absolute temperature in kelvins, n is the number of electrons exchanged, and [red] indicates that concentrations or pressures of all products on the reduced side of the equation are to be multiplied together; likewise, [ox] indicates that concentrations or pressures of reactants on the oxidized side are to be multiplied. The oxidized side is the left-hand side of the equations in Table 2-1. Concentrations or pressures, shown inside square brackets, are always given in molar (M) units for solutions and *atm* for gases. A concentration of one is used for solids and water; all pure solids have an invariant activity of unity, as does H_2O when it is present in such excessive amounts that it remains unchanged by the reaction.

For 25°C (77°F) the Nernst equation can be rewritten.

$$E = E° - \frac{0.0592}{n} \log \frac{[\text{red}]}{[\text{ox}]} \tag{2-13}$$

It is important to understand that the equation given here only corrects standard *reduction* potentials—not oxidation potentials, not total cell potentials.

- For example, if zinc metal is immersed at equilibrium in water with a Zn^{2+} ion concentration of only $10^{-6}M$, its electrode potential at 25°C (77°F) is

$$E_{Zn} = E°_{Zn} - \frac{0.0592}{n} \log ([Zn]/[Zn^{2+}])$$

$$E_{Zn} = 0.762 \text{ V} - \frac{0.0592}{2} \log (1/10^{-6}) = -0.940 \text{ V}$$

so that the metal is even more anodic (corrodible) in water than in a concentrated solution of its ions, since its potential is more negative.

- To complicate matters more, here is an example of correcting the potential of

$$O_2 + 2H_2O + 4e^- = 4OH^- \qquad E° = +0.401 \text{ V} \qquad (2\text{-}14)$$

to the more realistic conditions of air and water, where $p_{O_2} = 0.2$ atm and $[OH^-]=10^{-7}M$ for pH 7.

$$E = E° - \frac{0.0592}{n} \log ([OH^-]^4/p_{O_2}[H_2O])$$

$$E = +0.401 \text{ V} - \frac{0.0592}{4} \log [(10^{-7} M)^4/(0.2 \text{ atm}) (1)] = 0.805 \text{ V}$$

The concentration of OH^- is taken to the fourth power because four OH^- ions are the products on the reduced side of the equation, while the value of n is 4 because four electrons are reacted.

- For another example involving pH, consider correcting the hydrogen electrode to nonstandard conditions

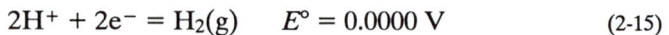

$$2H^+ + 2e^- = H_2(g) \qquad E° = 0.0000 \text{ V} \qquad (2\text{-}15)$$

$$E = E° - \frac{0.0592}{2} \log \{(p_{H_2})/[H^+]^2\}$$

$$E = 0.0000 \text{ V} - \frac{0.0592}{2} \log p_{H_2} - 0.0592 \text{ pH}$$

since pH is defined as

$$pH = \log \frac{1}{[H^+]} \qquad (2\text{-}16)$$

Note that it is possible for a solution pH to be negative if the $[H^+]$ is greater than $1M$, as it is in strong acids.

*Cell Potential

The cell reaction consists of an anode reaction and a cathode reaction. While in some cases more than one anode reaction or more than one cathode reaction may be operative, almost always one anode and one cathode predominate. The cathode (cat) reaction is a reduction (red), so the standard emf, as corrected by the Nernst equation, is the true half-cell potential. But at the anode the reaction is an oxidation (ox), just the reverse of a reduction, which means that the sign of its potential must be reversed.

$$E_{cell} = E_{red}^{cat} + E_{ox}^{anode}$$

$$E_{cell} = E_{red}^{cat} - E_{red}^{anode} \tag{2-17}$$

As stated previously, a spontaneous reaction, such as corrosion, must result in a reduction in Gibbs energy, so ΔG must be negative. Thus it follows from Equation 2-1 that E_{cell} must be positive if corrosion occurs, or a mistake has been made somewhere.

- For the corrosion of copper in acid at standard conditions,

$$E_{cell} = E_{red}^{cat} - E_{red}^{anode}$$

$$E_{cell} = 0.000\ V - (+0.34\ V) = -0.342\ V$$

a reaction that will not work unless an external voltage is applied.
- However, an aerated acid *could* corrode copper at standard conditions.

$$E_{cell} = E_{red}^{cat} - E_{red}^{anode}$$

$$E_{cell} = 1.229\ V - (+0.342\ V) = +0.887\ V$$

*Polarization

The +0.887 V calculated above is the initial driving force for corrosion (under standard conditions). In reality, as the anode and cathode reactions proceed, both anode and cathode immediately begin shifting potential until they have about the same potential, which will be somewhere between the original cathode half-cell potential (1.229 V) and the original anode half-cell potential (+0.342 V). The shift in potential is called "polarization" and always occurs to both the anode and the cathode in every corrosion cell.

*Concentration Cells

A corrosion cell can be set up if a metal is in contact with two different concentrations of a solution. The metal contacting the higher concentration of cathode reactant becomes the cathode. A concentration cell often develops in a tight crevice where corrosion gradually alters the solution composition.

- Suppose a tank contains unstirred acid with pH 0 near an inlet pipe and pH +1 elsewhere. This concentration difference sets up a potential difference and, hence, a corrosion cell, even though the tank metal is the same at both places. The metal in pH 1 acid is the anode because the difference in potential between the two places is

$$\Delta E = E_{pH = 0} - E_{pH = 1}$$

$$\Delta E = 0.0000 \text{ V} - [0.0000 \text{ V} - 0.0592 \text{ (pH)}] = +0.0592 \text{ V}$$

The pressure of the H_2 bubbles forming on metal cathode areas is assumed to be near atmospheric pressure in both locations, so it is omitted from the calculation.

"Positive" and "Negative" Electrodes

If the corrosion cell is considered as a source of power, as with a battery, electrons flow from the anode so the anode must be negative. Therefore, an automobile lead storage battery has the anode post marked with a negative sign. But on the other hand, if the corrosion cell has power applied to it, as an electroplating tank has, the cathode where metal will plate out must be connected to the negative line of the direct current (dc) power source, so it is the cathode that is negative.

The anode is negative for the external circuit of the cell (electron flow), but it is positive for the internal circuit (ion flow). Consequently, it is ambiguous to refer to "positive" and "negative" electrodes in corrosion processes.

2.4 CORROSION PRODUCTS AND PASSIVITY

The products of corrosion may be soluble or insoluble compounds or a gas (usually H_2). Soluble compounds usually increase corrosion by increasing the conductivity of the electrolyte. Insoluble compounds may precipitate on the anode surface, on the cathode, or somewhere in between, depending on the relative mobilities of the ions produced at the anode and cathode. If negative ions produced at the cathode (e.g., OH^-) reach the anode before the metal ions move away from it, the precipitate tends to form on the anode.

A layer of corrosion products on the metal, even if it is as porous and imperfect as rust scale, will slow down corrosion by slowing the diffusion of the reactants to the cathode sites and by slowing diffusion of metal ions from the anode. Consequently, if insoluble products form on the metal, the corrosion rate usually decreases with time, as shown in the idealized curves of Figure 2-4.

Passivation

A number of reactive metals that come in contact with corrosive environments can abruptly turn extremely corrosion resistant due to a phenomenon called "passivation" (see Figs 2-4c and 2-5).

- Iron in concentrated nitric acid does not appear to corrode at all. No corrosion is observed even if the acid is gradually diluted. But, if the metal surface is then scratched, a layer of gas bubbles sweeps across the surface and the dramatic production of hydrogen and nitrogen oxides gives strong evidence of corrosion.

In passivating, a metal reacts to form a layer of corrosion products so thin that it is invisible and so complete a barrier that it slows corrosion by several orders of magnitude. The passive film is at most only a few atoms thick, largely composed of amorphous oxides and hydroxides of the metal. The stainless steels, which have been especially developed to passivate in strongly oxidizing environments, form a passive film composed mainly of Cr_2O_3 containing Fe and $-OH$ groups within its structure.

2.5 CLASSIFICATIONS OF CORROSION

Corrosion takes on different appearances depending on the metal, the corrosive environment, the nature of the corrosion products, and all the other variables, such as temperature, stresses on the metal, and the relative velocity

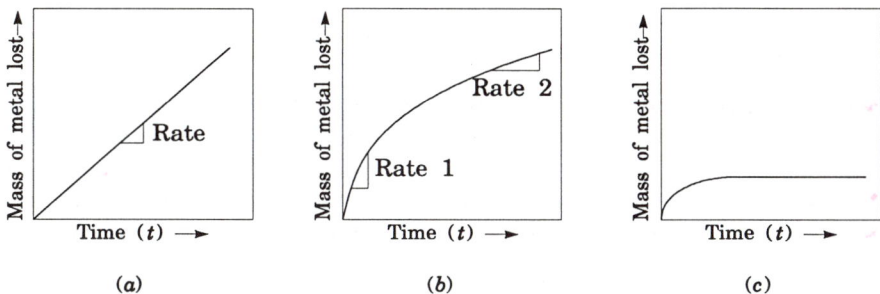

FIGURE 2-4. Theoretical change in corrosion rate with time: (a) constant rate if corrosion products are soluble or gaseous; (b) rate decreases with time if insoluble corrosion products form on the metal surface; and (c) rate drops sharply if the metal passivates.

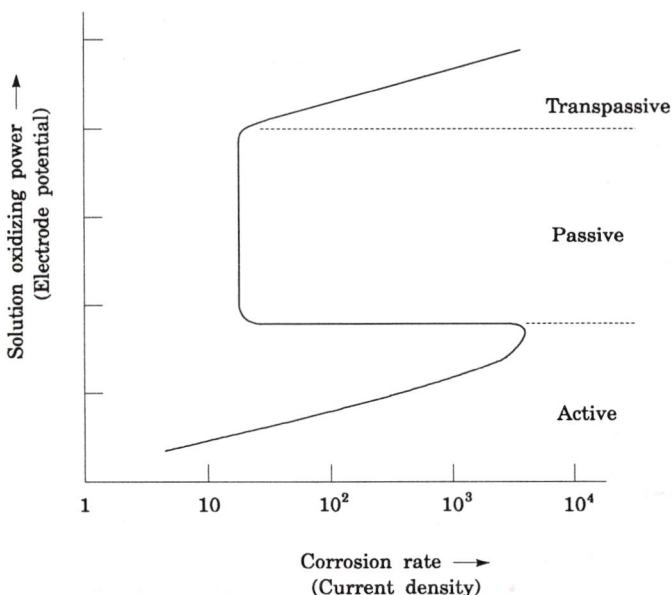

FIGURE 2-5.   Corrosion of a passivating metal in solutions of increasing oxidizing power.

of the metal and the environment. It is unrealistic to arbitrarily divide all corrosion into a number of specific types of corrosion as if they were unrelated.

Instead, it seems better to classify corrosion problems into the three general divisions of mechanisms which, as it turns out, are also the general divisions of severity, from Damage to Danger to Disaster. These correspond to the mechanisms of (1) chemical corrosion, (2) electrochemical corrosion, and (3) corrosive–mechanical interaction.

Chemical Corrosion

Uniform attack and the reaction of gases with metals, although actually occurring by electrochemical processes at a microscopic or atomic scale, are included in the general classification of *chemical corrosion* because no clear separate identification of the components of an electrochemical cell (anode, cathode, electrolyte, and electronic connection) can be made. In addition, some corrosion processes are definitely not electrochemical: the dissolution of metals in organic solvents, in liquid metals, and in some molten salts. These processes are also classified as types of chemical corrosion.

Reactions in organic solvents can involve a direct electron rearrangement between the metal atom and the reactive species of the solvent without the electrochemical steps of oxidation and reduction occurring. However, the appearance of the corrosion can be quite similar to all the various types of electrochemical corrosion. That is, chemical corrosion in organic solvents can produce uniform attack, crevice corrosion, pitting, erosion–corrosion, stress–corrosion cracking, and so on (see Fig. 2-6).

The methods of combating corrosion in organics are essentially the same as for electrochemical corrosion, except for the electrical methods of cathodic and anodic protection. Usually, these electrical methods cannot be used because of the high electrical resistivity of the organic liquid.

Data on the dissolution of solid metals by liquid metals and molten salts are given in Section 7.4. Gas–metal reactions, which are often referred to as

FIGURE 2-6.   Example of corrosion of metal (bronze pump impeller) in organic liquid. [Reprinted with permission from C. P. Dillon, Ed. *Forms of Corrosion: Recognition and Prevention*, NACE Handbook 1, National Association of Corrosion Engineers, 1982, Fig. 5.5.4.1, p. 86. Copyright © by NACE.]

"oxidation" regardless of the gas involved, are covered in detail in Chapters 14 and 15.

Uniform Attack

Uniform attack is by far the most common type of corrosion, but at the same time the least serious! As the metal corrodes it leaves a fairly smooth surface that may or may not be covered with corrosion products. A typical example would be the atmospheric corrosion of an old galvanized steel barn roof. Once the zinc galvanizing has corroded off, large areas of the steel quickly become heavily rusted, and while holes appear in only a few spots at first, all of the remaining steel is paper thin. An example of uniform corrosion is shown in Figure 2-7.

Uniform attack is considered a type of chemical corrosion because the corroding metal is serving as both the anode and the cathode. While the anode area is obvious in aqueous environments, since the entire surface of the metal is corroding, no separate cathode is identifiable. However, oxidation cannot occur without a corresponding reduction; thus the same metal

FIGURE 2-7.   This ship ran aground near the mouth of the Columbia River 60 years earlier.

surface must also be providing sites for the cathode reaction. The cathode sites are regions on the surface that are temporarily coated with a thicker layer of corrosion products or regions that are in contact with solution that is momentarily more concentrated in the reducible reactant. These cathode areas obviously move around constantly, because no region remains protected for long if the attack is uniform.

With uniform corrosion the engineer can estimate the life of the equipment and can plan for replacement on a regular schedule. Corrosion rates for thousands of metal–corrosive combinations are available in handbooks,[3] along with the variables of temperature, concentration, and so on. If published data cannot be found, simple laboratory tests will establish the corrosion rates of candidate materials in the real or simulated environmental conditions. Consequently, while uniform attack corrodes more tons of metal than all other types of corrosion, this attack creates only minor problems in most cases, because proper engineering can combat this corrosion before it becomes serious.

Electrochemical Corrosion

In electrochemical corrosion all the elements of an electrochemical cell are present and identifiable. The electrolyte is normally an aqueous solution in which electrical current is carried by migrating positive and negative ions, while the external connection between the anode and the cathode allows electron transfer from anode to cathode. The anode corrodes but the cathode does not. Consequently, attack is localized at the anodes while cathodes support that attack.

Where anodes and cathodes are caused by differences in metallurgical composition or structure, the important types of corrosion are galvanic corrosion, dealloying, and intergranular corrosion. If anodes and cathodes occur at different places on the metal because the environment differs in concentration, the most important types of corrosion are crevice corrosion and pitting (see Fig. 2-8).

Engineers plan for uniform attack, for which they can make reasonable estimates of corrosion rates. These engineers usually do not expect attack to be localized, with some metal corroding and some metal, not only not corroding, but actually intensifying the attack on the rest of the metal. Localized attack then creates the danger of extremely rapid corrosion concentrated at a few anodic places. How concentrated the attack will be cannot be predicted before variables, such as the cathode/anode area ratio, are

---

[3.] See the list of references at the end of this text.

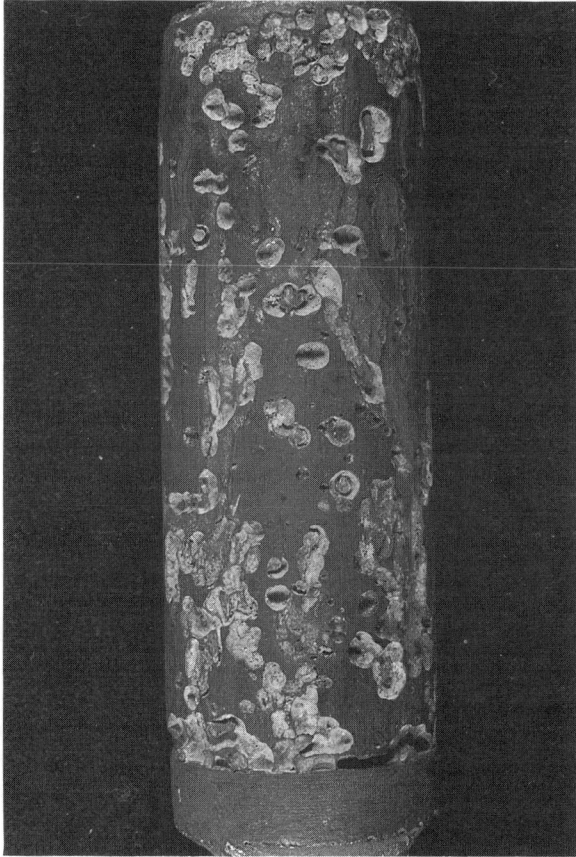

FIGURE 2-8.   Localized attack (pitting) of a sucker rod coupling from an oil well.

established by the corrosion cell. Note that it is not uncommon for the attack to be two or three orders of magnitude more intense than uniform attack.

Corrosive–Mechanical Interactions

Synergistic combination of corrosion with mechanical stresses creates a much higher level of danger than corrosion alone. For these situations the possibility exists for unanticipated, sudden failure of a component. This situation, in turn, may cause a machine to self-destruct, a chemical plant to burn (as in Fig. 2-9), a boiler to explode, and so on. The important corrosive–mechanical interactions include erosion–corrosion, stress–corrosion cracking, hydrogen damage, and corrosion fatigue.

FIGURE 2-9.   A small steam separator fractured in this gas plant. One man died and one was burned severely.

*2.6 POURBAIX DIAGRAMS

In 1938 Dr. Marcel Pourbaix presented his potential–pH diagrams to illustrate the thermodynamic state of a metal in dilute aqueous solutions. The advantages of depicting all the thermodynamic equilibria in a single, unified diagram was immediately evident to scientists and engineers, especially in corrosion, hydrometallurgy, and electrochemistry.

The axes of the diagram are the key variables that the corrosion engineer can control. The vertical axis shows the metal–solution potential, which can be changed by varying the oxidizer concentration in the solution or by applying an electrical potential to the metal. The horizontal axis shows the pH of the solution. The diagrams are divided into regions of stability, each labeled with the predominant species present.

For regions where metal ions are stable, the boundaries are usually drawn for equilibrium concentrations of $10^{-6}M$, chosen to show the limits of corrosion where soluble corrosion products would be barely detectable. However, in a specific corrosion situation, where solution concentrations are known to be greater than $10^{-6}M$ or the temperature is not 25°C (77°F), the diagram can be redrawn to fit the real conditions.

The potential–pH diagram for the iron–water system is shown in Figure 2-10. The *dotted lines* on the diagram show the theoretical limits of the stability of water. The upper line shows where O_2 should be generated on an anode and the lower line shows where H_2 should be given off at a cathode. Between the two dotted lines water is stable, so this is the important region in aqueous reactions. However, the actual stability range for water is usually much greater than the diagram indicates; water does not decompose as readily on most metals as it does on an ideal platinum surface.

Aside from the stability limits for water, Pourbaix diagrams have three different types of lines.

1. *Horizontal lines*, independent of pH. The equilibrium does not involve hydrogen ions. For example, the boundary between the Fe^{3+} and Fe^{2+} regions is for the equilibrium

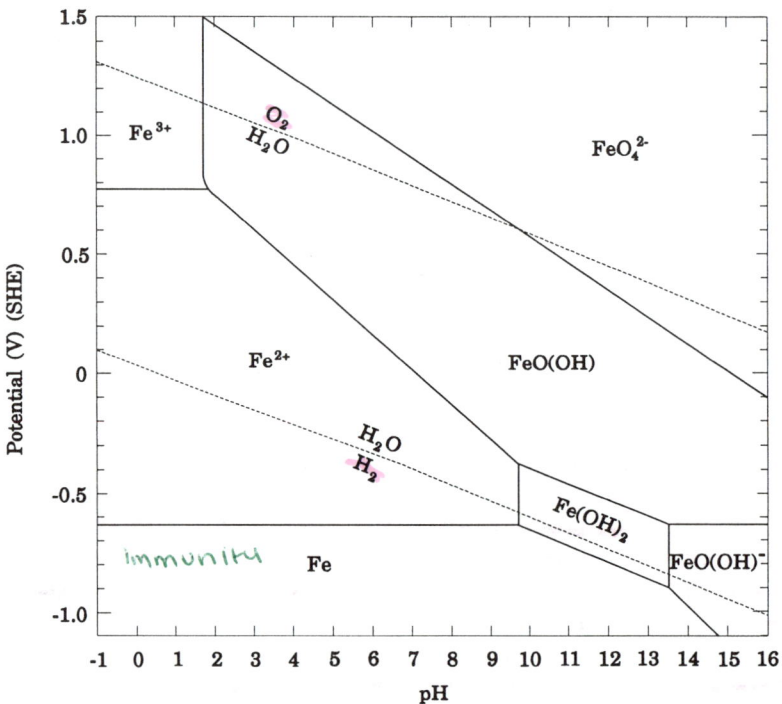

FIGURE 2-10.   Pourbaix diagram for the Fe-H2O system at 25°C (77°F) for $10^{-6}M$ activities of all metal ions.

$$Fe^{3+} + e^- = Fe^{2+} \qquad (2\text{-}8)$$

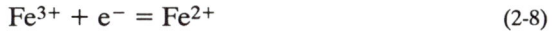

with the Nernst equation giving

$$E = E^\circ - \frac{RT}{F} \ln \frac{[Fe^{2+}]}{[Fe^{3+}]} \qquad (2\text{--}18)$$

2. *Vertical lines*, not involving oxidation or reduction. The boundary between the Fe^{2+} and $Fe(OH)_2$ regions shows the equilibrium

$$Fe^{2+} + 2H_2O = Fe(OH)_2 + 2H^+ \qquad (2\text{-}19)$$

Iron remains in the +2 valence state with no electrons exchanged, so the reaction can take place at any potential, positive or negative.

3. *Sloping lines*, involving both hydrogen ions and electrons. The boundary between Fe^{2+} and $FeO(OH)$ regions represents the equilibrium

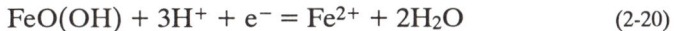

$$FeO(OH) + 3H^+ + e^- = Fe^{2+} + 2H_2O \qquad (2\text{-}20)$$

The Nernst equation (Eq. 2-13) will give the mathematical equation for this line.

$$E = E^\circ - 0.177pH - 0.0592 \log [Fe^{2+}]$$

To aid in understanding corrosion, Pourbaix diagrams can be divided into domains of immunity, corrosion, and passivation. The immunity domain is the region where the metal is the most stable species. The corrosion domains include all regions where soluble corrosion products (metal ions) are the most stable. And the passivation domains include all regions where the most stable species are insoluble corrosion products (oxides, hydroxides, and the like).

Operating in the passivation domain does not insure that corrosion will be acceptable. The insoluble corrosion products may or may not form a protective barrier between the metal and the solution. With steel, for example, the scale is loosely adherent at pH 6 but strong and tenacious above pH 8. Pitting of metals is especially probable at high potentials in the passivation domain near a transpassive corrosion region. The domains of passivation offer the possibility (or the hope?) of low corrosion rates but no guarantee.

In the corrosion domains, the corrosion rate increases as operating conditions get farther from boundaries of immunity and passivation.

- How can the corrosion of steel in water be stopped? At pH 7 the steel develops a potential of -0.5 V, just inside the Fe^{2+} region (see Fig. 2-10). One method is to apply a negative potential to the steel to bring it down into the Fe immunity region of the diagram (cathodic protection). A second method could be to increase the pH so that a solid corrosion product of $Fe(OH)_2$ or $FeO(OH)$ forms on the steel surface to passivate it. Or, in a third method, the steel potential can be made positive enough to get it out of the Fe^{2+} region (anodic protection).

The potential–pH diagrams for several important metal–water systems are illustrated in Figure 2-11.

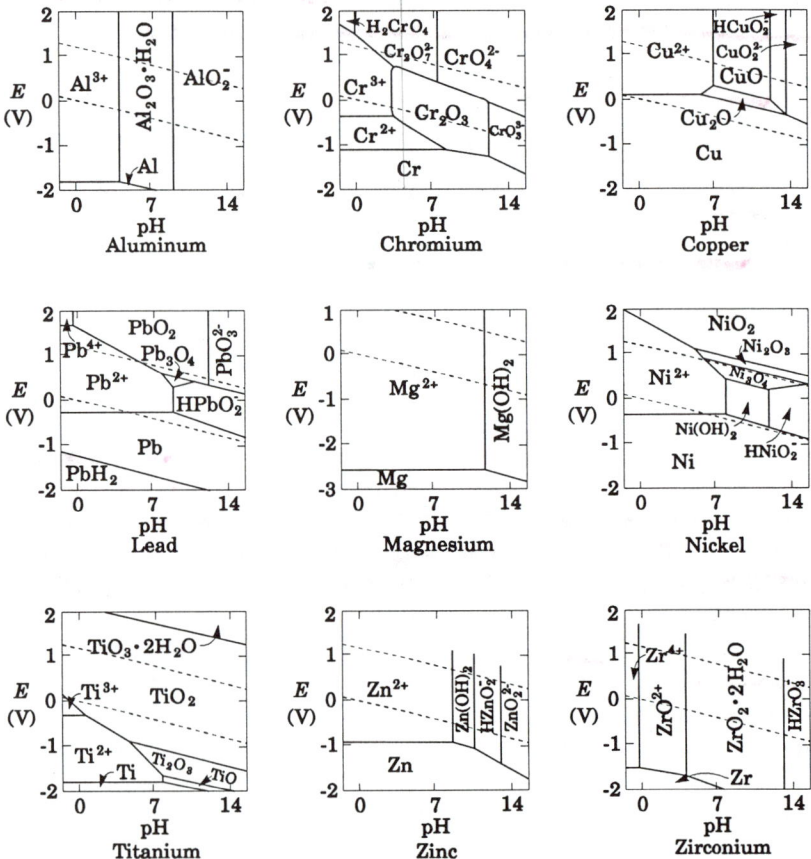

FIGURE 2-11.   Potential–pH diagrams for common metal–water systems. Metal ions are $10^{-6}M$, temperature is 25°C (77°F).

In spite of their importance, Pourbaix diagrams have their limitations.

a. The diagrams are drawn for the equilibrium situation, and practical problems are often far from equilibrium.
b. The diagrams for pure metals in simple solutions have been worked out but they have to be modified for alloys and to include effects of other ions that may be present in real-life solutions.
c. Metal in a region of passivity may or may not be well protected, depending on the degree of perfection of the layer of solid corrosion product.
d. The pH on the diagrams is not the pH of the bulk solution but the pH of solution in contact with the metal, in crevices, pits, under loose rust, and so on.
e. The most important data about corrosion, the corrosion rates, are not shown.

To overcome these difficulties, potential–pH diagrams are now being made from a series of potentiodynamic polarization scans at different pH values. The potential of an alloy in a complex environment is gradually increased as the current density is measured. Breaks in the polarization curve, corresponding approximately to the lines on the equilibrium Pourbaix diagram, show the transitions from regions of immunity to corrosion to passivation. Even zones of imperfect passivation (pitting) can be mapped, along with contours showing corrosion rates. These pseudo-Pourbaix diagrams give a much better understanding of short-term corrosion and can be constructed very quickly, although they do not show protective films that might develop over long periods of time. Figure 2-12 shows how such a diagram can be constructed.

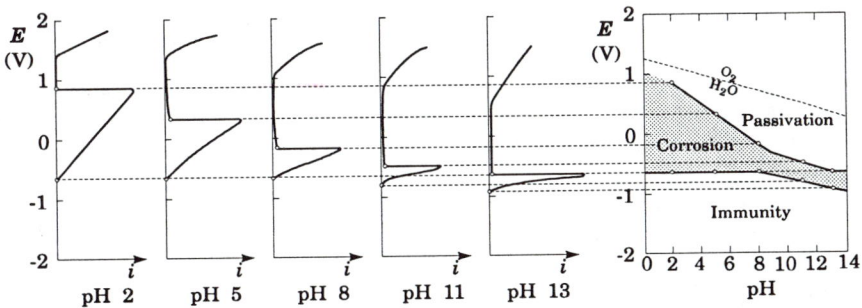

FIGURE 2-12. E–log I polarization curves and experimental potential–pH diagram for iron in chloride-free solutions of various pH values.

2.7 CORROSION RATES

The extent of corrosion is commonly measured either of two ways. In uniform attack, the mass of metal corroded on a unit area of surface will satisfactorily describe the damage. However, if attack is localized, the amount of metal removed *on average* over the entire surface is meaningless. The depth of penetration, whether by uniform attack, pitting, or whatever, gives a much better description of almost any type of corrosion except cracking. (A crack of any length is a warning of imminent disaster.)

Weight Loss

The corrosion rate in uniform attack (*only*) can be given as mass per area per time. (mass/area · time). The most common unit of weight loss has been milligrams per square decimeter per day (mdd), although numerous other metric, English, and bastard units have also been employed. In the International (SI) System of units now being used, weight loss is expressed as grams per square metre per day ($g/m^2 \cdot d$), which can be obtained from milligrams per square decimeter per day by simply dividing by 10. To have a very approximate idea of magnitudes, it is worth remembering that corrosion rates in the *order* of 1 $g/m^2 \cdot d$ are usually satisfactory.

- I have even seen corrosion rates in "grams per square foot per fortnight" published.

Depth of Penetration

The corrosion rate, whether for uniform or localized attack, is very often given as depth of penetration per unit time. For localized attack, such as pitting, the penetration usually refers to the depth of the deepest pit observed, since that one is likely to be the most dangerous. In the past, the most common unit was probably mils per year (mpy, where 1000 mils = 1 in.) or similarly, inches per year (ipy). The SI unit is millimetres per year (mm/y), which is a unit that is very easy to visualize mentally, whereas a mil does not mean much to most people. You can convert from one system to the other by remembering that 40 mpy = 1 mm/y, which is quite a severe attack. Conversion factors to change from traditional units to SI units are given in Table 2-2.

Acceptable Rates

What rates are acceptable? The answer to this question depends very much on what is corroding. On a ship, if the anchor corroded at 1 mm/y (40 mpy)

TABLE 2-2   Conversion of Common Corrosion Units to SI Units

To Convert from	To	Multiply by*
mpy	mm/y	0.0254
mpy	g/m²·d	0.0695d
ipy	mm/y	25.4
ipy	g/m²·d	69.5d
mdd	mm/y	0.0365/d
mdd	g/m²·d	0.100

*Note: d = metal density in grams per cubic centimetre (g/cm³) or megagrams per cubic metre (Mg/m³).

it would not disturb anyone, but corrosion of 0.01 mm/y (0.4 mpy) on the bearing in the ship's compass would be intolerable.

However, for general engineering work the following rules of thumb will prove useful:

<0.1 mm/y (4mpy)   = good corrosion resistance
0.1–1 mm/y (4–40 mpy) = may be satisfactory if greater corrosion can be tolerated
>1 mm/y (>40 mpy)   = usually excessive corrosion

Calculating Corrosion Rates

In most cases the engineer cannot calculate corrosion rates the way he/she can calculate equilibrium potentials with the Nernst equation; the rates are a function of so many variables that the only way to find them is to measure them. However, in cases where the anode and cathode are separated, if the current flowing between them can be measured, the corrosion rates can be calculated from the corrosion current.

According to Faraday's law,

$$\text{Corrosion rate} = i_{corr}/nF \qquad (2\text{-}21)$$

with i_{corr} being the corrosion current density in amperes per square metre (A/m²), F is 96,490 C/mol · e⁻, and n is in moles of electrons per mole of metal corroded. Since coulombs = amperes × seconds (C = A × s) the rate then comes out as moles of metal per square metre per second, which converts easily to the usual grams per square metre per day (g/m² · d), by multiplying by the atomic weight of the metal and 86,400 s/d.

- If iron corrodes in seawater to produce a corrosion current (a discharge of Fe^{2+} ions) of 0.3 A/m², the

$$\text{Corrosion rate} = i_{corr}/nF$$

$$= \frac{0.3\ \text{A}}{\text{m}^2}\ \frac{(55.85\ \text{g/mol Fe})\,(86400\ \text{s/d})}{\dfrac{2\ \text{mol e}^-}{\text{mol Fe}}\,(96{,}490\ \text{A}\cdot\text{s/mole}^-)}$$

$$= 7.5\ \text{g/m}^2\cdot\text{d}$$

- To convert the weight loss to depth of penetration, divide by the metal density, which for Fe is $7.86 \times 10^6 \text{g/m}^3$.

$$\text{Corrosion rate} = (7.5\ \text{g/m}^2\cdot\text{d})\,(1\ \text{m}^3/7.86 \times 10^6\ \text{g of Fe})\,(365\ \text{d/y})$$
$$= 0.35\ \text{mm/y}$$

It is an interesting coincidence worth remembering that for many common engineering metals (Fe, Al, Cu, Ni, etc.) the corrosion current density in amperes per square metre(A/m^2) is approximately the same as the corrosion rate in millimetres per year (mm/y).

Change in Corrosion Rate With Time

If insoluble corrosion products form on the metal, they tend to interfere with the electrode reactions. Even if they do not form a fully passive, protective film they slow diffusion of reactants to the cathode or diffusion of solvated metal ions away from the anode. Consequently, the corrosion rate usually slows down with time, as shown in Figure 2-4b. For this reason, when you state a corrosion rate you should also give the period of time over which it was measured. For example, "a corrosion rate of 0.35 mm/y in a 3-month test." Figure 2-4b shows the danger of extrapolating a short-term rate to longer times.

STUDY PROBLEMS

2.1 Aluminum cathodically protects an offshore drilling rig (steel) in seawater. (Neither metal passivates.) If *standard* conditions could develop at 25°C (77°F), how large would the driving force for corrosion be in volts?

2.2 From the standard reduction potential for mercury, calculate the standard Gibbs energy change for the ionization of mercury.

$$2\text{Hg} \rightarrow \text{Hg}_2^{2+} + 2\text{e}^-.$$

2.3 The Gibbs energy under standard conditions for the reaction $\text{Fe}_3\text{O}_4 \rightarrow 3\,\text{FeO} + \tfrac{1}{2}\,\text{O}_2$ can be expressed by the equation

$\Delta G°$ (kJ/mol) = 312.4 - 0.1252T, where T is the temperature in kelvins. (a) Is the reaction spontaneous at room temperature (298 K)? Show your proof. (b) At what temperature will the reaction change to spontaneous or nonspontaneous? (Any temperature above 1700 K would be inexact because FeO melts.)

2.4 An alkaline dry cell battery has a voltage of 1.5 V. The anode reaction in the cell is $Zn \rightarrow Zn^{2+} + 2e^-$. (a) Express the driving force for the cell reaction in joules. (b) Assuming standard conditions at 25°C (77°F), what would be the half-cell potential of the cathode reaction?

2.5 Automobiles in a company parking lot were severely corroded when acid vapors condensed in a thin film of moisture on the steel surfaces. What are the most probable anode and cathode reactions for the corrosion?

2.6 Electroplating waste solution consisting of 5% $CuSO_4$ and 10% H_2SO_4 at room temperature has seriously attacked Type 410 stainless steel drains and holding tank. What are the relevant anode and cathode reactions?

2.7 An undersea research chamber made of stainless steel (Fe−Cr alloy) is suspended in the Pacific by a galvanized (Zn coated) carbon steel cable. Explain the electrochemical processes occurring and describe where they are occurring.

2.8 Ferric chloride crystals ($FeCl_3 \cdot 6H_2O$) stored in an aluminum bin evidently picked up water from the air because an orange-red syrup quickly ate through the bin and ran over everything. Write the most probable anode reaction for corrosion, the most probable cathode reaction, and explain where or what the anode and cathode are.

2.9 Some people complain that pineapple juice has a bad taste if it sits in the open tin can for 1 or 2 days. Can you explain this from your knowledge of corrosion?

2.10 Steel gas pipe was laid parallel to (but not touching) copper water lines in the same trench in a new housing development. Within 6 months the gas was leaking everywhere, as evidenced by the general odor and the dead grass over the lines. (a) How could an electrolytic corrosion cell have been set up? (b) Assuming a cell *was* set up between Fe and Cu, identify: the anode, the cathode, the electrolyte, the anode reaction, and the cathode reaction.

2.11 Aluminum and steel pipes carrying city water were fastened together with an electrically insulating connector. Everything was painted with a good layer of aluminum paint so that no exterior corrosion occurred, but internally the steel corroded badly. (a) How could an electrolytic corrosion cell have been set up? (b) Assuming a cell *was* set up between Al and Fe, identify: the anode, the cathode, the electrolyte, the anode reaction, and the cathode reaction.

2.12 If silver and copper are connected in an electrolyte, which will be oxidized? Which will be reduced?

2.13 If lead and cadmium are connected in hydrochloric acid, which will be oxidized? Which will be reduced?

2.14 If you put two strips of copper in a 1 M solution of $CuSO_4$ (deaerated) and connect a 1.5 V dry cell battery between them (a) which one will corrode and (b) what is the driving force for corrosion (in volts)?

*2.15 A magnesium juicer with a nickel-plated nozzle squeezes orange juice (pH 3.3) at 25°C (77°F) and 1 atm. Calculate the voltage of the resulting corrosion cell, making reasonable assumptions for missing data.

*2.16 A bronze (Cu alloy) fitting is fastened to a steel ship splashed with seawater ($\sim$ 3.5% NaCl). Calculate the potential of the electrochemical cell, making reasonable assumptions about activities.

*2.17 What potential would magnesium metal have in an aqueous solution of $10^{-6}M$ $MgCl_2$ at pH 5 and 25°C (77°F)?

*2.18 Copper is immersed in $10^{-4}M$ HCl at 35°C (95°F) and 1 atm. At 35°C (95°F) $E°_{Cu} = +0.338$ V. Calculate the driving force for corrosion (in volts).

*2.19 What is the value of the corrosion current density, i_{corr}, for zinc corresponding to a corrosion rate of 0.05 ipy?

*2.20 Use the Ni-H_2O Pourbaix diagram to answer the following questions: (a) What are the regions of passivity? (b) In what potential range can the metal be cathodically protected at pH 10? (c) How will decreasing the potential from 1.0 to 0 V at pH 3 affect the corrosion?

*2.21 From the Cr/H_2O Pourbaix diagram (a) is chromium theoretically stable in water? Explain. (b) Chromium immersed in water stays bright and shiny. Does that agree with this diagram? Explain.

*2.22 From the Ni/H_2O Pourbaix diagram (a) what are the corrosion regions of the diagram, (b) can nickel be cathodically protected at pH 7? Explain. (c) Can nickel be anodically protected at pH 7? Explain.

*2.23 On the Fe/H_2O Pourbaix diagram, why is the line separating the Fe and Fe(OH)$_2$ regions a *sloping* line?

2.24 Calculate a conversion factor for converting corrosion of zinc expressed as ipy to the units of grams per square metre per day (g/m$^2 \cdot$ d).

*2.25 What is the potential of the hydrogen electrode at pH 7, 1 atm, and 40°C (158°F)?

*2.26 If 1 in.^{2} of iron surface produces a current of 1 mA, what is its corrosion rate?

*2.27 If the corrosion current is 1.000 mA from a sample of copper with a surface area of 1.000 cm^2, what is the corrosion rate in millimetres per year (mm/y)? Corrosion product is CuO.

3

*Electrochemical Corrosion Theory

The driving force for corrosion is the potential difference developed by the corrosion cell

$$E_{cell} = E_{red}^{cat} - E_{red}^{anode} \qquad (2-17)$$

However, the cell potential does not correctly predict the corrosion rate, and it is the corrosion rate that is the essential determiner of a metal's suitability in a corrosive environment. Logically, if the cell potential is small, the corrosion rate will be low. On the other hand, a large cell potential does not necessarily mean that the metal must corrode badly. It may passivate, for example, and corrode at an extremely low rate.

Corrosion kinetics, the rates of the electrode reactions, are related to the thermodynamic driving force that is measured by the cell potential. This relationship depends on several factors, all connected with the "polarization" of the electrodes in the cell.

The term "polarization" refers to a shift in potential caused by a flow of current. An anode increases its potential as more current flows from it into the electrolyte, while the cathode's potential decreases as current flows onto it. Both electrodes in the cell polarize until they reach essentially the same potential; the corrosion potential. Polarization is also often called "overvoltage," a term commonly used in commercial electrochemical processes, such as electroplating, to describe the additional voltage that must be applied to overcome the polarization of the electrodes. An understanding of the causes of polarization is essential to an understanding of corrosion.

33

3.1 EXCHANGE CURRENT DENSITY

An electrode *at equilibrium* with its environment has no net current flow to or from the surface of the metal. Actually, a "dynamic equilibrium" is established in which the forward and reverse reactions are both occurring, but at equal rates. If the forward reaction is

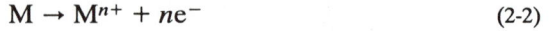

$$M \rightarrow M^{n+} + ne^- \tag{2-2}$$

with positive current (M^{n+} ions) flowing from the electrode, an exactly equal current flows back onto the surface as the metal plates back on

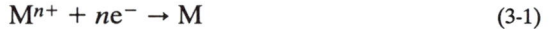

$$M^{n+} + ne^- \rightarrow M \tag{3-1}$$

Thus the net current flow is zero. The oxidation process is exactly reversed by its corresponding reduction process.

In terms of the current,

$$I_{ox} = I_{red} \tag{3-2}$$

In this equation I is the current in amperes (A) and the subscripts ox and red refer to oxidation and reduction. This equilibrium current, either I_{ox} or I_{red}, is called the *exchange current* I_0. The exchange current cannot be measured directly but can be found by extrapolation, as is shown in Figure 3-1. The exchange current may be extremely small but it is not zero.

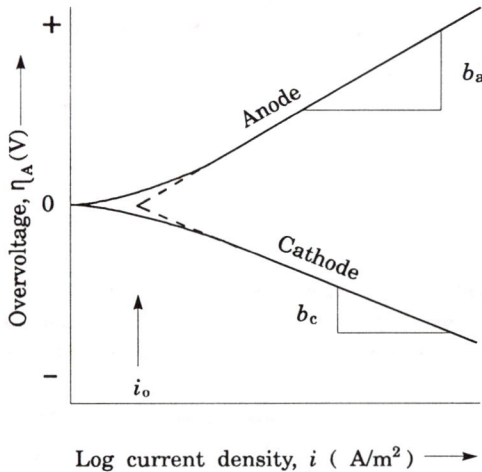

FIGURE 3-1.   Experimentally measurable anodic and cathodic polarization curves showing activation polarization. Tafel slopes are b_a and b_c. Exchange current density is i_0.

Often it is more convenient to use the *exchange current density* i_0 expressed as amperes per square metre (A/m^2), rather than the current in order to eliminate the variable of electrode size. The exchange current density is a direct measure of the electrode's oxidation rate or reduction rate at equilibrium.

$$\text{rate}_{ox} = \text{rate}_{red} = i_0/nF \quad (\text{mol/m}^2 \cdot \text{s}) \qquad (3\text{-}3)$$

The term i_0 is a function of the reaction, the concentration of reactants, the electrode material, the temperature, and the surface roughness. Typical examples of exchange current densities for a variety of reactions and electrodes are given in Table 3-1.

Note that a Pt surface makes the H$_2$ reaction extremely easy, but not the O$_2$ reaction. The i_0 values for corroding and plating metals are obviously *not* in the same order as their standard electrode potentials $E°$.

3.2 ACTIVATION POLARIZATION

All electrodes, both anodes and cathodes, undergo activation polarization when current flows. A slow step in the electrode reaction is responsible for the shift in electrode potential. If activation polarization is the main cause of polarization of the electrodes the corrosion is said to be *reaction controlled*.

Electrode reactions are not one-step processes in which everything hap-

TABLE 3-1   Approximate Exchange Current Densities at 25°C (77°F)

Reaction	Electrode	Solution	i_0 (A/m^2)	Reference
$2H^+ + 2e^- = H_2$	Al	1 M H$_2$SO$_4$	10^{-6}	Parsons 1959
$2H^+ + 2e^- = H_2$	Cu	0.1 M HCl	2×10^{-3}	Bockris 1953
$2H^+ + 2e^- = H_2$	Fe	1 M H$_2$SO$_4$	10^{-2}	Bockris 1953
$2H^+ + 2e^- = H_2$	Ni	1 M H$_2$SO$_4$	6×10^{-2}	Bockris and Reddy 1970
$2H^+ + 2e^- = H_2$	Pb	1 M HCl	2×10^{-9}	Bockris 1953
$2H^+ + 2e^- = H_2$	Pt	1 M H$_2$SO$_4$	8	Bockris and Reddy 1970
$2H^+ + 2e^- = H_2$	Ti	1 M H$_2$SO$_4$	6×10^{-5}	Bockris and Reddy 1970
$2H^+ + 2e^- = H_2$	Zn	1 M H$_2$SO$_4$	10^{-7}	West 1970
$O_2 + 2H_2O + 4e^- = 4OH^-$	Pt	0.1 M NaOH	4×10^{-9}	Bockris and Reddy 1970
$O_2 + 2H_2O + 4e^- = 4OH^-$	Au	0.1 M NaOH	5×10^{-9}	Parsons 1959
$Cu^{2+} + 2e^- = Cu$	Cu	Sulfate	4×10^{-1}	West 1970
$Fe^{2+} + 2e^- = Fe$	Fe	Sulfate	2×10^{-5}	West 1970
$Ni^{2+} + 2e^- = Ni$	Ni	Sulfate	2×10^{-5}	West 1970
$Pb^{2+} + 2e^- = Pb$	Pb	Perchlorate	8	West 1970
$Zn^{2+} + 2e^- = Zn$	Zn	Sulfate	3×10^{-1}	Bockris and Reddy 1970

pens simultaneously. Instead, adsorption (ads) of reactants onto the electrode surface may be followed by electron transfer, then readjustment of valence electrons between neighboring adsorbed species or perhaps between an adsorbed atom and a molecule in the electrolyte, and finally desorption of the products.

For reduction of H^+ on an iron cathode the slow step appears to be

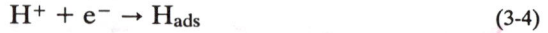

$$H^+ + e^- \rightarrow H_{ads} \tag{3-4}$$

while on some other metals the next step in the reaction seems to be even slower.

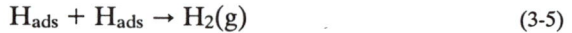

$$H_{ads} + H_{ads} \rightarrow H_2(g) \tag{3-5}$$

The reactants must have extra energy, an *activation energy*, in order to proceed to form products. The electrode can be activated only by shifting its potential, that is, by polarizing. The metal ions coming off the anode can be driven off faster if the anode becomes more positive, and similarly, the reactant species in the electrolyte can receive electrons more rapidly from the cathode if the cathode becomes more negative.

The Tafel equation describes the activation polarization, η_A, caused by a flow of current.

$$\eta_A = \pm\, b \log (i/i_0) \quad \text{(in volts)} \tag{3-6}$$

where i is the current density and b is a constant called the Tafel slope. The sign on the right-hand side of the equation is positive for anodes and negative for cathodes. A graph of η_A versus $\log i$ gives a straight line with a slope of $\pm b$, as shown in Figure 3-2. It should be noted that the Tafel slopes for oxidation and reduction are generally not equal, although they are usually both in the range of 0.05–0.15 V for each 10-fold change in current, or roughly 0.10 V per decade.

- *Example:* How much will nickel have to polarize to corrode at a current density of 1 A/m² (roughly 1 mm/y)?

- *Answer:* Use the Tafel equation, assuming a Tafel slope of 0.10 V per decade and the i_0 from Table 3-1.

$$\eta_A = +\, b \log (i/i_0) = +(0.10 \text{ V}) \log[(1 \text{ A/m}^2)/(2 \times 10^{-5} \text{ A/m}^2)]$$

$$= \underline{0.47 \text{ V}}$$

Figure 3-2 is only a theoretical diagram because in reality the Tafel equation does not hold within ±50 mV of the equilibrium potential. At

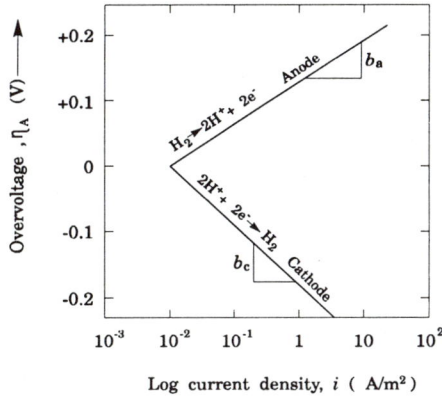

FIGURE 3-2. Theoretical activation polarization diagram for the hydrogen reaction. Tafel slopes for the anode and cathode reactions are indicated as b_a and b_c, respectively.

small overvoltages no electrode can be completely an anode or a cathode; to some extent the oxidation and reduction processes both occur on the electrode surface. The measurable, net current decreases to zero if the overvoltage decreases to zero, as shown in Figure 3-1. This figure also illustrates how the exchange current density i_0 can be found. An electrode can be polarized enough to put it in the Tafel (straight-line) region so that the line can be extrapolated back to the equilibrium potential where $\eta_A = 0$ and $i = i_0$.

3.3 CONCENTRATION POLARIZATION

While all electrodes undergo activation polarization some electrode processes are *diffusion controlled*; that is, they undergo concentration polarization in addition to activation polarization.

The anode is rarely controlled by diffusion. The cathode process, however, is often diffusion controlled because reactants in the environment (the electrolyte) must diffuse to the cathode surface in order to pick up electrons.

• Would a diffusion-controlled cathode reaction be more likely in a concentrated acid or a dilute acid?

In a concentrated acid the concentration of hydrogen ions is extremely high so that diffusion distances are short, making diffusion occur so fast that it does not slow the reaction rate.

Especially in solutions where O_2 reduction is the primary cathode reaction (Reaction 2-5), the cathode must polarize severely as more current is forced upon it. That is, the cathode must become much more negative to attract more reactant species to its surface. The reactant becomes depleted in the vicinity of the cathode, requiring additional reactant to diffuse through this depleted zone.

At some point, however, when diffusion distances become so long that increasing negative polarization cannot attract any more reactant, a limiting current density, i_L, is reached. This situation is illustrated in Figure 3-3.

- For many years the "four-minute mile" was an unobtainable dream for runners until dedicated training broke that barrier. But *some* limit does exist ($\sim$ 3.6 min?), which cannot be beaten no matter how large a prize is offered.

The concentration polarization, (η_C,) is defined by the equation

$$\eta_C = \frac{RT}{nF} \ln (l - i/i_L) \qquad (3\text{--}7)$$

which shows that diffusion control plays no significant part until the cathodic current approaches the limiting current density i_L within about an order of magnitude.

When the corrosion rate is controlled by the limiting current density, Faraday's law (Eq. 2-21) becomes

$$\text{Corrosion rate} = i_L/nF = DC_b/x$$
$$(\text{m}^2/\text{s}) \, (\text{mol/m}^3)/(\text{m}) = \text{mol/m}^2 \cdot \text{s} \qquad (3\text{-}8)$$

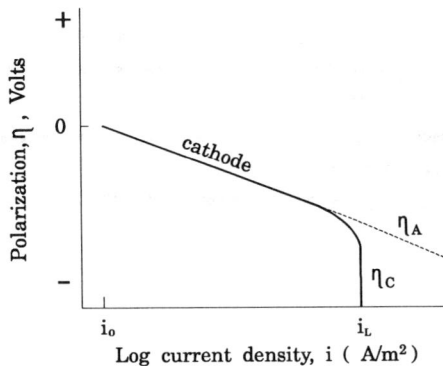

FIGURE 3-3.   Activation polarization, η_A, plus concentration polarization, η_C, of a cathode reaction. Limiting current density is indicated by i_L.

in which D is the diffusivity of the reacting species, C_b is the concentration of the species in the bulk of the solution beyond the depleted zone, and x is the thickness of the depleted zone and is the distance the species must diffuse. The three major variables that affect the limiting current density are temperature, concentration, and velocity.

Increasing the temperature increases the diffusivity D *exponentially* and thus can have a tremendous effect on the corrosion rate. In the case of diffusion of oxygen, however, its solubility decreases with increasing temperature, which tends to counteract its increased diffusivity.

- "The race is not always to the swift nor the battle to the strong, but that's the way to bet," is a cynical view of humanity but very true for corrosion.

Increasing the relative velocity, or flow rate, of the liquid corrodent over the metal surface reduces x, the thickness of the depleted zone and, therefore, decreases the diffusion distance and increases the diffusion rate. In theory, the limiting current density is related to the local flow rate (v) by

$$i_L \propto v^m \qquad (3\text{-}9)$$

where m is a constant that theoretically equals 0.5 for laminar flow or 0.8–0.9 for turbulent flow.

- In reality, corrosion products, surface deposits, or protective coatings will greatly reduce the sensitivity to flow rate.

Schematically, the effects of the major variables are shown in Figure 3-4. Increasing temperature, concentration, or flow rate increases the limiting current density.

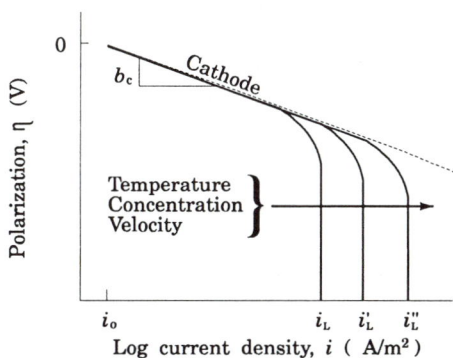

FIGURE 3-4.   Increasing temperature, concentration of corrodent, and velocity all increase the limiting current density of a cathode process controlled by concentration polarization.

3.4 RESISTANCE POLARIZATION

An additional overpotential, the resistance polarization η_R, is required to overcome the ohmic resistance of the electrolyte and any insoluble product film on the surface of the metal. This overpotential is defined by Ohm's law as

$$\eta_R = IR \tag{3-10}$$

where I is the current and R is the resistance, in ohms (Ω), of the electrolyte path between anode and cathode and is directly proportional to the path length. In typical corrosion processes the anodes and cathodes are immediately adjacent to each other so that resistance polarization makes only a minor contribution to the overall polarization, as indicated in Figure 3-5.

Resistance polarization drops to zero instantaneously if the current is turned off, whereas η_A and η_C decay gradually with time. Consequently, for systems such as cathodic protection in soils, where anode and cathode are separated by a long, high-resistivity electrolyte path, the engineer can measure resistance overpotential by momentarily interrupting the protection current.

3.5 POLARIZATION DIAGRAMS

The diagrams of potential versus log current density that have been used to illustrate these types of polarization have a significance that is broader than merely illustrating the electrical behavior of the anode and cathode. The potential of the anode metal is a measure of its corrodibility, as illustrated by the standard emf series (Table 2-1); that is, the more negative the metal's

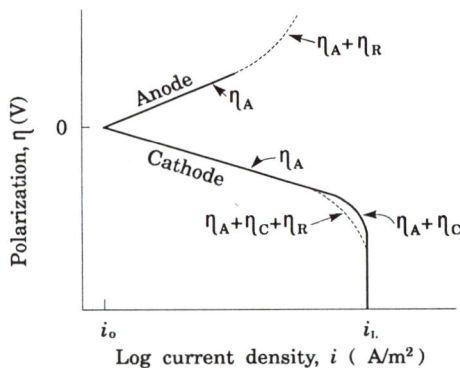

FIGURE 3-5.   Polarization curves for anode and cathode reactions showing contributions for activation, η_A, concentration polarization η_C, and resistance polarization η_R.

potential the more easily corroded it is. Consequently, when the anode polarizes in the positive direction it is becoming less corrodible as it opposes a shift away from equilibrium.

At the cathode the environment reacts. The higher the cathode potential the more corrosive the environment, or the greater is its oxidizing power. Therefore, since the cathode polarizes in the negative direction, the environment is becoming less aggressive. Obviously, polarization of both cathode and anode saves the metal from immediate destruction.

- Very few processes in nature work in your favor, but polarization is one that does.

Polarization of both anode and cathode continues until they reach the same potential: the corrosion potential. At the corrosion potential, the current density coming from the anode is the corrosion current density. This situation is best illustrated by putting the polarization curves for both anode and cathode on the same diagram, such as in Figure 3-6.

The corrosion potential (E_{corr}) is established by the intersection of the cathode and anode polarization curves. Since the two electrodes are likely adjacent to each other on the same piece of metal their potentials could hardly be very dissimilar.

The cathodic polarization curve in Figure 3-6 shows some concentration polarization in addition to activation polarization, since this is often the situation for the cathode process. If the intersection were on the vertical end of the cathode line, the corrosion current density (i_{corr}) would also be the limiting current density (i_L).

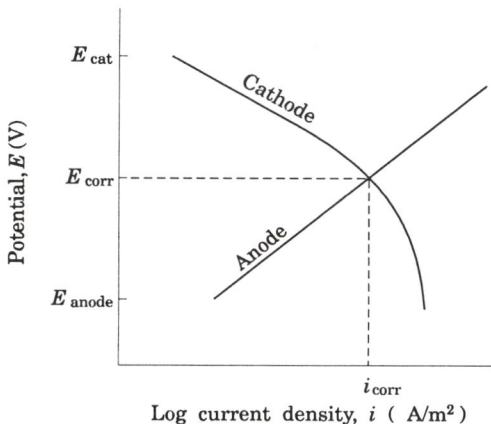

FIGURE 3-6.   Polarization diagram for a corrosion cell where the cathode reaction is partially diffusion controlled.

The corrosion potential lies somewhere between the equilibrium potentials of the anode and the cathode, but not precisely half-way between them. If E_{corr} is close to the equilibrium potential of the anode (E_{anode}), then the cathode has polarized much more than the anode and the reaction is said to be under "cathodic control."

Two Reduction Reactions

To illustrate a more complicated corrosion situation, take the case of a metal corroding in an environment that has two possible cathode processes. Here the two cathode curves are added because the *total* current to the cathodes must equal the total current from the anode. For example, a metal corroding in acidified ferric chloride would have the polarization diagram shown in Figure 3-7.

The two cathode reactions involved are reduction of ferric ions to ferrous (Reaction 2-8) and the reduction of hydrogen ions in the acid to hydrogen gas (Reaction 2-6). A third possible cathode reaction is shown on the diagram, the reduction of metal M^+ ions as they plate out on the metal, but it could only occur at potentials far below E_{corr}. Likewise, the other oxidations, $Fe^{2+} \rightarrow Fe^{3+} + e^-$ and $H_2 \rightarrow 2H^+ + 2e^-$ cannot occur either, because their potentials are more noble than E_{corr}, so they were omitted from the diagram to avoid confusion.

At the corrosion potential the contribution of each cathode reaction can be evaluated. Point 2 on the diagram shows the current density due to ferric

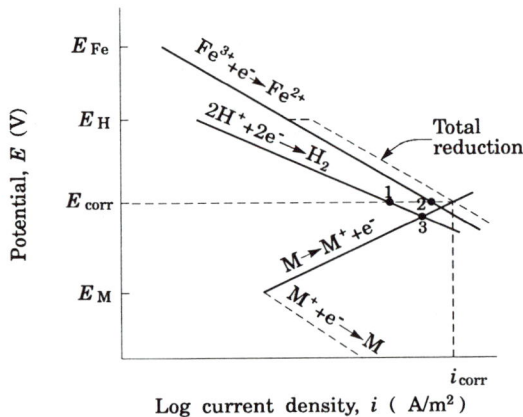

FIGURE 3-7.   Polarization diagram for corrosion of a metal in acidified ferric chloride. (1) H^+ reduction when Fe^{3+} is also present, (2) Fe^{3+} reduction, and (3) H^+ reduction in the absence of Fe^{3+}.

ion reduction and point 1 shows the contribution of the H^+ reduction. Note that fewer H_2 bubbles will be generated in acidified $FeCl_3$ (point 1) than if the metal were corroding in the acid alone (point 3).

Separate Anode and Cathode

When the anode and cathode are separate pieces of metal it is unlikely that they will be the same size. In this case the total anode current (not current density) equals the total cathode current and the polarization diagram must show the logarithm of the current, not current density. Figure 3-8 shows, in a general way, the polarization diagram for zinc and steel corrosion in the atmosphere that would result if the zinc coating on galvanized steel is damaged.

Both steel and zinc serve as cathodes, although most of the cathode reaction is occurring on the zinc. However, the corrosion potential of the couple is below the potential at which iron could corrode, so only zinc corrodes. In doing so it cathodically protects the steel.

It is worth noting also that for the uncoupled metals, the corrosion rate of zinc in the atmosphere is much less than for steel. Figure 3-8 would seem to suggest otherwise if you compare points 1 and 2, but because of zinc's much larger surface area, its corrosion current density is less than steel's.

- If the coating does not corrode slower than the base metal, why bother with a coating?

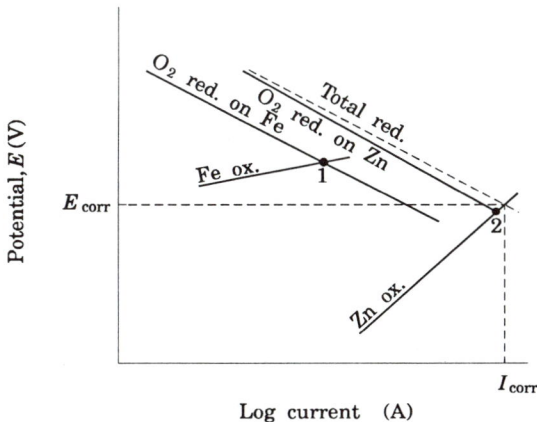

FIGURE 3-8.   Polarization diagram for corrosion of zinc and iron in the atmosphere. Point 1: Corrosion of iron not coupled to zinc. Point 2: Corrosion of zinc not coupled to iron.

Passivating Anode

For metals that form a passive film in their environments, the anode polarization curve takes on an S-shaped appearance, as shown in Figure 3-9. As the metal polarizes, the current density increases in the active region until a passive film forms, dropping the current density several orders of magnitude. A further increase in polarization potential has very little effect on the current density until the transpassive region is reached where the current density again begins increasing rapidly with increasing potential. In the transpassive region the metal ions in the passive film are being oxidized to a higher valence, which damages the film and reduces its protective character. Also, $O_2(g)$ may be generated on the anode surface at high potentials, damaging the film.

Figure 3-9 shows four cathode lines for increasingly oxidizing conditions. Line A will produce active corrosion, line C will passivate the metal, and line D will force it into the transpassive region. Line B, however, is not so clear-cut; it intersects the anode curve at three places: (1) in the passive region, (2) at the active–passive transition, and (3) in the active region. We can immediately dismiss intersection 2, which is unstable; the passive film is either rapidly forming or rapidly disintegrating at this point. If the film is in the process of forming, it moves the situation to intersection 1. If the film is disintegrating it quickly comes off and moves the intersection to 3.

Whether the corrosion is at point 1 or point 3 depends on the metal's prior history. If the metal has just been placed in the corrosive environment, its potential follows the anode curve as it polarizes from its equilibrium potential

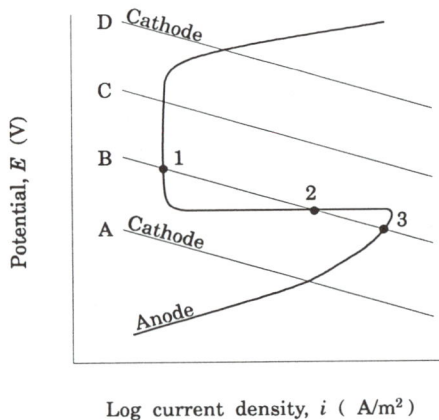

FIGURE 3-9.   Polarization diagram for an active–passive metal. Cathode lines A–D indicate increasingly oxidizing environments.

up to point 3. However, the environment is not quite oxidizing enough to get it over the active peak to the passive state, point 1. It stays at point 3, corroding severely, but never quite severely enough to form the passive film.

On the other hand, when a metal passivates under cathodic conditions C, it might remain passive at point 1 even after the environment becomes less oxidizing, dropping the cathodic condition to B. The metal might continue passivated at point 1 for some time, perhaps for days, until by accident or chance the passive film is damaged. Then, the film is swept off, the potential drops to point 3, and rapid active corrosion occurs. A change in environmental conditions from C to B, therefore, is insidious because of the danger of a later sudden, catastrophic increase in corrosion at a time when no changes are being made in operating conditions.

• Try not to get paranoid about this.

STUDY PROBLEMS

*3.1 Titanium, a passivating metal, corrodes severely in a molten salt bath when connected to a small piece of platinum but undergoes essentially no corrosion when connected to a large piece of platinum. Illustrate these two conditions on a single polarization diagram.

*3.2 An aluminum canoe corrodes at a very high rate in a certain swamp. (a) Draw the polarization diagram for the situation, labeling every line. (b) Indicate corrosion potentials and exchange current densities.

*3.3 An aluminum canoe corrodes at a very low rate in a certain lake. (a) Draw the polarization diagram for the situation, labeling every line. (b) Indicate corrosion potentials and exchange current densities.

*3.4 The maximum corrosion rate of aluminum in nitric acid at 25°C (77°F) occurs at 41% acid. Sketch the polarization diagram that would illustrate this situation.

*3.5 Draw polarization curves for corrosion of an active–passive metal, such as aluminum in dilute nitric acid, in which the reaction is diffusion controlled.

*3.6 Sketch polarization curves to illustrate a situation in which increasing the solution velocity decreases the corrosion rate.

*3.7 With the aid of a sketch, show clearly how corrosion of an active metal might *not* be affected by connecting it to a more noble metal in an electrolyte.

*3.8 (a) Sketch the polarization curves for corrosion of steel in the earth, assuming that the corrosion process is controlled by diffusion. (b) Label the curves completely. (c) Indicate how the corrosion rate can be found from the graph.

*3.9 With polarization curves, show how they could be used to describe the corrosion situation where iron coupled with nickel corrodes worse in dilute hydrochloric acid than iron coupled with lead. None of the metals passivates.

*3.10 The corrosion rate of copper in freshwater is almost directly proportional to the water velocity. Draw the polarization diagram to illustrate this situation.

*3.11 Show with a polarization diagram how decreasing the pH might decrease the corrosion of stainless steel in sulfuric acid.

*3.12 The Dome of the Rock, a Moslem mosque, is the most eye-catching building in Jerusalem. Its dome is capped by aluminum impregnated with gold. If Israel had a wet climate, a corrosion cell might be set up. Illustrate on one polarization diagram the situations for corrosion of aluminum with and without the gold.

*3.13 Draw a graph of polarization curves for the situation where nickel is corroded in an acid with a strong oxidizer added.

*3.14 Explain how you could calculate the amount of H_2 generated by the corrosion described in Problem 3.13.

*3.15 Sketch polarization curves to explain how an active–passive metal has extremely low corrosion in aerated solution, but in nonaerated solution the metal sometimes is satisfactory but sometimes corrodes very severely.

*3.16 If the Tafel slope is already known for the oxidation process, the corrosion rate can be determined from two measurements. What are they?

*3.17 Someone has said, "The Tafel equation is one of the very few solid links between thermodynamics and kinetics." Explain.

*3.18 Show how you could determine the corrosion potential of iron in river water if you knew the appropriate exchange current densities and Tafel slopes. In addition, specify which exchange current densities and Tafel slopes you would need.

*3.19 A nickel electrode has an exchange current density of 10^{-5} A/m^2 in dilute sulfuric acid. What is its oxidation rate at equilibrium, expressed as grams per square metre per day? The rate at which nickel ions plate out on the electrode at equilibrium?

*3.20 A copper electrode has an exchange current density of 2×10^{-5} A/m^2 in a sulfate solution. What is its oxidation rate expressed as grams per square metre per day? At what rate do copper ions plate out on the electrode at equilibrium?

*3.21 Compare the rates of hydrogen generation on platinum and on zinc at 25°C (77°F) in 1 M H_2SO_4 at equilibrium. Which is a better cathode metal?

4

Metallurgical Cells

Corrosion can change from uniform corrosion to a localized attack because of differences in the metal or because of variations in the environment. This chapter deals with the metallurgical variables and the types of corrosion resulting from them.

The important principle to remember is that corrosion attacks inhomogeneities in the metal. Real metals contain many inhomogeneities, most of them deliberately put in to achieve high strength. A perfectly pure metal crystallized in a perfect crystal structure would be both incredibly strong and highly corrosion resistant, but no such metal exists.

4.1 METAL PURITY

Commercially pure metals typically contain a few tenths of 1% of impurities. These small amounts of extraneous atoms do not appear to have much effect on corrosion in natural environments such as soil and water, but they do affect the protective nature of the oxide scale in atmospheric corrosion.

In steels, the normal impurities do not change the corrodibility of the metal if the aqueous environment is between pH 4 and 13.5, but in acids, the sulfur and phosphorus increase attack by making $H_2(g)$ generation easier. Copper in steel also increases corrosion in acids, but improves atmospheric corrosion resistance. The so-called "weathering steels" achieve exceptional atmospheric resistance by alloying with a few tenths of a percent of Cu, Ni, and Cr. Figure 4-1 illustrates a typical application of weathering steel.

In recent years it has been found that extremely high-purity metals are extraordinarily resistant to corrosion. For example, 99.998% Al corrodes at only one thirty thousandth the rate of commercial 99.2% Al.

FIGURE 4-1.   Bridge girder of high-strength, low-alloy weathering steel. (Courtesy of Charles Bradford.)

High-purity ferritic stainless steels now available are refined to extremely low carbon and nitrogen contents to give them a corrosion resistance that rivals the very best.

4.2 CRYSTAL DEFECTS

Although metals are crystalline the crystal structures are not perfect. In addition to impurity atoms, they all contain many atom vacancies, that is, sites where atoms should be present but are not. The vacancies permit some diffusion of atoms within the crystal. In addition, all real crystals contain linear flaws called "dislocations," where atoms are crowded too closely on one side of the line and are packed too loosely on the other side. These dislocation lines give the metal ductility but also increase diffusion and corrode more rapidly than the surrounding crystal, as shown in Figure 4-2.

Cold Work

Cold work (rolling, hammering, drawing, etc.) is usually done at room temperature, although it can be at any temperature below the metal's recrystallization temperature. While cold work strengthens the metal at very little cost, it greatly increases vacancy and dislocation densities. Impurities can then concentrate at the dislocations to create very localized corrosion cells that contribute to pitting. In acids, cold working can increase H^+ adsorption sites at dislocations on the metal surface. Cold work also introduces local internal stresses that make the metal more susceptible to stress-corrosion cracking. Figure 4-3 shows the surface of cold-worked brass after it has been polished and etched with acid.

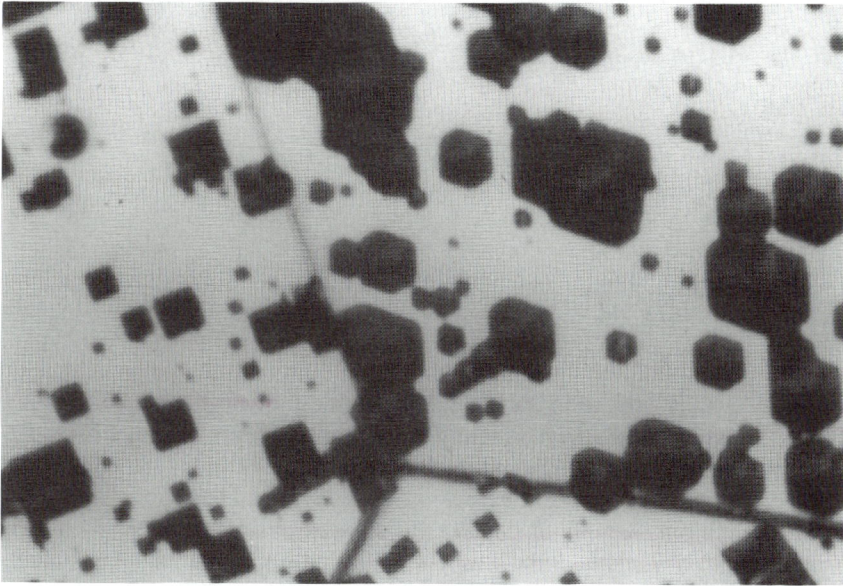

FIGURE 4-2.   Etch pits in 3% silicon steel. Ferric sulfate solution has eaten out the distorted structure around dislocation lines where they intersect the metal surface. Magnification 2750×. [Reprinted with permission from *Metals Handbook*, 8th ed., Vol. 8, ASM International, Materials Park, OH, 1973, Fig. 3056, p. 113.]

FIGURE 4-3.   Corrosion of brass cold-worked 60%. Magnification 320×.

4.3 GRAIN STRUCTURE

Grain Size

The metal "grains," the individual crystals, should be extremely small to give the best toughness, but that means that the grain boundaries, which are narrow regions of mismatch between the grains (see Fig. 4-4), will be numerous and take up an appreciable fraction of the metal surface.

Atoms at the grain boundaries are easily corroded because they are not bonded as strongly as atoms within the grains. More importantly, the grain boundaries serve as collecting sites for impurity atoms that do not fit well inside the crystals. At moderate temperatures diffusion is much more rapid along the boundaries than within the grains, so atoms collect more rapidly and form precipitates at the boundaries. All these inhomogeneities localize the corrosion attack.

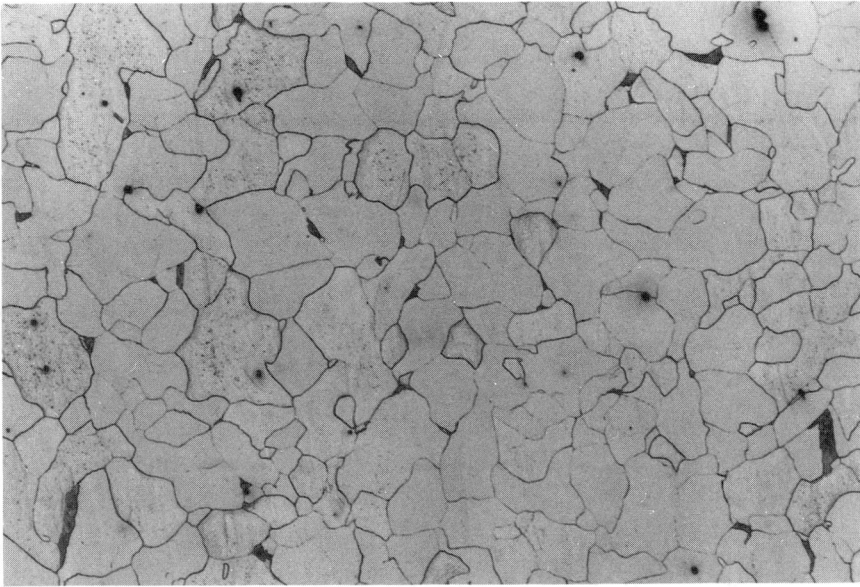

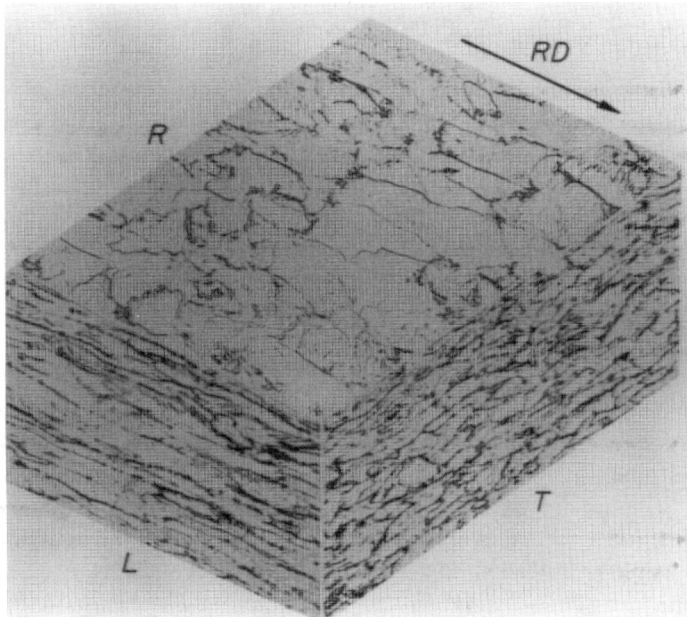

FIGURE 4-4. Photomicrographs of metal grain structures. (*a*) Grain boundaries in annealed iron (200×). (*b*) Rolled surface (*R*), longitudinal (*L*), and transverse (*T*) sections of low-carbon steel cold-rolled 65% (490×). *RD* is the rolling direction. [Reprinted with permission from J.T. Michalak *Metals Handbook*, 8th ed., Vol. 8, ASM International, Materials Park, OH, 1973, Fig. 3373, p. 220.]

Grain Shape

Cold work severely distorts the shape of the metal grains; rolling, for example, flattens the grains and elongates them in the rolling direction as in Figure 4-4b. A transverse cut of the same metal shows that the grains have been particularly flattened in the vertical, or "short transverse" direction, but the grain boundaries are not lined up the way they are in the long transverse or the longitudinal (rolling) directions.

A transverse cut through cold-worked metal exposes much more grain boundary area than a longitudinal cut does. Consequently, corrosion attack on the grain boundaries in a transverse cut can be much greater. Also, cold work tends to align certain crystal directions with the direction of working and forces more of the closely packed atomic planes to lie parallel to the metal surface. The close-packed planes have the strongest bonding between atoms and, consequently, are more corrosion resistant than planes perpendicular to them that would predominate in a transverse cut.

High Temperatures

Hot working (shaping the metal above its recrystallization temperature) anneals the metal as rapidly as it deforms so the grains stay fairly uniform in shape and orientation, and have a fairly fine grain size on the surface. Any slag inclusions get stretched out in stringers parallel to the working direction, creating glassy, noncorroding areas.

Castings normally solidify from the outer walls inward, with the grains elongated perpendicular to the surfaces. Impurities do concentrate in the grain boundaries, and, in addition, alloys may undergo some "coring," where the various metals in the alloy partially separate during freezing. This procedure can produce adjacent grains with somewhat different chemical compositions, and thus different corrosion potentials. Castings should be annealed to homogenize them before they are put into corrosion service; unfortunately, this is often not done.

Welds are particularly inhomogeneous (Fig. 4-5). Filler metal is usually chosen to be slightly cathodic to the base metal to prevent preferential attack of the weld. However, as the weld solidifies, the center of the weld cools more slowly than the edges, giving the center a coarser grain size. The heat-affected zone (HAZ) just adjacent to the weld does not melt but does recrystallize to a very fine grain size if it has been cold worked. The differences in grain size combined with the high-temperature diffusion of impurities to the grain boundaries concentrate corrosion attack on the HAZ.

FIGURE 4-5.   Variation in grain size and shape across an arc weld in pressure-vessel steel plate (7×). [Reprinted with permission from *Metals Handbook*, 8th ed., Vol. 8, ASM International, Materials Park, OH, 1973, Fig. 2983, p. 77.]

4.4 SOLID SOLUTION ALLOYS

Interstitial solid solution, in which small nonmetallic atoms (H, C, N, etc.) squeeze into the spaces between the larger atoms of the metal crystals, profoundly affects the mechanical properties of metals. The most notable of these alloys is the solution of carbon in iron to make steel. Surprisingly, interstitial atoms do not affect the corrosion resistance of commercial alloys greatly.

- It is true that high-carbon steel corrodes a little more than low-carbon steel in seawater, but the difference is slight; all plain-carbon steels corrode at about 0.1–0.2 mm/y (5–10 mpy) in quiet seawater. Rates like these mean that regular painting and cathodic protection are necessary to preserve a ship indefinitely, such as the Queen Mary.

Substitutional solid solution, such as chromium atoms replacing some of the iron to make a stainless steel, may greatly improve the corrosion resistance. Alloying a third element, nickel, can improve it even more. When enough alloying atoms replace atoms of the host crystal, the substituting metal imparts some of its own properties to the alloy.

• Alloying 12% Cr with iron makes a stainless steel. But 6% Cr is not one-half as good. With anything less than 10% Cr the iron alloy corrodes much like ordinary carbon steel, because it does not passivate with a complete Cr_2O_3 film.

4.5 GALVANIC CORROSION

Galvanic, or two-metal, corrosion is the best known of all types of electrochemical corrosion. Experienced engineers are well aware that connecting dissimilar metals in a corrosive environment will cause rapid destruction of one of the metals, so they do not often make that mistake, unless the corrosive environment shows up uninvited.

The Galvanic Series

The standard emf series (Table 2-1) lists the reactivity of pure metallic elements in solutions of their own ions, but real corrosion situations are never that ideal. Corrosion can be analyzed much better with a practical "galvanic series" that lists real commercial alloys in an identical, real environment. Table 4-1 shows a galvanic series for seawater. Potentials are not given because they change with time, and because they depend on the amount of dissolved O_2, salts, and so on in the water. (Not even seawater is homogeneous!)

Note that the order of metals in the galvanic series is not exactly the same as in the standard emf series; for example, compare the positions of aluminum and zinc. Also, observe that most of the stainless steels may be either passive or active in seawater. It is this instability of the passive film in chloride solutions that makes stainless steels so susceptible to crevice corrosion.

The galvanic series in other environments will be slightly different than that for seawater. For example, if aluminum is connected to galvanized steel, the aluminum would be the anode in acids but the cathode in basic solutions, or in seawater.

TABLE 4-1   Galvanic Series in Seawater

Noble end	Graphite
best *corrodable* *cathode*	Platinum
	Hastelloy C
	Titanium
	Alloy 20 stainless steel
	316, 317 Stainless steel, passive
	Monel Cu-Ni alloys
	302, 304, 321, 347 Stainless steel, passive
	Silver
	Nickel
	Silver solder
	Mill scale on steel
	Lead
	430 Stainless steel, passive
	90Cu-10Ni
	410, 416 Stainless steel, passive
	Bronzes
	Pb-Sn Solder
	Copper
	Tin
	Brasses
	316, 317 Stainless steel, active
	Nickel cast iron
	302, 304, 321, 327, 410, 416, 430, active
	Low-alloy steel
	Mild steel, cast iron
	Aluminum alloys
most corrodable -anode	Zinc
Active end	Magnesium

Coupling two metals that are very far apart in the galvanic series makes a corrosion cell if they are put in an electrolyte, such as seawater. The more noble metal is the cathode and the active metal corrodes as the anode. However, for metals that are near each other in the galvanic series, if their potential difference is small then corrosion may not be too severe.

• Would you advise using lubricants containing graphite aboard seagoing aluminum boats? Check the galvanic series.

The metals in Table 4-1 are ranked according to their corrosion potentials, but the corrosion current (or corrosion rate) that flows between two coupled metals is not shown. The corrosion rate depends on not only how active (corrodible) the anode metal is, but also on how readily the cathode reaction proceeds on the surface of the more noble metal.

The Area Effect

Cathode reactions are usually much more sluggish than anode reactions. For reaction at the cathode surface, first a surface site must become vacant, then a nearby reducible ion or molecule must adsorb at the empty site, one or more electrons must be transferred from the metal to the adsorbed species, and usually the now-reduced species must desorb to make room for more reactant. Consequently, corrosion cells do not work well if the cathode area is so small that few surface sites are available.

The cathode/anode area ratio is a critical factor in determining how severe the corrosion will be in a galvanic cell. A large cathode intensifies attack on a small anode. Therefore, if galvanic corrosion cannot be avoided, it may be made negligible if the cathode/anode area ratio is small enough.

- Steel plates fastened with copper rivets corrode in seawater at an acceptable rate (see Fig. 4-6), but copper plates fastened with steel rivets concentrate all attack on the rivet heads. Failure would be rapid and sudden.

Consequently, an important rule to remember for galvanic corrosion is never to paint the anode metal without also painting the cathode. It seems logical to paint only the metal that will corrode, but this assumption is not correct. Paint will reduce the anode area to microscopically small regions at the bottom of the pores in the paint film. Couple these little anode areas to an unpainted cathode that is thousands of times larger and either holes will be drilled through the anode metal or, more likely, rust will quickly form under the paint and strip it off.

- A large warehouse was constructed with a painted structural steel frame and aluminum panels for walls and roof. Condensation on the cold metals loosened the paint and caused rusty water to drip on the contents stored below. The warehouse owner demanded that the steel be sandblasted and repainted properly, while the contractor insisted that the original paint had been applied correctly. They called in a corrosion expert who told them to paint the aluminum panels. This finding seemed ridiculous, because the aluminum was not corroding. However, it stopped the galvanic corrosion of the steel by reducing the cathode/anode area ratio.

The Distance Effect

In environments with low electrical conductivity, such as freshwater, galvanic corrosion occurs only near the connection point where the two metals make contact. The high resistivity of the solution confines the cell current to the

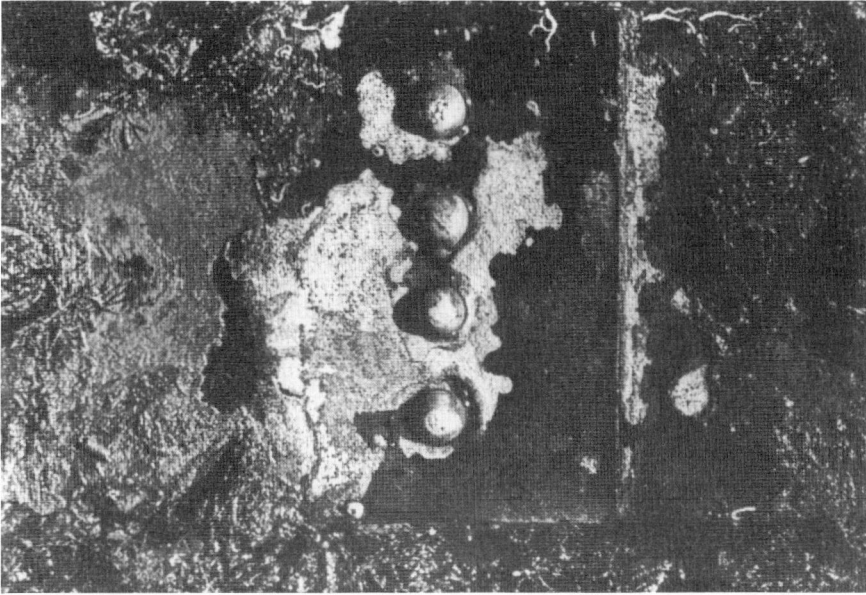

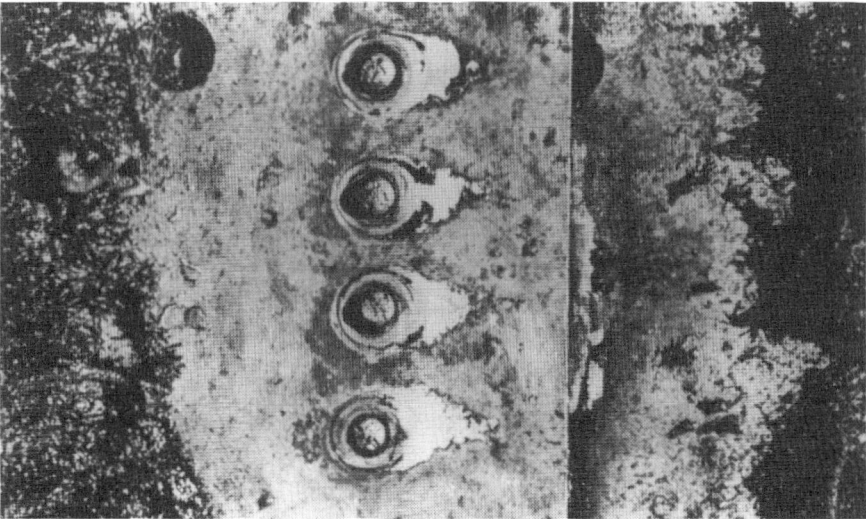

FIGURE 4-6. The area effect in galvanic corrosion. (*a*) Steel plates with copper rivets in seawater 15 months. (*b*) Copper plates with steel rivets, same environmental conditions. [From *Corrosion in Action*, The International Nickel Company of Canada, Ltd., Toronto, 1955, Fig. 41, p. 18. Courtesy of Inco Alloys International.]

region around the junction, causing a deep ditch to form in the anode metal where it touches the cathode. Attack in a narrow band on only one side of a metal junction confirms a diagnosis of galvanic corrosion. This "distance effect" offers another way to combat galvanic corrosion: space anode and cathode far enough apart and galvanic corrosion will virtually cease even though the metals are still electrically connected by an external conductor. In effect, that is what painting both the anode and the cathode does, by increasing the electrical resistance through the fine, winding paint pores.

- Many old houses with steel water pipes have been modernized with additional plumbing made of copper. To prevent galvanic corrosion the plumbers often put insulated connectors between the two kinds of piping (Fig. 4-7). Building codes, however, require the plumbing to be electrically continuous for grounding purposes so electricians fasten external metal straps across the insulated couplings. Surprisingly, this procedure does not cause severe galvanic corrosion. The insulated spacer between the two pipes separates them enough so that the water's resistance prevents the exchange of much current. (Also, old rusty steel is only about 0.2 V more reactive than copper so the driving force is not great.)

Preventing Galvanic Corrosion

In addition to all the usual techniques used to slow or stop corrosion, the following is a list of some of the ways to combat galvanic corrosion specifically.

1. Make all components out of the same metal.
2. Select metals close to each other in the galvanic series.
3. Electrically insulate the two metals from each other.
4. Separate the two metals with a middle piece that can be replaced or has a potential intermediate between the other two.

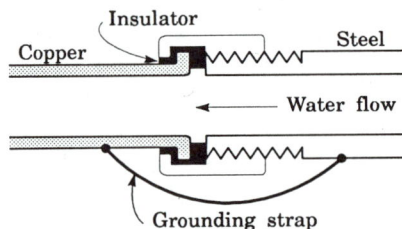

FIGURE 4-7.   Insulated connector between steel and copper pipes, shorted by grounding strap.

5. Do not apply a coating with any porosity to the more corrodible metal in the couple without also coating the more resistant metal.
6. Avoid a small anode connected to a large cathode.
7. Extend the distance between the metals in the electrolyte.
8. Avoid contaminating one metal with the corrosion products of another (e.g., $Al + Cu^{2+}$).
9. Use galvanic corrosion to protect the cathodic metal.

This last point needs explaining. Galvanic corrosion can sometimes be exploited to provide cathodic protection for critical parts.

- In heat exchangers the thin-walled tubes, made of corrosion-resistant metals, such as copper or bronze, are often connected to heavy steel tube sheets. The steel corrodes badly, of course, but protects the tubes which must not leak. In this situation the tubes serve only as cathodes because electrons are streaming into them from the steel.

*Mixed Potential Approach to Galvanic Corrosion

A polarization diagram can illustrate how galvanic corrosion increases the corrosion rate of an active metal, as shown by Figure 4-8.

Metal A is reactive, and by itself will corrode at a rate proportional to I_A. Metal B, which is quite noble, corrodes only at a rate proportional to I_B if not coupled to A. When A and B are connected, the cathode reaction can take place on both metals (indicated by the dashed line labeled "Total red"), but in actuality almost all the cathode reaction is occurring on B.

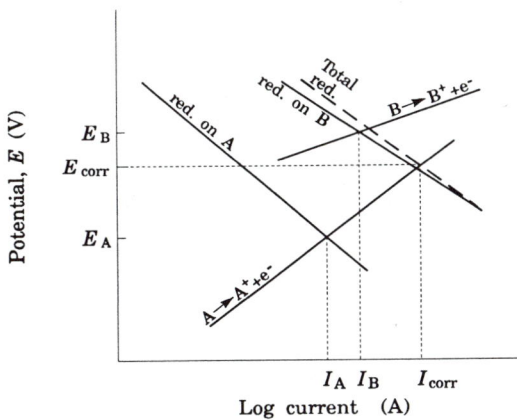

FIGURE 4-8.   Polarization diagram of galvanic corrosion for an A–B couple.

The galvanic couple corrodes at a rate proportional to I_{corr} at the potential E_{corr}. Note that at this potential metal B cannot corrode at all; the line for oxidation of B does not extend down to E_{corr}. Only metal A corrodes, and at a rate much greater than if it were not fastened to B. That is, $I_{corr} \gg I_A$.

The increase in corrosion potential caused by galvanic coupling can occasionally be put to good use if it helps passivate the reactive metal. Figure 4-9 shows such a situation. Uncoupled, the corrosion of metal A is proportional to I_A. Coupling to metal B raises the corrosion rate of A to I_B, but coupling to C passivates A and lowers its corrosion to I_C. Titanium can be passivated in this manner by coupling it to platinum in seawater.

The area effect can also be illustrated by a polarization diagram. Figure 4-10 shows metal A corroding alone (corrosion current I_A), and when connected to a small cathode B (corrosion current I_B, or to a large cathode B' (current I'_B,). The only difference between the reduction reactions on B and B' is the much larger exchange current on the much larger cathode.

Observe that the relative amount of anodic or cathodic control of the corrosion process is indicated by the corrosion potential. As the cathode reaction becomes easier, and the corrosion is less likely to be limited (controlled) by the cathode reaction, the corrosion potential increases.

- Nothing is more sensitive to contaminating ions than a beer-drinker's palate. Steel fermentation vats at one brewery were coated inside with baked phenolic, which was often damaged when sediment was scraped out. Replacement vats had bare stainless steel bottoms welded to phenolic-coated, carbon steel sidewalls. This large cathode/small anode arrange-

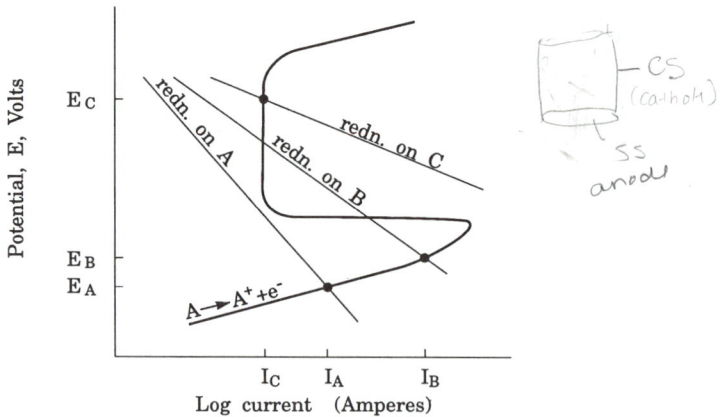

FIGURE 4-9.   Polarization diagram of a passivatable metal galvanically corroded if coupled to metal B but passivated if connected to metal C.

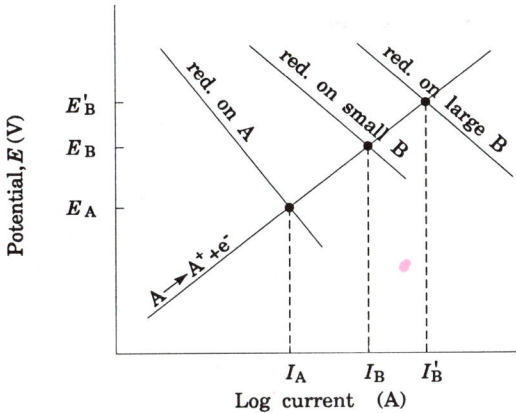

FIGURE 4-10. Effect of galvanically coupling a corrodible metal (A) to cathodes of various sizes.

ment perforated the steel in a 2-in. band above the stainless steel (the distance effect). What was worse, bitter ferrous ions contaminated the beer.

4.6 DEALLOYING

The corroding of one metal from an alloy, while leaving the other components in place, is called dealloying. This damage is a type of galvanic corrosion on a microscopic scale that can occur in alloys of elements separated widely in the galvanic series. It leaves the alloy in a very weakened condition so that stresses can easily initiate stress-corrosion cracking, or corrosion fatigue, or even simple mechanical fracture due to overloading.

Dealloying is also called "selective leaching" or "parting." The process occurs only with certain alloys (described later), only under certain conditions, and is usually very slow (years). Because of the possibility of sudden fracture, dealloying can be a very serious problem.

Dezincification of Brass

Brass is an alloy of copper (noble) with up to 40% zinc (very reactive). With brasses containing greater than 15% Zn, zinc atoms can corrode out and leave the copper behind. The metal color changes from yellow to the pinkish color of copper, so the type of corrosion is unmistakable. Dezincification proceeds from the metal surface inward. Attack may be uniform over the surface, or it may be localized, especially under a deposit.

Stagnant solutions, high Cl^- concentrations, high O_2 or CO_2, and deposits of any kind (lime, bacterial colonies, dirt, seaweed, etc.) provide excellent conditions for dezincification. An example of dezincification is shown in Figure 4-11.

- Dezincification of brass screws in wet wood, the usual situation with boats, can let the screws break suddenly. They always break under high stresses in rough waves, of course.

The mechanism in most cases involves the simultaneous dissolution of both zinc and copper ions in the brass, immediately followed by reduction and replating of the copper. The resulting structure has the same external form and dimensions as the original brass; screws still have their slot and their threads, for example. But the metal has almost no strength because 15–40% of its atoms are missing. It has no ductility and fractures under stress rather than deforming.

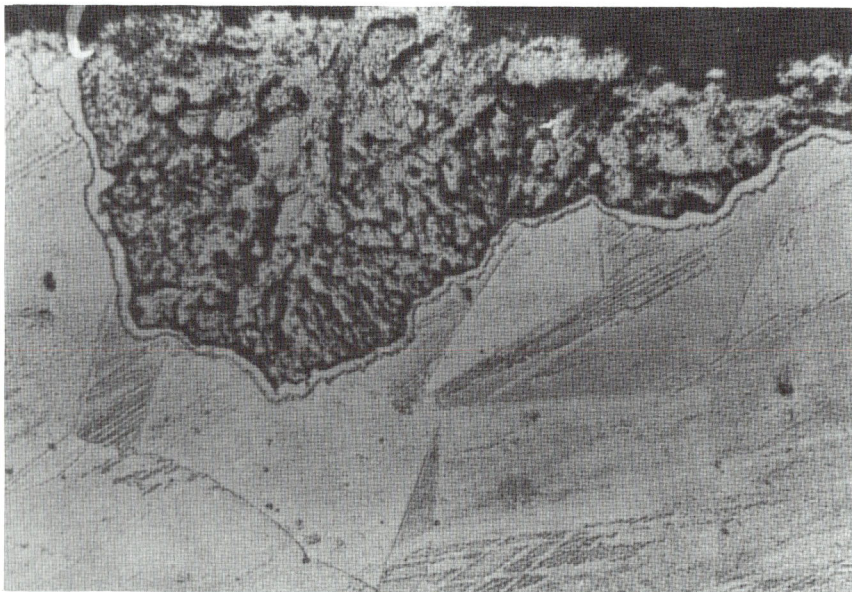

FIGURE 4-11.   Dezincification of 70Cu-30Zn cartridge brass in $1M$ NaCl, 79 days. [Reprinted with permission from C. P. Dillon, ed., *Forms of Corrosion: Recognition and Prevention*, NACE Handbook 1, National Association of Corrosion Engineers, 1982, Fig. 7.5.2.1, p. 101. Copyright © by NACE.]

Dealloying of Other Alloys

A number of copper alloys besides the brasses may suffer dealloying. Among them are aluminum bronze (90Cu-10Al), which sometimes undergoes "dealuminification" in acids or seawater, and occasionally the cupronickels (10–30% Ni), tin bronzes, and silicon bronzes.

Alloys of gold or platinum with more reactive metals, such as silver, nickel, or copper, may be dealloyed in strong oxidizing acids. High-nickel alloys, although extremely corrosion resistant, occasionally undergo dealloying of Cr, Fe, Mo, or W in molten salts.

Graphitic Corrosion of Gray Cast Iron

Gray cast iron is an ancient alloy that has been used for centuries for underground water pipes, even though it is extremely brittle. Its microstructure consists of a matrix of pearlite or ferrite, usually, with graphite flakes distributed throughout (see Fig. 4-12).

In saltwater, dilute acids, and some soils, especially those with sulfate-reducing bacteria, the iron is slowly corroded by galvanic action with the

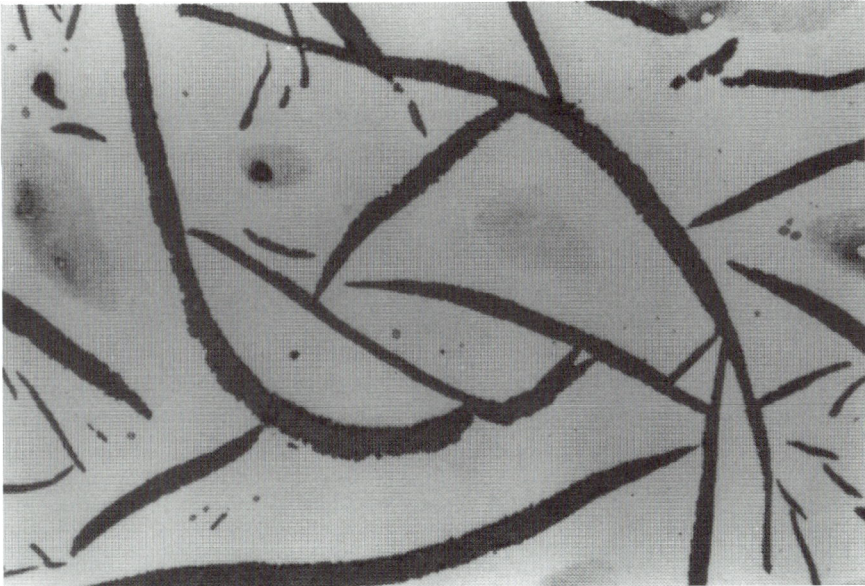

FIGURE 4-12.   Microstructure of gray cast iron showing graphite flakes (160×). [Reprinted with permission from J. C. Moore in *Metals Handbook*, 8th ed., Vol. 8, Taylor Lyman, Editor. ASM International, Materials Park, OH, 1973, Fig. 2993, p. 87.]

graphite until the microstructure becomes rust with the graphite flakes still distributed as before. This structure is so weak it can be cut with a knife (Fig. 4-13).

• "Graphitized" pipe looks like rusty pipe. Its shape is unchanged; even the manufacturer's trademark cast into the pipe can still be read. If it has not been broken, the pipe can still carry water at internal pressures of up to around 3.5 MPa (500 psi), and packed in soil it may give good service. But being weak and brittle, it will break from frost heave, from water hammer, or any other mechanical stress. Many a building has burned to the ground because the sudden change in water pressure for fire fighting has broken the water mains.

At the present time most of the major cities in North America have come to the realization that they must rapidly replace their cast iron water and sewer systems at a cost that will run into the billions of dollars.

• One test for graphitic corrosion is the famous Clang and Clunk Test: Hit the pipe with a hammer. If the metal goes "clang," it is good. If it goes "clunk," or breaks, it is not.

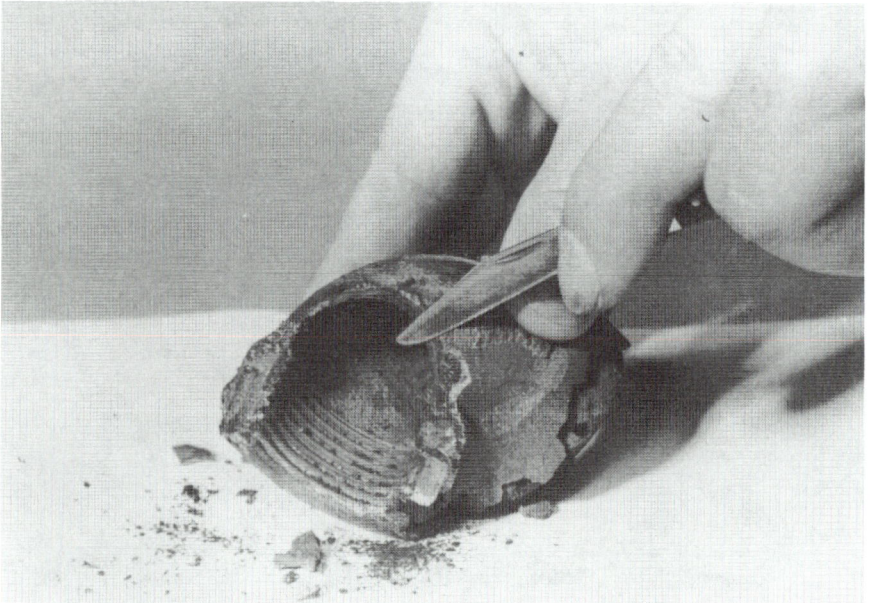

FIGURE 4-13.   Graphitic corrosion of cast iron elbow. [From *Corrosion in Action*, The International Nickel Company of Canada, Ltd., Toronto, 1955, Fig. 6. p. 6. Courtesy of Inco Alloys International. General Supervision by F. L. LaQue]

4.7 INTERGRANULAR CORROSION

Galvanic corrosion at the grain boundaries, although on a microscopic scale, is caused by compositional differences between the grain boundaries and the metal adjacent to the boundaries. The boundaries are collection sites for a great number of ill-fitting atoms that would distort the crystals if left inside, as indicated in Figure 4-14. Diffusion is rapid along the boundaries so that new phases can form rapidly, and the new phases do not require additional energy to produce new surfaces as they would within the crystals.

Sensitization of Stainless Steels

Stainless steels held in the temperature range of about 500–800°C (900–1500°F) often "sensitize" so that they become very susceptible to intergranular attack in certain corrosive environments. The surface then will just crumble (sometimes called "sugaring") as the individual grains fall out. The metal also becomes very susceptible to stress-corrosion cracking and corrosion fatigue. A typical example is shown in Figure 4-15.

Much of the carbon in a stainless steel resides in its grain boundaries where it can react with the chromium in the alloy. At low temperatures diffusion is too slow to allow much reaction, but at around 500°C (900°F) chromium begins diffusing into the grain boundaries to precipitate the carbide $Cr_{23}C_6$ within the boundaries.

These precipitates themselves have no appreciable effect on corrosion, but a narrow area next to the boundaries has been seriously depleted of chro-

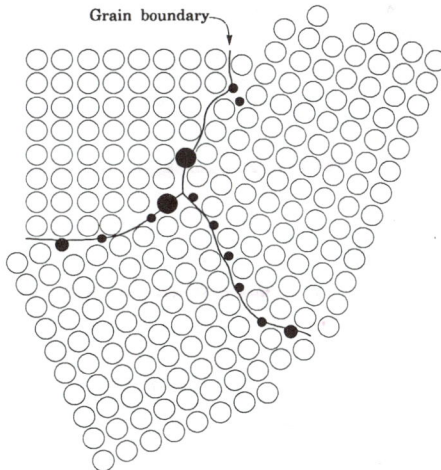

FIGURE 4-14.   Schematic diagram of grain boundaries separating three metal crystals. (Dark atoms are impurities.)

FIGURE 4-15.   Sensitized 304 stainless steel elbow exposed to 30% nitric acid at 90°C (200°F). [Reprinted with permission from R. James Landrum, *Fundamentals of Designing for Corrosion Control*, National Association of Corrosion Engineers, 1989, Fig. 3.74, p. 147. Copyright © by NACE.]

mium, thus setting up a microscopic galvanic cell between the grains, which contain perhaps 18% Cr, and the boundary regions where the chromium content may approach zero.

- Stainless steel heat exchanger tubes were softened with a propane torch to bend them. Within a few weeks of operation they all failed at the bends. *Cause:* Charging with carbon plus heating in the critical temperature range for sensitizing.

  The problem frequently occurs when austenitic stainless is welded. The weld will cool rapidly but the heat affected zone (HAZ) may sensitize, especially if the plate is thick, because the HAZ cools through the sensitizing temperature range slowly. An intergranular corrosion attack on the sensitized HAZ is often referred to as "weld decay."

- All zinc is supposed to be removed from the weld area of galvanized steel before welding, to ensure both a sound weld and a healthy welder, but anyone who knows welders knows that the removal may be pretty slipshod. If a weld of galvanized steel to stainless steel is contaminated with zinc, all grain boundaries around the weld become loaded with zinc, ensuring intergranular corrosion at horrendous rates. An explosion and fire that destroyed an entire chemical plant was traced to the welding of galvanized steps to the side of a storage tank.

Prevention of Sensitization

First, it should be noted that some environments do not cause intergranular corrosion even if the stainless steel is sensitized. For these environments, notably high-temperature gases, no special precautions need to be taken in welding, stress relieving, and the like.

To prevent sensitization of stainless steels:

1. Perform any heat treatment above the sensitization temperature, followed by quenching or cooling as rapidly as possible.
2. Clean metal surfaces of all oil, grease, or paint before heating or welding.
3. Avoid welding, if practical.
4. Arc weld; never use an acetylene torch.
5. When casting stainless, do not use molds with oil or polymer binders.
6. Use low-carbon "L" grades, such as 304L, that contain 0.03%C max, and thus practically never sensitize.
7. Use stabilized grades of stainless steel, such as type 321 containing Ti, or type 347 containing Nb (Cb) and Ta. These strong carbide-forming elements tie up all the carbon as TiC or NbC, leaving none for Cr to react with.

Knife-Line Attack

Welded, thin sheet of *stabilized* stainless steel occasionally suffers corrosion attack right at the fusion line on both sides of the weld. The attack is so sharply localized that it looks as if the weld were cut out with a knife.

The TiC and NbC form only in the temperature range of 870–1150°C (1600–2100°F); at higher temperatures they decompose. The stabilizing carbides dissolve as the metal melts, but the cooling may be so rapid after welding that they have no time to re-form. Instead, the $Cr_{23}C_6$ carbides form at lower temperatures, thus sensitizing the stainless steel along the fusion line.

The cure is to reheat the welded sheet to approximately 1065°C (1950°F), where $Cr_{23}C_6$ decomposes and the NbC or TiC forms. After the chromium has had time to diffuse back into the grains, the cooling rate is unimportant. The low-carbon grades do not suffer knife-line attack and thus can be substituted for stabilized grades if necessary.

End-Grain Attack

In extremely corrosive environments a metal sheet or pipe may be resistant on all surfaces parallel to its rolling or extrusion direction but suffer deep attack on transverse surfaces. The rolling has lined up nonmetallic inclusions into thin stringers, which are only exposed to the corrosive at the ends of the plate or pipe. Microscopic galvanic cells are set up between the stringers and the surrounding metal, resulting in deep pits into the cut ends. Figure 4-16 shows a badly corroded example of end-grain attack.

Stainless steels are especially susceptible to end-grain attack because the metal may go transpassive down inside the pits, setting up an even more serious galvanic cell.

- A stainless steel distillation column for concentrated nitric acid showed little corrosion except at the ends of riser pipes and bubble caps, where end-grain corrosion had eaten its way in. In fact, many of the bubble caps entirely disappeared.

FIGURE 4-16.   End-grain attack on a carbon steel valve handle in SO_2 fumes.

End-grain attack can be prevented by a seal weld on transverse-cut surfaces. Melting the end grain is sufficient; adding filler metal is unnecessary (see Fig. 4-17).

- In a paper mill a drum filter had a screen made of copper alloy sheet with many small holes. Although the sheet thickness remained unchanged, the screen had to be scrapped when the round holes enlarged into ellipses in the direction that the sheet had been rolled.

Exfoliation

Exfoliation, also called lamellar or layer corrosion, is a type of intergranular corrosion that causes the metal to swell up like flaky pastry. When the metal grains have been highly elongated in platelet shape by cold work, and the grain boundaries are electrochemically different from the grains, exfoliation can occur.

The problem occurs most often with high-strength aluminum alloys but is also observed with the cupronickels. Extrusion can build up a high dislocation density at the grain boundaries as it elongates them, making the boundary regions more corrodible than the bulk of the grains. Age hardening, especially underaging, creates precipitates in the boundaries that make the adjacent metal more anodic.

Intergranular corrosion produces voluminous solid corrosion products that split the metal grains apart and force them outward to leave a laminated,

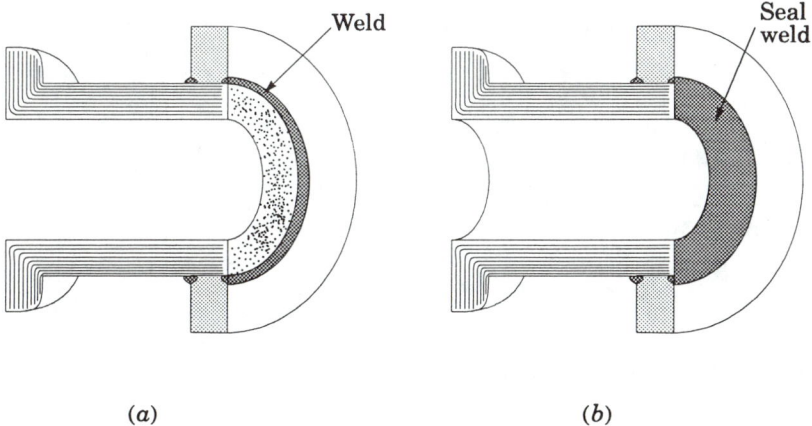

Weld

Seal weld

(a)   (b)

FIGURE 4-17.   Sketch of a seal weld to prevent end-grain attack.

flaky, or blistered appearance that is easily detected in its early stages. The splitting of the metal at the grain boundaries is not increased by applied stress and does not lead to stress-corrosion cracking.

Exfoliation can be prevented by slightly overaging aluminum alloys to give a more uniform distribution of precipitate throughout the grains. A recrystallization anneal will produce equiaxed grains that are not susceptible to exfoliation.

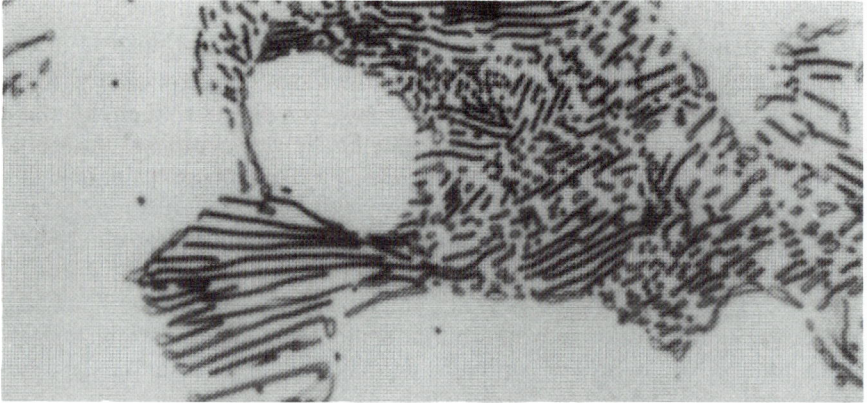

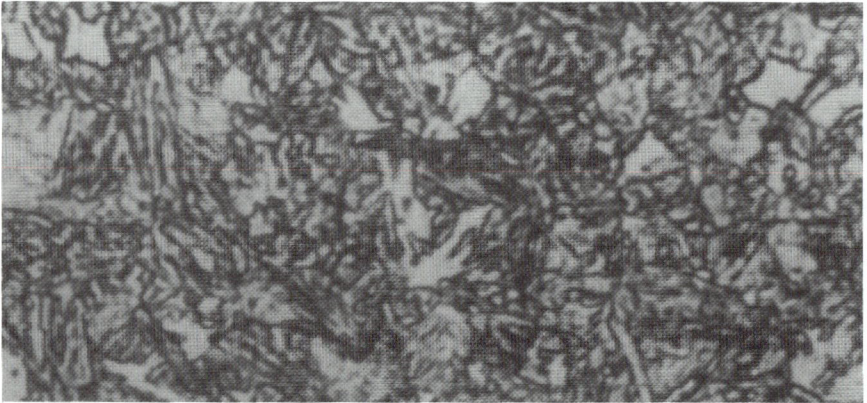

FIGURE 4-18 (a,b).   Microstructures of steel. (*a*) Pearlite, 2350×; (*b*) tempered martensite, 1310×; and (*c*) spheroidite, 2380×. [Reprinted with permission from *Metals Handbook*, 8th ed., Vol. 7, ASM International, Materials Park, OH, 1972, Figs. 217, 220, 232, pp. 30, 31. Taylor Lyman, Editor.]

4.8 CORROSION OF MULTIPHASE ALLOYS

Alloys with two or more phases are generally not as corrosion resistant as solid solution alloys. Since each phase has a different corrosion potential, distinct microscopic anodes and cathodes are established on the metal surface.

Pearlitic steel (Fig. 4–18*a*), the usual form of steel if not heat treated, is a good example of a two-phase alloy. Consisting of alternating layers of nearly pure ferritic iron and the more noble iron carbide, the steel rusts preferentially at the ferrite.

Heat treatment to strengthen alloys commonly produces a finely divided second phase dispersed throughout the crystal structure of the primary phase.

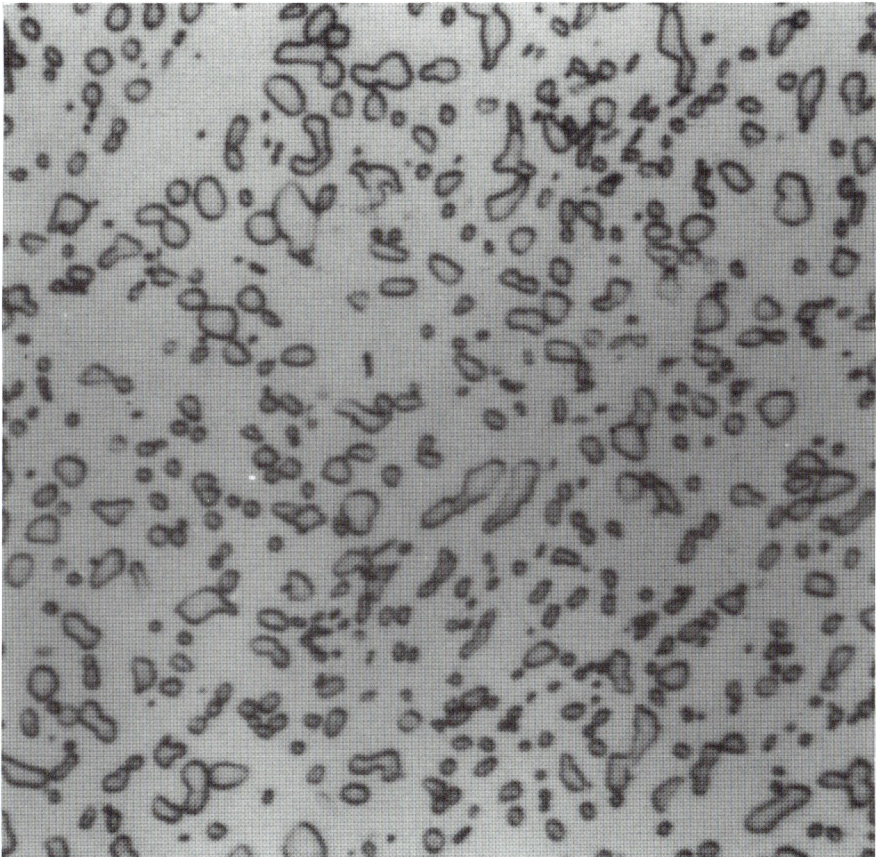

FIGURE 4-18 (c).   (*Continued*)

While very small precipitates do not offer much surface area for the cathode or anode process, still they do increase the corrosion rate above the single-phase, nonheat-treated alloy.

Distribution of a second phase varies with the way the metal is shaped: rolled, cast, extruded, or whatever, and also with the temperature where the second phase forms. The most serious situation that can develop is formation of the anodic phase in a continuous network, as sometimes happens at the grain boundaries. Preferential corrosion along the network creates sharp notches that intensify stresses to initiate a crack.

Ringworm Corrosion

Heat-treated steel pipes and tubing sometimes corrode in a band or ring around the pipe near the threaded or flanged ends. The pipe has been heat treated to tempered martensite (Fig. 4-18*b*) or bainite, which are reasonably corrosion resistant with extremely fine second-phase precipitates. In addition, the ends of the pipes have been reheated for upset forging to flange

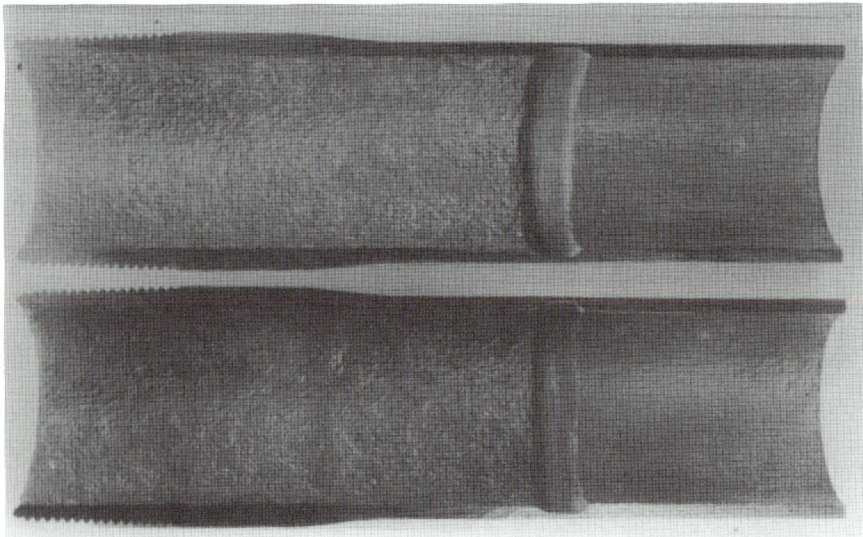

FIGURE 4-19.   Ringworm corrosion of oil well tubing. [Reprinted with permission from *Corrosion Control in Petroleum Production*, TPC Publication No. 5, National Association of Corrosion Engineers, 1979, Fig. 2.19, p. 15. Copyright © by NACE.] NACE Group Committee T-1 on "Corrosion Control in Petroleum Production."

them or to thicken them for threading. These ends are normalized to a very fine pearlite, still reasonably corrosion resistant.

At a short distance from the ends, however, the metal became only hot enough to transform the structure to spheroidite (Fig. 4–18c). Spheroidite has a ferrite matrix with large spheres of iron carbide (large cathode areas). The resulting corrosion confines itself to the spheroidite region and eats out the ring around the pipe. Figure 4–19 shows an example of oil well tubing with ringworm corrosion.

FIGURE 4-20.   Variation in corrosion of a stainless steel spring from a sulfuric acid pump.

Regions of local stress

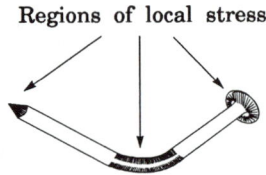

FIGURE 4-21.   Anodic regions revealed in the corrosion of a nail. The shank acts as a cathode.

4.9 STRESS CELLS

An electrochemical cell develops locally where a metal is stressed. Elastically strained metal is anodic to annealed metal and, in addition, the stress may damage protective film. An example is the spring in Figure 4–20 where the center of the spring, being strained the most, also corroded the most.

Plastic deformation produced by cold work increases dislocation density greatly as well as leaving residual internal stresses in the metal. Corrosion attacks the deformed surfaces, although the corrosion potential may not shift greatly (−40 mV is typical) because both anodic and cathodic processes are stimulated. Figure 4–21 shows localized corrosion of the cold-formed head and point of a nail.

- Stolen automobiles often have their engine numbers ground off the block and different numbers stamped on. Police labs treat the numbers with an acid that brings back the original numbers by corroding the cold-worked metal under the original stamping.

STUDY PROBLEMS

4.1   A water well was designed with a stainless steel screen on the bottom end of a carbon steel pipe that was primed and coated near the connection. Explain how corrosion will be affected.

4.2   Rivets of Al–5% Mg have fractured in a ship superstructure made of aluminum alloy. All metal was painted, but corrosion is assumed to have played some role in the failures. It is suggested that the rivets be replaced with stainless steel rivets of the same strength. Comment.

*4.3   Draw polarization curves showing how increasing the size of the cathode in a galvanic corrosion cell would affect the corrosion rate.

4.4   Plaster board was nailed on the walls of a new house, with a nail piercing a copper water pipe in the wall. Two years later a large leak developed. What did the distance effect and area effect have to do with the corrosion?

4.5   A stainless steel weld repair corroded in the HAZ even though the welder brushed dirt and loose paint from the metal surface before arc welding. The welding rod had a corrosion potential 100 mV higher than the pipe. Service temperature was 215°C (420°F). What was the probable cause of corrosion? Explain.

4.6   Briefly explain the following observations. (a) Cold-swaged tubing shows very slightly lower corrosion resistance than annealed tubing. (b) Tubing swaged only part of its length shows much lower corrosion resistance than annealed tubing.

4.7   An aluminum boat hull was assembled with high-strength steel rivets. The rivets were then painted with a paint consisting of 92% zinc powder. (Aluminum is cathodic to steel in fresh water.) Where would corrosion occur? Was painting the rivets a good idea? Do you have a better idea?

4.8   Which would last longer for piping aerated water: stainless steel pipe welded with a carbon steel rod or carbon steel pipe welded with a stainless steel rod? Explain.

4.9   One manufacturer claims, "Almost any kind of welding method is suitable for stainless steel except oxy-acetylene welding." Explain.

4.10 A zinc-coated steel bolt with a brass washer was used to fasten two aluminum parts. Immersed in water containing some ammonia, the brass corroded slightly, the zinc corroded away almost completely, and the aluminum was slightly darkened around the bolt but was uncorroded. What were the anode and cathode reactions?

4.11 Steel wool scouring pads stored on the edge of a stainless steel sink corrode after one use. To prevent corrosion could you (a) use a pad of copper turnings, (b) replace sink with a more resistant metal, (c) heat the pad to dry it, (d) insulate pad from sink, or (e) galvanize the steel wool? Explain your answer in each case.

*4.12 Draw the polarization curves for galvanic corrosion (solid lines). On the same graph (with dotted lines) draw the polarization diagram that shows the effect of painting the corroding metal so that only 1% of it contacts the solution.

4.13 A stainless steel manufacturer states, "Aluminum fastenings should not be used with stainless steel." Could you use stainless steel fastenings with aluminum? Explain.

*4.14 Draw the polarization diagram for a galvanic corrosion situation before and after the cathode has been painted.

*4.15 A condenser with titanium tubing and bronze (mainly Cu) tube sheets has severe corrosion of the tube sheets by the cooling water. Construct a polarization diagram to illustrate this problem, showing every anode and cathode process involved.

*4.16 If a corrosion potential were listed for iron in the galvanic series, would it be higher, lower, or about the same as the standard equilibrium potential in the series? Why?

*4.17 A steel brace on an off-shore drilling rig is held in place with four small stainless steel bolts at 15 m (50 ft) below the water line. If the surface area of the brace were doubled, how would the corrosion rate be affected? Assume that corrosion is due to galvanic action only. Draw a polarization diagram to illustrate the situation.

4.18 Type 304 stainless steel was selected for a chemical process using HNO_3 with 0.1% HCl. Corrosion was not bad except at areas adjacent to *some* welds. Why only *some* welds?

4.19 A stainless steel pressure vessel containing HNO_3 and smelter slag was sealed with a lead O-ring that leaked within 48 h because of corrosion of the lead. Against the advice of the corrosion engineer, a gold O-ring was substituted. It worked fine. Why?

4.20 Brass radiators soldered with lead–tin solder have had corrosion problems. A machinist suggests switching to a silver solder. Discuss.

4.21 A stainless steel distillation tower was found to have sharp ditches running along the sides of the weld beads at the upper (hot) end of the tower. To prevent recurrence, what changes would you advise in (a) welding technique, (b) metal, and (c) heat treatment.

5

Environmental Cells

One drop of water sitting on a metal surface creates an electrochemical cell. The metal may be approximately uniform in composition but the water drop is not, and that is all that is required to start localized attack. The environment needs only to vary in concentration, velocity, or temperature to set up local anode and cathode areas on the metal. The cathodes, of course, will be where the environment reacts best.

5.1 CORROSIVE CONCENTRATION

Oxygen Concentration Cells

The most common cathode reaction of all is the reduction of oxygen from the air.

$$O_2 + 2H_2O + 4e^- \rightarrow 4OH^- \qquad (2\text{-}5)$$

The drop of water mentioned above sets up an electrochemical cell because the surface dissolves oxygen from the air, while farthest from the air the oxygen concentration is lowest. In this way an oxygen concentration cell develops, with the metal at the circumference of the drop being the cathode and the metal in the center the anode. This situation is depicted in Figure 5-1.

Any place that the metal contacts solution having an exceptionally high oxygen concentration will be a cathode, and any place where the oxygen is especially low will tend to be anodic, if it is not located too far from the cathode. The distance effect is always important in electrochemical cells, as is the cathode/anode area ratio.

- An inspection of old highway culverts by the Wyoming Highway Department found that most corrosion occurred in the center of the culvert pipes

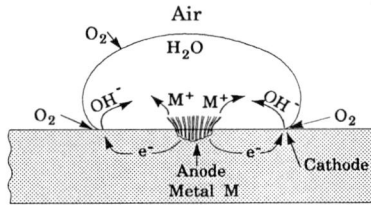

FIGURE 5-1.   Diagram of corrosion of metal under a water droplet.

under the paving centerline, where the oxygen concentration in the soil was at its lowest.

Other Concentration Cells

While oxygen concentration cells are the most common, any cathode reactant that varies in concentration on the metal surface can create a corrosion cell. A variation in pH, for example, can establish a cathode region where pH is low, provided it is low enough (perhaps <pH 4) to make hydrogen ion reduction (Reaction 2-6) an important cathode process.

If corrosion is caused by a strong oxidizer, the cathodes will tend to be located at regions of high oxidizer concentration. Oxidizers commonly used in the chemical industries include O_2, nitric acid (HNO_3), hot concentrated H_2SO_4, hydrogen peroxide (H_2O_2), and solutions containing ferric ion (Fe^{3+}), chromate (CrO_4^{2-}), dichromate ($Cr_2O_7^{2-}$), permanganate (MnO_4^-), or perchlorate (ClO_4^-).

- Offshore steel platforms have two oxygen concentration cells on each leg. The first is in the splash zone and just below the water surface. Corrosion is moderate at lower depths. The second concentration cell is at the mud line where the O_2 concentration is much higher in the water than in the mud (see Fig. 5-2).

Critical Humidity

Any significant corrosion in gases at temperatures below 100°C (212°F) actually is almost always electrochemical because an extremely thin film of moisture condenses on the metal surface to serve as the electrolyte. Other pollutants in the gas (SO_2, salts, etc.) dissolve in the moisture to increase its conductivity and in some cases offer alternative cathode reactions. Temperature fluctuations and surface contamination allow moisture to condense at relative humidities well below 100%.

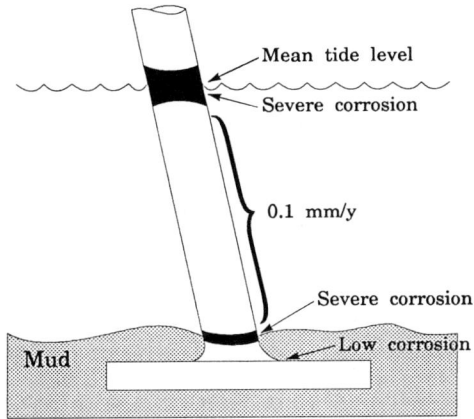

FIGURE 5-2.   Oxygen concentration cells on legs of offshore steel structure.

• Steel corrosion in an SO_2 polluted atmosphere is negligible below 60% relative humidity but is over 100 times higher at 90% humidity.

5.2 VELOCITY

The relative velocity between metal and environment can profoundly affect the corrosion rate. Either metal or environment can be moving: the metal in the case of a boat propeller, or the environment in the case of a solution flowing through a pipe.

Going from stagnant conditions to moderate velocities may lower corrosion. Oxygen concentration cells can be prevented by distributing a more uniform environment through the system. If inhibitors have been added, they also can be distributed more evenly and, therefore, may be more effective. In addition, moderate velocities can prevent suspended solids from settling out and creating crevice corrosion situations under the sediment. A more uniform environment also reduces the possibility of pitting.

On the other hand, increasing velocity may increase the supply of reactant (usually O_2) to the cathodes. Because the diffusion of the reactant is often the rate-controlling (i.e., slowest) step in the whole corrosion process, the corrosion rate of an *active* metal commonly increases with increasing velocity, until the velocity gets so high that diffusion is no longer rate controlling. This situation is illustrated in Figure 5-3*a*.

For metals that can *passivate,* increasing velocity could increase corrosion until conditions become oxidizing enough to form a passive film. From that point on, velocity has virtually no effect unless it becomes so great that it

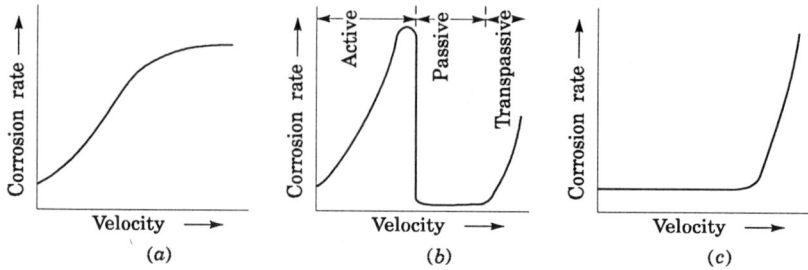

FIGURE 5-3.   Effect of velocity on corrosion rate. (*a*) Diffusion-controlled corrosion of an active metal with soluble corrosion products. (*b*) Active–passive metal. (*c*) Metal protected by a thick scale of corrosion product.

sweeps off the passive film (see Fig. 5-3*b*). But take note that while passive films are so thin that they are invisible, they are also tough enough to withstand any reasonable velocity.

For metals that are protected by a thick layer of corrosion product, the corrosion rate may be satisfactory at low velocities but above a critical velocity the protective layer will be eroded away (see Fig. 5-3*c*).

• The critical velocity for copper in seawater is only 0.6–0.9 m/s (2–3 ft/s), but admiralty brass, a copper alloy developed particularly for seawater, is good up to 1.5–1.8 m/s (5–6 ft/s).

For the rather uncommon corrosion processes under *anodic control*, where corrosion occurs as fast as the metal atoms can detach themselves from the surface, the velocity of the solution has practically no effect on corrosion.

5.3 TEMPERATURE

An old rule of thumb is that increasing the temperature 10°C (~20°F) doubles the corrosion rate. This approximation gives some idea of the exponential effect that temperature can have on corrosion, although this rule can be misleading in some situations. Increasing temperature increases reaction rates, diffusion rates, and the rate of dissolution of gases in water. It also increases the ionization of water, which improves the ionic conduction and lowers its pH.

• The first thing a corrosion engineer does after he dies is to go to the Devil and plead, "For God's sake, lower your operating temperature."

For *passivated* metals the protective film breaks down and repairs itself much more rapidly as temperature increases. It becomes more easily dam-

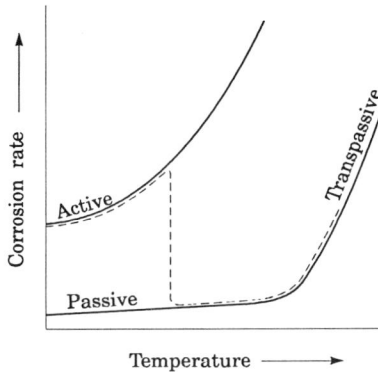

FIGURE 5-4.   Effect of temperature on the corrosion rate of active metals, passive metals, and active–passive metals (dotted line).

aged by Cl^- ions and thus more susceptible to pitting. At some point the film breaks down entirely and the metal becomes transpassive, corroding at a very high rate (see Fig. 5-4).

Some metals, notably zinc in water, are active at low temperatures but passivate at high temperatures, as shown by the dotted line in Figure 5-4. An *active–passive* metal may change the corrosion situation quite abruptly.

Increasing temperature reduces the solubility of gases in water. Where O_2 reduction is rate controlling the corrosion may actually decrease as temperature increases. Thus, steel corrodes hardly at all in boiling water.

Although temperature can affect corrosion tremendously it is not always controlled as it should be. Higher temperatures mean higher production rates and higher efficiencies so temperatures are often pushed to the very limit or a little beyond.

• A new hydrofluoric acid plant designed to use concentrated H_2SO_4 at 120°C (250°F) showed high corrosion rates of the carbon steel equipment right from start-up. A renowned corrosion engineer was called in and spent several hours observing the production by looking over the operators' shoulders. He then informed the astounded engineers that the operators were actually running at 165°C (325°F). As Yogi Berra has said, "You can observe a lot just by watching."

*5.4 POLARIZATION CURVES

Figure 5-5a shows that increasing the *concentration* of the corrosive (the cathode reactant) increases the corrosion rate of an active metal. The reactant could be dissolved oxygen gas, an acid, or some other corrosive.

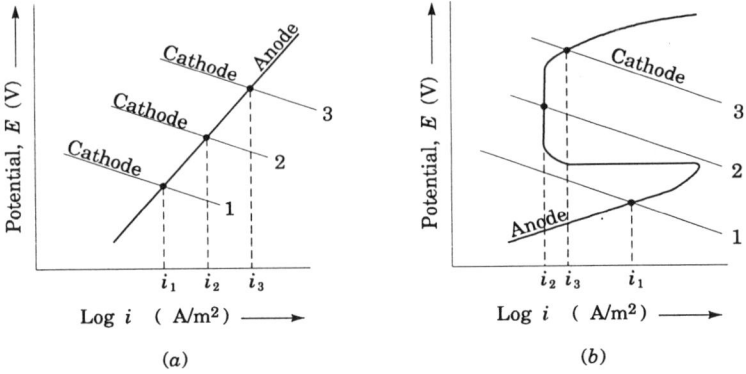

FIGURE 5-5.   Effect on the corrosion current density when concentration of reactant is increased from 1 to 3. (a) Active metal and (b) active–passive metal.

Figure 5-5b for an active–passive metal shows that a low concentration of oxidizer (line 1) might be insufficient to passivate, a higher concentration (line 2) could passivate, but too high a concentration (line 3) could make the metal transpassive, increasing its corrosion rate again. Passivating inhibitors use this principle to change from a high corrosion rate in the active condition to low corrosion with sufficient oxidizer added.

Increasing velocity reduces the concentration polarization of the cathode, thus increasing the corrosion of an active metal (Fig. 5-6a), and possibly passivating an active–passive metal (Fig. 5-6b).

Increasing the temperature has a couple of important effects on polarization curves. First, the exchange current densities increase, shifting the curves to the right. Second, the Tafel slopes decrease because polarization is re-

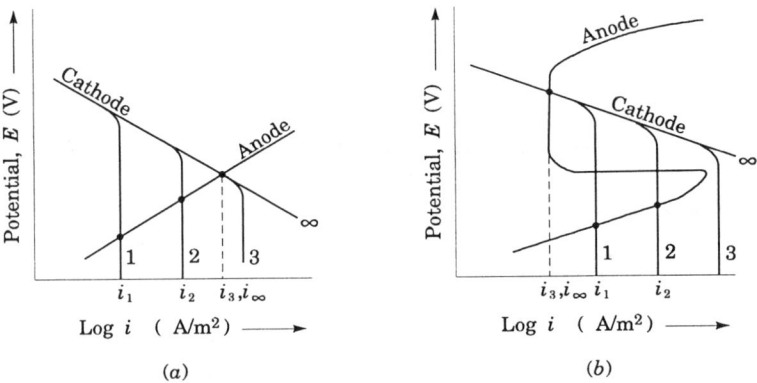

FIGURE 5-6.   Effect of increasing flow velocity on (a) an active metal and (b) an active–passive metal.

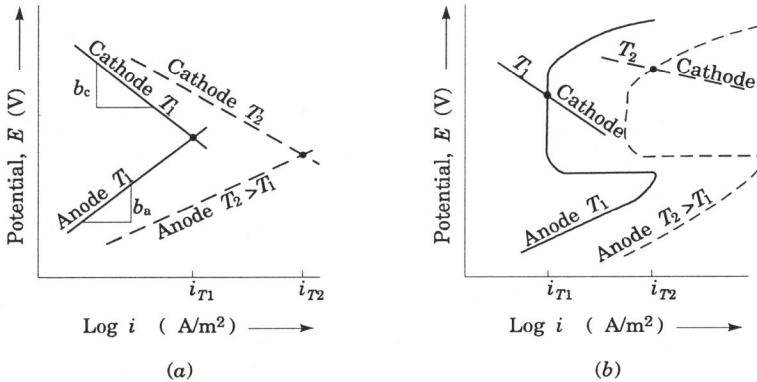

FIGURE 5-7. Effect of temperature on polarization curves of (*a*) an active metal and (*b*) an active–passive metal. Solid lines are the lower temperature T_1, dotted lines are the higher temperature T_2.

duced. The result is much greater corrosion, as Figure 5-7 shows. The anodic polarization curve of an active–passive metal has a much narrower passive potential range at high temperatures so that the corrosion is likely to become transpassive, but even if the passive film remains, the corrosion rate (current density) is high.

5.5 CREVICE CORROSION

Crevice corrosion is caused by a concentration cell in which oxygen, usually, or some other oxidizer, is depleted inside the crevice. The crevice can be created in three different ways.

1. A tight metal–metal joint that is just loose enough to allow moisture to penetrate, such as in the threads of nuts and bolts, under rivets, between lapped joints, in wire screens where the wires cross, or between the tubes and the tube sheets in boilers (see Fig. 5-8 and 5-9).

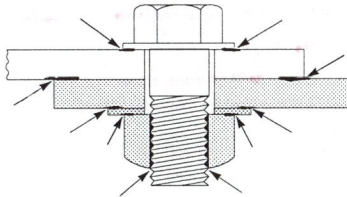

FIGURE 5-8. Example of crevice corrosion of a bolted joint. The threads in the nut and bolt will be the first to fail.

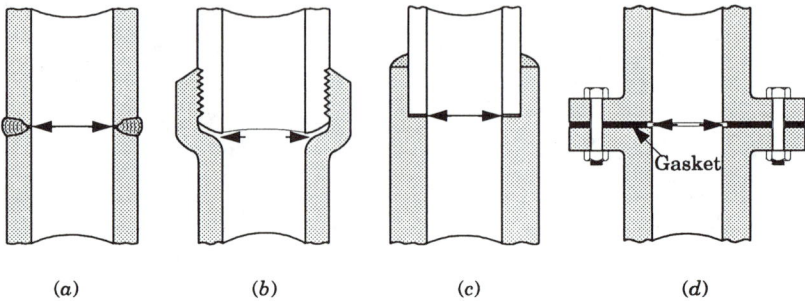

FIGURE 5-9.   Crevices created in pipe connections by (*a*) incomplete penetration of a butt weld, (*b*) a threaded joint, (*c*) a socket connection and (*d*) a badly fitting gasket. Arrows show location of crevices.

2. Metal contacting absorbent nonmetallic material, such as fiber gaskets, foam insulation, porous paint, valve packing, or wood.
3. Metal surfaces where deposits have formed. These deposits may be dirt, rust, bacterial colonies, precipitated salts, or the like.

• Automobiles used to have bolted-on fenders to make replacement easier. Canvas gaskets between fenders and body prevented squeaking, but once moisture got through the paint to the canvas, the fenders tended to fall off.

When liquid is flowing outside, crevices must be quite tight (in the order of a micron) to keep the solution inside stagnant. Chloride ions speed the development of corrosion in the crevice, but even so, a long incubation period precedes any localized attack. Within the crevice the metal commonly looks like it has pitted, as Figure 5-10 shows on a stainless steel flange. Stainless steels are particularly susceptible because they become active within the crevice and passive outside, developing a large potential difference between anode and cathode areas. Crevice corrosion often initiates stress-corrosion cracking or corrosion fatigue.

When corrosion begins, the crevice corrodes uniformly just as the metal outside the crevice does, but in time the O_2 within the crevice has all reacted and no more O_2 enters because the crevice is so narrow. The cathode reaction continues outside the crevice, however, with electrons being supplied from the anode within the crevice. Corrosion outside actually decreases because of the electron current coming from the crevice.

The production of positive metal ions attracts anions, particularly Cl^-, into the crevice to form metal chloride. These soluble chlorides hydrolyze, reacting with water according to the reaction

$$MCl + H_2O \rightarrow MOH + HCl \tag{5-1}$$

FIGURE 5-10.   Crevice corrosion of stainless steel pipe flange under a composition gasket. [Reprinted with permission from E. V. Kunkel, "Selecting Gaskets to Limit Corrosion of Stainless Steel Bolted Joints in a Chemical Plant," in *Corrosion*, Aug. 1954, Vol. 10, No. 8, Fig. 1, p. 260. Copyright © by NACE.]

where M is the metal that has corroded.

Hydrochloric acid lowers the pH in the crevice and, in ionizing, also releases the Cl^- ions to react with more metal ions. The high acidity breaks down passive films, increasing the dissolution of the metal.

- One petroleum refinery, notorious for its corrosion troubles, had a problem in the coolers that handled the light ends because H_2O and HCl vapors were condensing in the tubes. The engineer added ammonia (NH_3) to neutralize the condensate but great amounts of ammonium chloride (NH_4Cl) crystals precipitated in the tubes causing crevice corrosion throughout. New tubes cost $800,000.

Chloride ions are not the only anions that migrate into the crevice, of course; other aggressive ions, such as thiosulfate ($S_2O_3^{2-}$) can produce the intense attack. However, if hydroxide ions, which migrate even faster than chloride ions, are present, they help to shut down the attack by tying up the metal ions as insoluble metal hydroxides.

Crevice corrosion can be prevented in several ways.

1. Eliminate crevices: weld butt and lap joints with a continuous weld, not a skip weld; use nonabsorbent gaskets, such as Teflon; caulk seams.
2. Keep the metal clean: remove deposits; filter out solids; avoid intermittent flow that might allow debris to settle out.
3. Add alkalies to neutral chloride solutions.

- A 30-in. cast iron pipe buried near a coal storage yard quickly developed a leak. The work crew that dug it up found that the soil contained 17 g/L of H_2SO_4 that had leached from the coal. They patched the leak by welding a ½-in.-thick steel sleeve completely around the pipe. A year later the pipe was leaking again (steel is anodic to cast iron). They dug up the pipe, patched it, and coated the area with coal tar enamel reinforced with fiberglass cloth. They installed a drain tile to remove the acid and they added an impressed current cathodic protection system with a graphite anode. Within another year they had to install a Duriron anode to replace the graphite, which had dissolved in the acid. The next year the pipe started leaking again. When they dug it up they found a new problem: crevice corrosion caused by a wooden board left lying on the pipe in a previous excavation. They repaired the pipe once more and finally had no more trouble.

Filiform Corrosion

Filiform corrosion appears as fine filaments of corrosion products growing across the metal surface, usually under a protective film. A common example is the pattern of fine, red-brown lines of rust that form on the surface of painted or lacquered tin cans stored at high humidity for long times.

The filament is a crevice that is continuing to grow. Its head is the anode, with a low O_2 concentration and low pH. Its tail, consisting of precipitated corrosion products, serves as the cathode and follows along after the head. Cracks in the coating along the tail allow entry of O_2 and moisture.

Filiform corrosion occurs only where the relative humidity is continuously above 60%. While it affects the superficial appearance of the metal, it does not cause the metal to fail.

5.6 PITTING

- People use the term "pit" to describe any localized corrosion spot that forms a cavity. "Pits" can start from crevice corrosion, dezincification, galvanic corrosion of a reactive phase in an alloy, failure of a noble metal coating, corrosion by water droplets, erosion–corrosion, fretting, bacterial corrosion, or even the distinct mechanism of pitting.

Pitting causes more *unexpected* corrosion than any other form of attack. It is particularly insidious because the total amount of metal lost is small but perforation is rapid. Figure 5-11 shows an example of pitting inside a water line.

Pitting Mechanism

Real pitting initiates as a localized breakdown of a protective film on the metal. Breakdown may occur at a weak spot in the film, such as at a defect or impurity on the metal surface, or at a scratch. Alternatively, breakdown may occur because of a momentary change in solution concentration. Nor-

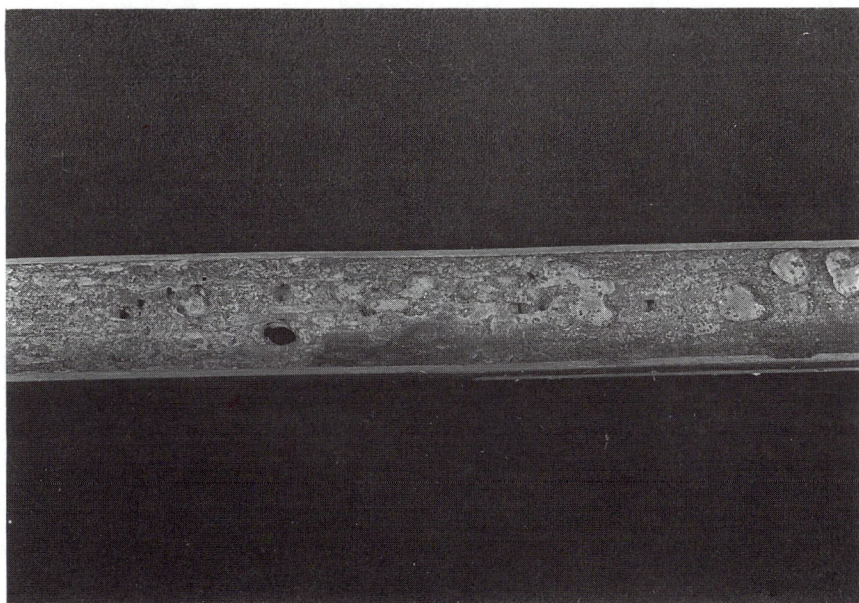

FIGURE 5-11.   Pitting of water pipe. The sharp-edged, flat-bottomed pits indicate dissolved CO_2 as the cause.

mally, breakdown is followed by rapid film repair, but if the damaging conditions persist long enough the film may not be able to repair itself and a pit is initiated. The active corrosion pit is surrounded by passive metal, establishing a small electrochemical cell with a large potential gradient.

After initiation, the mechanism of propagation is just the same as for crevice corrosion. In essence, the pitted metal has created its own crevice. The cathode reactant becomes depleted in the pit while the anode reaction continues. Migration of chloride and other aggressive ions into the pit permits metal salts to form and hydrolyze, making the solution in the pit very acidic. However, crevice corrosion nucleates more easily than pitting because pitting first has to make a crevice. Metals that pit are also susceptible to crevice corrosion but the converse is not necessarily true.

Pitting most commonly occurs in chloride solutions, especially if these solutions are stagnant or slow flowing. Stainless steels are most susceptible, with cold-worked metal the most susceptible and polished surfaces the least susceptible, because surface finish greatly affects the perfection of the pro-tective film.

- In many plants more pitting occurs during shutdown periods than during normal operation. During shutdowns stagnant solutions are sometimes allowed to sit in pipes, pumps, and vessels. In fact, it is often difficult to drain everything.

Pits usually grow in the direction of gravity. On the side walls of tanks they often grow down the walls in streaks as the dense solution of HCl eats its way down and gradually out to the surface.

Corrosion Rate

As conditions inside the pit keep increasing in acidity and Cl^- concentra-tion, the corrosion rate keeps increasing with time but the probability of repassivation also increases with time. It is very likely that the concentrated solution in the pit will be washed out by liquid flowing over the surface or by convection currents. The maximum pit depth d often obeys the equation

$$d = kt^n \tag{5-2}$$

where k and n are constants and t is the time. For aluminum in water, n is usually one-third, which indicates that most pits repassivate before they get very deep. For steel in soil, n can vary anywhere from 0.1 to 0.8 depending on the type of soil, its aeration, and its other characteristics.

Pitting is really an intermediate situation between no corrosion and uniform attack, so the metal comes close to having been the right choice for the job.

- But coming close only counts in horseshoes and hand grenades.

As with any form of localized attack, weight loss is *no* criterion of pitting damage, and the average pit depth is not much better. The maximum pit depth is the best indication, although this does vary with the amount of surface area examined. Some investigators calculate a pitting factor.

$$\text{Pitting factor} = \frac{\text{maximum pit depth}}{\text{average metal penetration}} \qquad (5\text{--}3)$$

where the average penetration is determined from the weight loss. Figure 5-12 shows how much the depth of pitting can vary. The depth of pitting usually decreases if the number of pits increases because of the reduction in cathode/anode area ratio.

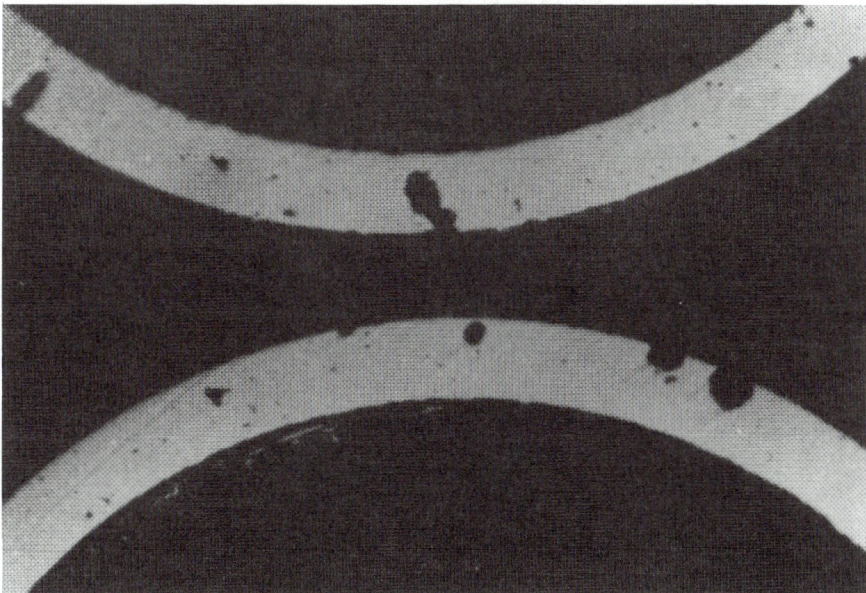

FIGURE 5-12.   Transverse sections of 304 stainless steel tubes from hydraulic systems of naval aircraft. Failures were from pitting. [Reprinted with permission from *Metals Handbook*, 8th ed., Vol. 10, ASM International, Materials Park, OH, 1975. Fig. 18(*a*), p. 193.] Howard E. Boyer, Editor.

• Aluminum parts with pits the diameter of a human hair were rejected in the Apollo space program.

Whether a pit is active or has repassivated can be found by adding a drop of distilled water and testing with pH paper. If the pit is more acidic than the surrounding unpitted surface, it is still active. In steel boilers, active pits can be identified by a red oxide (Fe_2O_3) cap covering black oxide (Fe_3O_4) deposits inside the pit.

Prevention of Pitting

Pitting can be prevented in many ways.

1. Reduce environment corrosivity: reduce Cl^- concentration, lower temperature, reduce acidity, and reduce oxidizer concentration.
2. Prevent stagnant conditions: increase flow rates and drain all equipment during shutdowns.
3. Add inhibitor, but in sufficient amounts to protect the entire surface.
4. Use cathodic or anodic protection.
5. Shot peen the surface. Shot peening increases the potential of steel about 100 mV, reducing pitting tendency.
6. Select a more resistant material. When chloride is present, type 304 stainless steel is often replaced by 316 or 317. Even more resistant are the nickel alloys Hastelloy G-3, Inconel 625, and Hastelloy C-22.
7. Use protective coatings: organic coatings, zinc-rich paints, and metallic zinc are commonly used.
8. Increase thickness of the metal. The time for a pit to penetrate aluminum varies as the cube of thickness; doubling the thickness increases the life by a factor of 8.
9. Passivate the metal. After cleaning, stainless steel is usually washed with 20% HNO_3 to give it a strong passive film.

5.7 MICROBIAL CORROSION

Localized corrosion by microbial growths has only recently been recognized as a serious industrial problem, but it is now appearing in such diverse places as the soil, inside pipelines, oil wells, in heat exchangers and condensers, and aircraft fuel tanks. Nearly all common engineering alloys are susceptible, except titanium with its incredibly resistant passive film.

A wide variety of bacteria, algae, and fungi have been identified as causing problems. They all require some water and most types require organic matter for food. In addition, most bacteria need oxygen for metabolic processes, although some types are anaerobic, growing only in the absence of oxygen. Typically, they grow best in temperate climates and at fairly neutral pH values, but exceptional types thrive at pH values as low as zero and temperatures from at least -10 to $+99°C$ (15–210°F). The corrosion usually takes the form of pits under the microbial colony, particularly at welds and their HAZs.

- Corrosion by sulfate-reducing bacteria was first discovered in Holland in 1934. Waterlogged soil with a neutral pH (the sea just pumped from it) corroded steel severely. But since the soil had no oxygen, no acid, and no oxidizing ions, it had no known cathode reaction!

All microbial colonies set up an *oxygen concentration cell*. Under their protective biofilm or slime the oxygen concentration is very low, creating an anodic region for crevice corrosion. This situation develops for both aerobic types, where the microbes consume the O_2, and anaerobic types, where bacteria shield themselves from air by growing a protective biofilm.

Sulfate-Reducing Bacteria. For steel in soils the most common bacterial corrosion is caused by various anaerobic sulfate-reducing bacteria (SRB). In their metabolic process these bacteria reduce sulfate in the soil to sulfide, consuming hydrogen in the process

$$SO_4{}^{2-} + 8H_{ads} \rightarrow S^{2-} + 4H_2O \qquad (5\text{-}4)$$

the reaction being catalyzed by an enzyme (hydrogenase) produced by the bacteria. These bacteria remove adsorbed hydrogen from the steel surface, "depolarizing" the cathode areas; that is, these bacteria speed up the cathode process, which is usually rate controlling. The sulfide produced in this reaction reacts with Fe^{2+} produced at the anode and forms FeS, a jet black precipitate on the metal. The FeS provides a cathode much more efficient than steel for H_2 evolution and, thus, further increases the corrosion. The attack may reach 2.5 mm/y (100mpy). Recent research strongly indicates that the SRB also generate an extremely corrosive metabolic product other than hydrogenase, but as yet this product has been unidentified. Although these bacteria are anaerobic and should not be able to grow in aerated soil, they are often found in the O_2 depleted regions under colonies of aerobic bacteria.

Acid-Producing Bacteria (APB). Acid producers, such as the sulfur-oxidizing types, are aerobic. These bacteria oxidize elemental sulfur or sulfide to H_2SO_4, forming up to 10 % H_2SO_4 as a waste product of their metabolism. They are often found in soils around sulfur storage facilities, in oil fields, and sewage disposal sites. Other bacteria make highly corrosive organic acids. The sulfuric or organic acids attack metals, concrete, and polymer coatings.

- We do not refer to sulfur-oxidizing bacteria by initials. In an oil field that term is usually reserved for the tool push (the boss).

Iron-Oxidizing Bacteria. Iron-oxidizing bacteria (IOB), also aerobic, feed on Fe^{2+} and exude Fe^{3+} through their biofilm, forming a crust of rust around themselves. The growing bacterial colony often bursts this crust, grows beyond it, forms another crust, and so on. The result will be "tubercles," or knoblike mounds of Fe_2O_3 rust (an example is shown in Fig. 5-13). The IOB cause crevice corrosion, accelerate the anode process, and some are now known to concentrate Cl^-, which makes them especially bad for stainless steels. The oxygen-starved regions inside the tubercules permit the growth of SRB also.

The presence of bacteria can often be detected by the unusual color of the corrosion products: the reddish-brown tubercles of IOB, the black iron sulfide of SRB, and bright green pits in copper alloys are examples. Test specimens inserted within an operating system may fail to detect bacteria because the colonies are very localized and often grow in the most inaccessible places. Pipelines must be "pigged" (scraped out) and the corrosion products examined. Iron sulfide can be immediately identified by a drop of HCl, which produces the characteristic rotten-egg smell of H_2S, but a precise identification of bacteria can only be made by microscopic examination of their size and shape. Field kits that test for SRB are now available; they require a 5-28 d incubation time, however.

Organic inhibitors, so effective in most types of corrosion, are of no use in bacterial corrosion; in fact, the bacteria feed on them. Even some organic coatings are damaged by bacteria. Biocides can be completely effective but must be selected to match the type of bacteria present without endangering the natural environment or animal life. Aldehydes, phenolics, chlorine and chlorine-containing compounds, and quaternary ammonium compounds are frequently used. Cathodic protection has often been said to be effective, but this is now being questioned. The increase in pH at the surface of cathodically protected metal might either inhibit or encourage microbial growth.

5.8 TEMPERATURE CELLS

Differential temperature cells may localize corrosion in such equipment as heat exchangers, boilers, immersion heaters, and reactor cooling systems, because they cannot have isothermal surfaces if they are to work. A hot spot in a boiler tube often corrodes severely because it is anodic to the large surrounding cooler area—the cathode/anode area ratio effect again.

In theory, the hot metal should be more reactive than the cool metal if both are active or both are passive, but in many cases the temperature

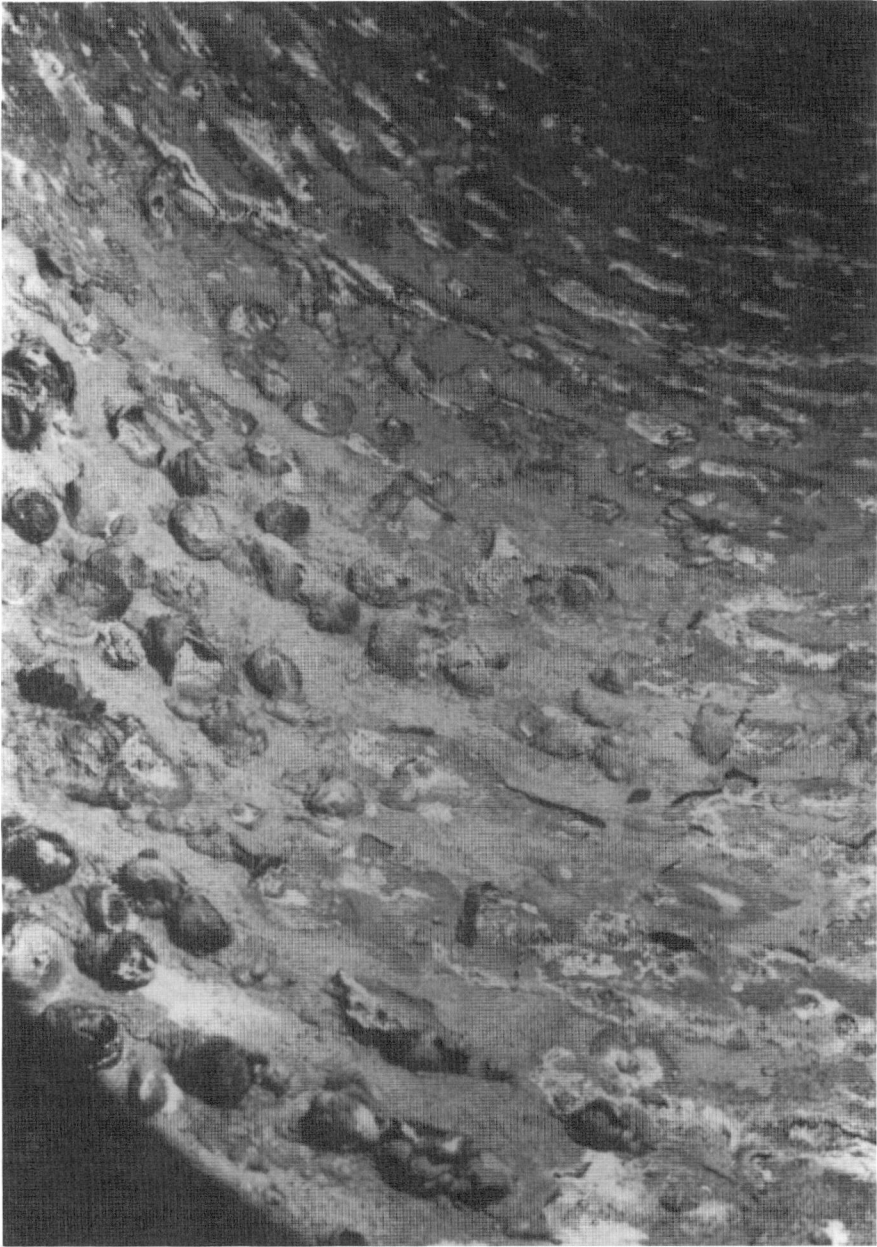

FIGURE 5-13. Rust tubercles formed by iron-oxidizing bacteria. [Reprinted with permission from *Corrosion Control in Petroleum Production*, National Association of Corrosion Engineers, 1979, Fig. 2.11, p. 12. copyright © by NACE.]

difference is no more than 75°C (135°F), which creates a potential difference of only a few millivolts. However, if surface films are affected, an active–passive cell can concentrate intense attack on the active metal, whether it is hot or cold.

- Most household hot water tanks were galvanized steel at one time. The zinc was anodic to steel in the 60°C (140°F) water. With the advent of automatic washing machines, water temperatures had to be raised to 80°C (175°F), which passivated the zinc and destroyed the steel by galvanic action. Glass-lined or Monel (Cu-Ni alloy) tanks are now used, or cathodic protection is applied.

Differential temperature cells persist for the life of the equipment and may lie dormant causing no trouble for years until some minor change in the environment affects the protective nature of the film on either the hot or cold metal. A higher conductivity, a slight increase in aggressive ion concentration, or whatever, can suddenly trigger a disastrous localized attack.

Bear in mind that the surface temperature of the metal is often hotter than the process fluid. The actual skin temperature is much more important than either the total amount of heat transfer or the temperature of the environment.

- Heat transfer situations can be tricky. For example, one engineer was chagrined to find that a steel heater that held boiling tar acid corroded four times as fast as a steel sample just immersed in the stuff.

5.9 CONDENSATE CORROSION

Localized corrosion occurs where moisture-laden hot gases contact a cooler metal surface and condense. This attack on cold metal, sometimes called "dewpoint corrosion," causes almost as much corrosion as on hot metal in many industries. Figure 5-14 shows the deep pits that can result.

Condensation is now recognized as a major factor in atmospheric corrosion but it has been a more obvious problem in the combustion of fossil fuels. Attack may occur, for example, where air leaks into a hot gas stream, where combustion gases contact the cold end of air heaters, where thermal insulation is poorly designed or maintained, and where flue gases condense in the top part of chimney stacks.

Condensate corrosion has its maximum intensity at the point where moisture first condenses, since that is the hottest region where electrochemical corrosion can occur. Attack is much more serious if SO_2 or SO_3 is present

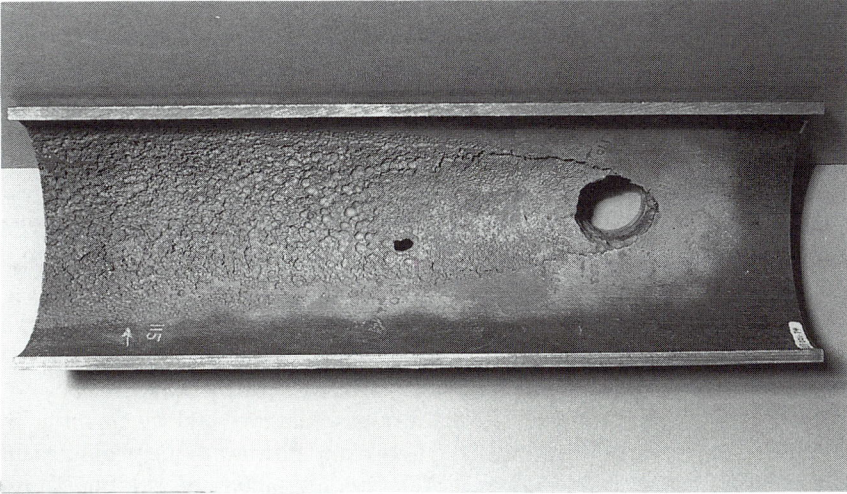

FIGURE 5-14.   Condensate corrosion at injection point on gas line.

in the gas. The H_2SO_4 that forms is both corrosive and hygroscopic, picking up more moisture from the gas stream. As little as 5 ppm of SO_3 raises the dewpoint by as much as 55°C (100°F).

Deep grooves can be cut in the metal if droplets of moisture run down the metal surface or are blown across the surface by the gas stream. Figure 5-15 shows such a situation.

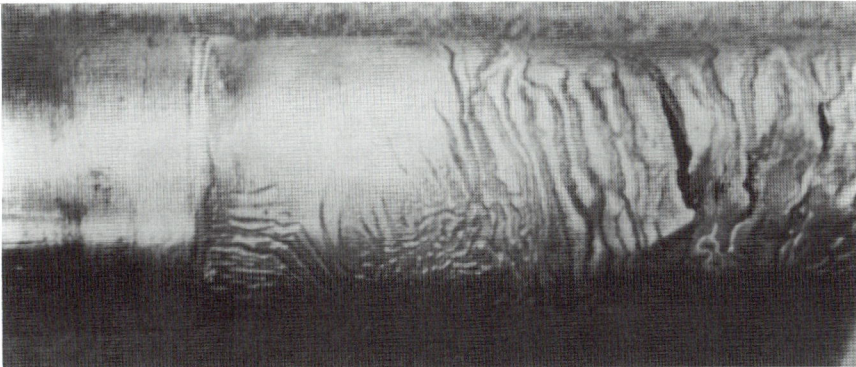

FIGURE 5-15.   Corrosion of an admiralty brass condenser tube by steam condensate containing high levels of NH_3 and O_2. The unattacked end was protected by the tubesheet. [Reprinted with permission from Barry C. Syrett and Ronald L. Coit in *Metals Handbook*, 9th ed., Vol. 13, Joseph R. Davis, Editor. ASM International, Materials Park, OH, 1987, Fig. 6(*a*), p. 988.]

5.10 STRAY CURRENT CORROSION

Electrical currents leaving their intended path and passing through soil, water, or other electrolyte will follow any high-conductivity metallic paths available. They corrode the metal where they leave it and return to the electrolyte. This corrosion, sometimes called "electrolysis," is basically independent of O_2 concentration and pH.

The currents may have left their intended path because of high-resistance electrical connections or poor insulation, or they may just pass through the electrolyte because it offers one alternate path to complete their circuit. Electrical currents do not choose only the lowest-resistance path if offered a choice. They divide inversely with the resistance and flow through all available paths, but they prefer metal because of its low resistivity.

Any source of direct current (dc) can release damaging stray currents. In times past electric trolley cars caused trouble, and nowadays the problem can often be traced to electric railway switch engines, electrical welding equipment, electroplating shops, mine haulage systems, cranes, breaks in high-voltage dc transmission lines, or impressed-current cathodic protection systems. Where cathodic protection is the source, the stray currents are usually termed "interference" (see Fig. 5-16).

Figure 5-17 illustrates how currents escaping from the rails of an electric railway can cause corrosion of a nearby buried pipeline. The damage is done where the current leaves the pipe as Fe^{2+} ions and returns to the soil. Figure 5-18 shows how welding repairs have caused serious corrosion of ships where the electrical generator has been grounded to the earth rather than to the ship.

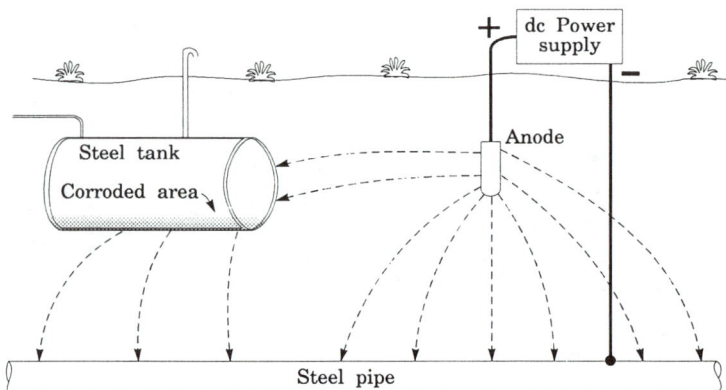

FIGURE 5-16. Stray current corrosion of a tank caused by an impressed current cathodic protection system.

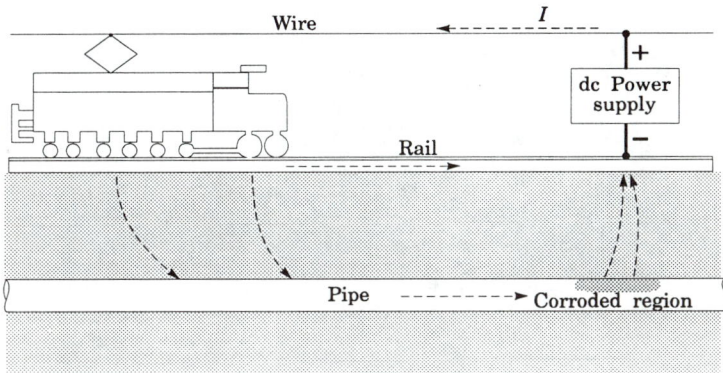

FIGURE 5-17. Stray current corrosion of a buried pipeline caused by an electric railway.

Alternating currents (ac) generally do much less damage than direct currents (perhaps 1% as much for steel), but they can damage passive films on such metals as stainless steel and aluminum by increasing the film porosity. High-frequency ac does very little damage because it plates back most of the metal ions on the half-cycle after it removes them.

Stray current corrosion often reveals itself as a fairly extensive corrosion region or a single, broad pit. The metal has a smooth surface and very little corrosion products. The surface looks as if it had been washed away but the environmental conditions do not involve high fluid velocity. Figure 5-19 shows examples.

The best way of preventing stray current corrosion is to eliminate the current at its source, if that is possible. The problems in Figures 5-17 and 5-18

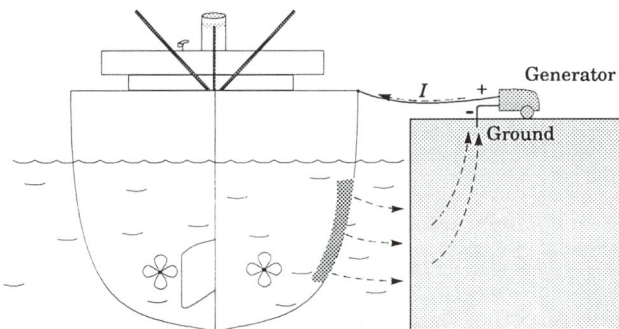

FIGURE 5-18. Stray current corrosion of the dock side of a ship hull caused by electric welding where return connections arc faulty.

FIGURE 5-19. Cast iron pipe from domestic water distribution system, graphitized and attacked locally by stray current. [Reprinted with permission from *Materials Protection and Performance*, Vol. 11, No. 10 October 1972, front cover. Edward C. Greco, editor. Copyright © NACE.]

could be corrected by eliminating the grounding to the earth and using a wire return instead. Good electrical connections, good insulation, and proper grounding are essential.

For a pipeline, inserting insulated couplings along the line makes it a much higher resistance path and thus reduces the amount of stray current picked up and discharged. In some situations it is possible to connect an electrical bond between the corroding structure and the current source at the point of worst corrosion. Most current, then, will flow through the bond rather than discharging into the electrolyte.

Another possible solution is to discharge the stray current from a sacrificial anode connected to the structure at the point of worst corrosion, replacing the anode when it corrodes away. If all else fails, corrosion can be prevented by bucking out the stray current discharge with an equal and opposing current flow onto the structure from an impressed-current cathodic protection system, which itself could cause interference on some other structure.

- One method that does *not* work, although it has been tried many times, is coating the metal where it is corroding. This technique reduces the anode area, causing all current to discharge through minute flaws in the coating and thus immediately drills holes through the pipe or tank. Stray current pick-up could be prevented if the *entire* structure were coated but that is usually impractical if the structure is already in place.

STUDY PROBLEMS

5.1 Briefly explain the *most probable* cause of corrosion where (a) rust blisters have formed inside a steel pipe carrying fresh water and (b) a stainless steel pipe carrying seawater at moderate velocity looks perfect except for one hole in it.

5.2 Name the general type, or types, of bacteria for which each of the following descriptions applies: (a) anaerobic, (b) black iron sulfide forms, (c) tubercules of rust form, (d) pH is reduced, (e) inhibitors are ineffective, and (f) concrete is damaged.

5.3 You are put in charge of the uranyl nitrate evaporation in a large uranium mill. The equipment is all type 304 stainless steel, which is usually resistant to the solutions, but pitting may occasionally occur. Which of the following possible precautions would help you avoid corrosion (and let you glow with pride)? (a) Caulk all lap joints, (b) remove salt layer from metal surfaces regularly, (c) decrease flow rates, or (d) add inhibitors selected by weight-loss lab tests. Explain.

5.4 Lake water for cooling is pumped to a nearby plant through a 6-in. steel pipe. It was inspected after 1 year of service and found to be in excellent shape. After 3 years it had numerous deep, internal pits arranged in a line along the bottom of the pipe. How would you solve the problem? Explain.

5.5 A pipe carrying river water is corroding badly only at low spots in the line. (a) What type of corrosion is probably occurring? Explain. (b) The obvious answer to the problem is, "Do not have any low spots." What would be the next-best answer?

5.6 You are put in charge of rum production in a distillery. The plant equipment is all anodized aluminum, which usually is resistant to rum, but pitting occasionally may occur. Which of the following possible precautions would help you avoid corrosion (and prevent staggering costs)? (a) Reduce the flow rate, (b) redesign equipment to avoid turbulent flow, (c) wire brush all internal surfaces regularly, or (d) replace any pitted aluminum with a noble metal that is considerably more resistant to rum. Explain.

5.7 You are put in charge of starch paste production. The plant equipment is all passive anodized aluminum, which is usually resistant to starch paste, but pitting may sometimes occur. Which of the following possible precautions would help you avoid corrosion (and let you stick around)? (a) Redesign equipment to avoid turbulent flow, (b) avoid very slow flow rates, (c) polish all internal surfaces to a bright, smooth finish, or (d) replace any pitted sections with a noble metal that is considerably more resistant to starch paste. Explain.

5.8 You are put in charge of the bleach section of a pulp mill. The 304 stainless steel piping is carrying 35°C (95°F) sodium hypochlorite (NaClO) solution, a strong oxidizer. Corrosion rate is quite low except at threaded joints, which are rusted and leaking. Which of the following possible remedies would help you stop the corrosion (and become the company's fair-haired child)? (a) Clean frequently, (b) caulk joints, (c) add inhibitor, (d) replace with type 304L. Explain.

5.9 Aluminum–magnesium alloys are resistant to seawater, except in certain circumstances, such as (a) when in contact with porous asbestos insulation and (b) when seawater contains iron rust (hydrated ferric oxide). Explain.

5.10 Condenser tubes made of 70Cu-30Zn brass often pit when seawater is used in them. (a) Where would you expect the worst pitting and why? (b) How might pitting be prevented?

5.11 A 316 stainless steel vessel (17Cr-12Ni-2Mo) was used successfully in a carbon black plant for several years. It contained a carbon-black slurry with carbonates, chlorides, and sulfates at pH 7 and temperatures from 200–500°C (400–900°F). After once being heated and boiled dry, with solids baked onto the surface, it failed in a matter of weeks. Discuss.

5.12 You are put in charge of perfume production for a soap company. The plant equipment is all anodized aluminum, which pits with some perfumes. Which of the following possible precautions would help you avoid corrosion (and come out smelling like a rose)? (a) Redesign equipment to avoid turbulent flow, (b) close crevices in lap joints by caulking, (c) remove suspended solids early in the process, or (d) add inhibitor selected by weight-loss lab tests. Explain.

5.13 You are put in charge of maintenance in a fertilizer plant. A type 316 stainless steel valve handling 45% phosphoric acid (H_3PO_4) has severe corrosion of the shaft only under rubber O-rings. Which of the following possible remedies would solve this problem (and help your career grow)? (a) Use a nonabsorbent gasket for the O-rings, (b) redesign the valve to drain completely when closed, (c) replace the shaft with a passivating metal, or (d) imbed O-rings in a plastic sealant. Explain.

5.14 Microbial corrosion is found on a pipeline in a soil with a pH 1. (a) What type of bacteria would most likely be responsible? (b) Would corrosion be worse in wet, heavy soils or well-aerated soils? Explain.

5.15 An aluminum pipe carrying crude oil with some brine and H_2S developed numerous small pinholes. Describe the mechanism of the corrosion.

5.16 Lab tests of 316 stainless steel in 60% nitric acid, boiling point 120°C (250°F) showed corrosion rates of 0.005 mm/y (0.2 mpy) at both 37°C (99°F) and 65°C (149°F), but 0.088 mm/y (3.4 mpy) at 99°C (210°F). Why would temperature have this effect?

5.17 A steel pipe carries naturally corrosive water to a stainless steel storage tank. The pipe has corroded badly, forming loosely adherent rust and the bottom of the tank is severely pitted. Suggestions?

Memos

(These real corrosion problems may have several solutions or they may have no completely satisfactory solution. The more helpful advice you can offer to Chet the better, but you will need to explain why he should take your advice.)

5.18 Memo To: Corrosion Engineer
From: Chet Bailey, Production Supervisor

In the old plant we've got a 6-in. 304 stainless pipe 100 ft long carrying light fuel oil with some water, ethylene dichloride, and hydrochloric acid at 150°F that has been in service 3 years. It's holding up pretty well except for two sections, both about 5 ft long that are badly pitted and leaking. I plan to have the bad sections cut out and new pipe welded in, but am supposed to check with you first, since you're supposed to be some sort of hot-shot expert in this corrosion thing. The entire system will be phased out in another year.

5.19 Memo To: Corrosion Engineer
From: Chet Bailey, Production Supervisor
One of our engineers in the Processing Section suggests substituting
50% H_2SO_4 in their operation instead of the 95% H_2SO_4 we've always
used. Briefly, his reasons are (1) yield will not be substantially changed;
(2) production will decrease 20%, but at the present time we have 50%
more capacity than we need; (3) lower cost of the dilute acid will achieve
a saving of $10,000/year; and (4) corrosion of the steel reaction tank
and piping will be reduced. The idea looks good to us but the manager
thought we should check with you on item 4, since you're supposed to
know something about corrosion.

5.20 Memo To: Corrosion Engineer
From: Chet Bailey, Production Supervisor

We have a problem with pitting of a 304 stainless steel pressure vessel
containing a 50% $Al_2(SO_4)_3$ solution, pH 2, at 225°F with rapid agitation. After we spot-weld the pits we will try cutting down on the stirring,
since I suspect it isn't needed anyway. Also, we might try raising the pH
to 11 in this vessel, although up to now we've been doing that in the
next stage. Do you have any other ideas? Now that we've got a hot-shot
corrosion engineer it's time you got to work. This is an almost-new
vessel and we've got to save it.

6

Corrosive-Mechanical Interactions

A synergistic interaction exists between corrosion and stress. The stresses may be applied directly through the metal as tension, torsion, or compression, or they may be applied to the metal surface through the environment, such as occurs with high-velocity liquids and impingement by suspended solids.

Stresses on the metal increase its corrosion. These stresses create stress cells of course, but more importantly they disrupt the surface film that has protected the metal to some extent. The resulting cell of filmed versus unfilmed metal increases and localizes the corrosion attack.

Localized corrosion concentrates the stress: intergranular corrosion, for example, creates a notch that acts as a stress raiser. Corrosion can even produce additional stresses by charging the metal with hydrogen that creates internal pressures, or by forming voluminous solid corrosion products that create a pressure on the metal in crevices.

In these circumstances the corrosion is often rapid, or if fracture is initiated, failure will be sudden. The possibility of a disaster is always present.

6.1 EROSION–CORROSION

The rapid movement of a corrosive liquid over a metal forms grooves, elongated pits, or waves on the surface, usually in a directional pattern. The flow damages any protective film and tends to undercut it, as the sketch in Figure 6-1 illustrates. The rapid flow facilitates delivery of the cathode reactant so that the corrosion tends to become reaction controlled.

Horseshoe-shaped pits form on some metals, with the open edge of the horseshoes pointing downstream, as shown in Figure 6-2. In very corrosive environments the surface is covered with one streaky pit on top of another

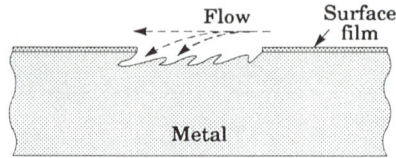

FIGURE 6-1.   Erosion–corrosion of metal surface showing undercutting.

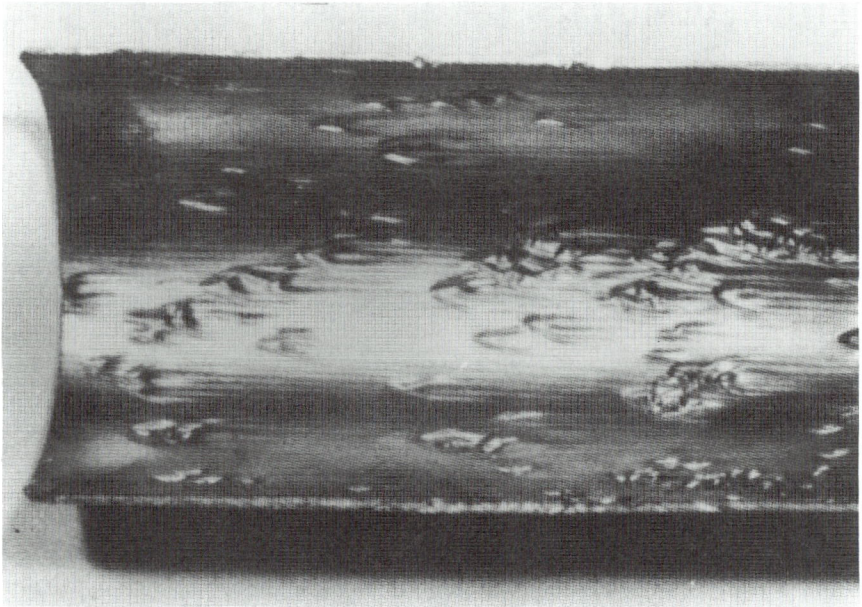

FIGURE 6-2.   Horseshoe-shaped pits in a copper pipe carrying river water. Flow was from left to right. [Reprinted by permission from W. G. Ashbaugh in *Metals Handbook*, 9th ed., Vol. 11, Kathleen Mills, Manager of Editorial Operations. ASM International, Materials Park, OH, 1986, Fig. 15, p. 189.]

(Fig. 6-3). At high velocities or turbulent conditions, some metals take on a wavy surface (Fig. 6-4).

Most metals are susceptible to erosion–corrosion, especially if they rely on a mechanically weak surface film for corrosion protection. Steel, with its weak rust layer, often undergoes erosion–corrosion. Although the passive layer on stainless steels is much stronger, it can suffer erosion–corrosion if the liquid velocity is high. Titanium, however, forms a passive film so protective that erosion–corrosion is seldom a problem.

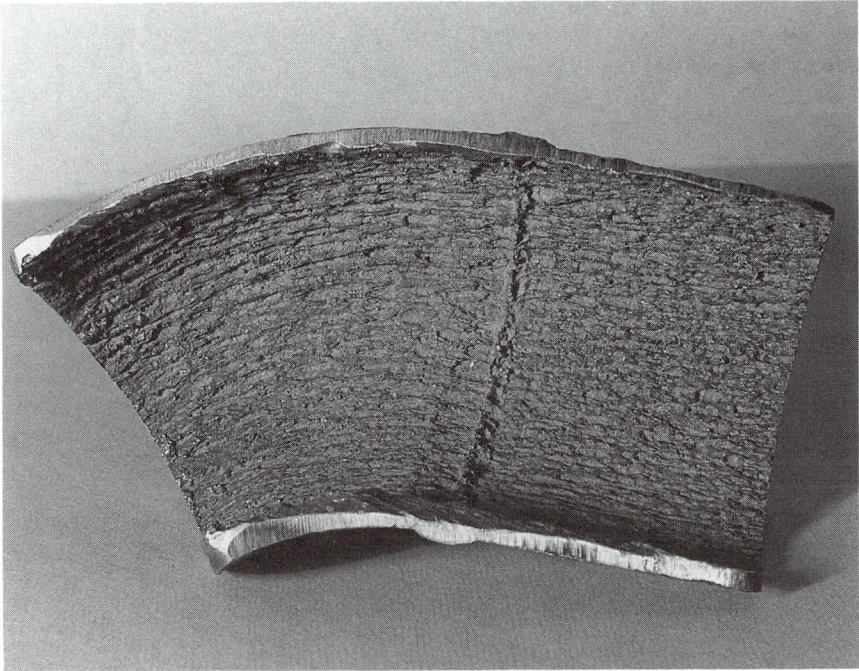

FIGURE 6-3.   Streaky pits of erosion–corrosion.

Strong, hard metals are more resistant to erosion and if they are also one-phase corrosion-resistant alloys they can perform well in erosion–corrosion situations. The single-phase 14% Si cast iron is excellent. Hardening processes that produce two-phase alloys generally make them less resistant to erosion–corrosion than alloys with solid solution hardening.

As one would expect, the problem is usually worse at high velocities, high temperatures, and high concentrations of the corrosive. However, if the corrosive environment strengthens the protective film, it sometimes happens that increasing velocity will reduce erosion–corrosion; this situation may occur with austenitic stainless steels in some environments.

Heat exchangers operate more efficiently at high velocities but the failure rate is often strongly related to velocity.

$$\text{Failure rate} \propto (\text{velocity})^n \qquad (6\text{-}1)$$

where n may be as high as 2.5. This equation, however, does *not* hold if the protective film is not damaged. Below a critical velocity the failure rate is zero.

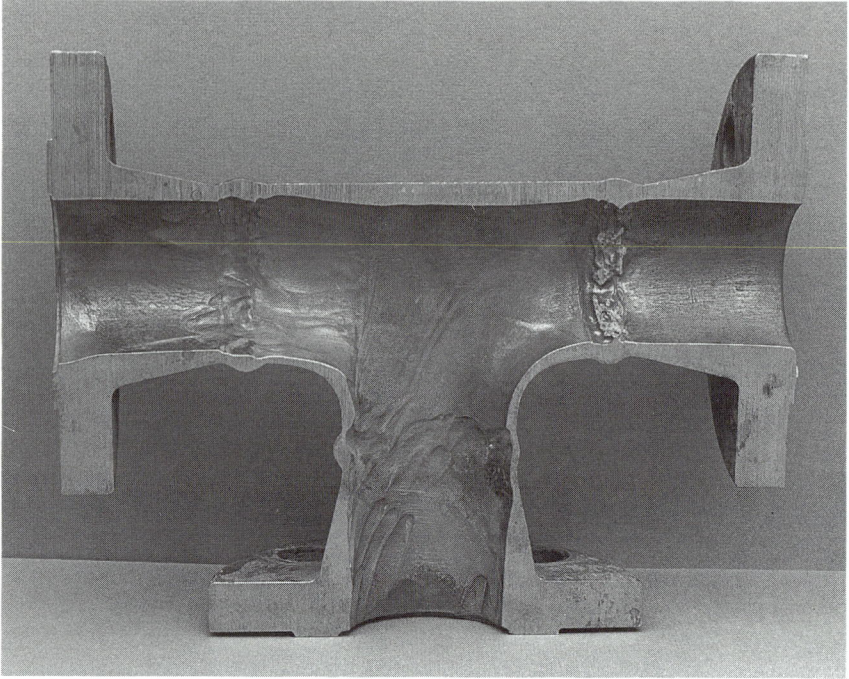

FIGURE 6-4.   Wavy surface, showing erosion-corrosion.

Environments that cause erosion–corrosion include gases, liquid metals, and organic liquids, as well as aqueous solutions. Liquid metals are particularly troublesome because of the high temperatures involved and their high density, which causes great mechanical wear when they are in motion.

Turbulence

Most erosion–corrosion problems occur in turbulent flow. The tubes in a heat exchanger, for example, often suffer erosion–corrosion only in the first few inches at the inlet end. Where the fluid has settled down to laminar flow the tubes look good.

- This "inlet-tube corrosion" can be combatted by extending the tubes a few inches beyond the tube sheet, by inserting short replaceable liners (ferrules) at the tube inlets, or even by reversing flow when the tubes are nearing the end of their life.

Turbulence is created by a change in pipe diameter, an abrupt change in flow direction, or an obstruction in a pipe, such as a weld bead protruding into the stream by $\frac{1}{8}$ in. or more. Erosion–corrosion then occurs just downstream from the weld, as illustrated in Figure 6-5.

- One of the great corrosion engineers, now retired, hated welders, especially welders who tried to do a perfect job. If they did not sensitize the metal they left big weld beads inside the pipes to create turbulence. He would clamp patches on a pipe until there was no room for another patch before he would call in a welder.

Impingement

Where the liquid is forced to make a sharp change in flow direction it strikes the metal surface. The metal will usually deform elastically but surface films may crack. If the liquid carries suspended solids, or even entrained bubbles, the damage is greatly intensified. About the worst erosion is caused by extremely fine silica sand with unworn edges, as is found in mining slurries.

Proper engineering can help you prevent erosion–corrosion in several ways.

1. Selecting strong, one-phase metals.
2. By coating the metal with a resilient or strong coating.
3. By decreasing the fluid stream velocity.
4. By avoiding turbulence; avoid abrupt changes in flow direction, avoid obstructions, make surfaces smooth.

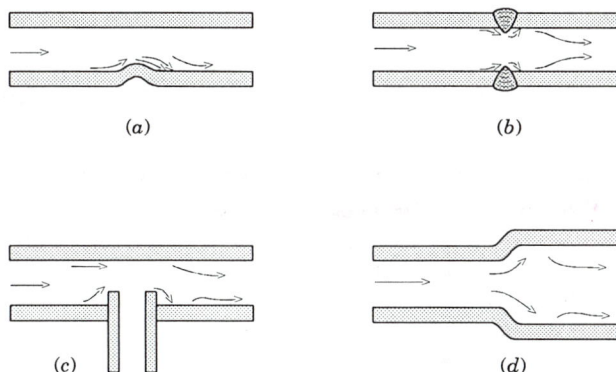

FIGURE 6-5. Erosion–corrosion from turbulence at (a) restriction in pipe, (b) protruding weld bead, (c) projecting pipe, and (d) change in diameter.

5. By increasing the thickness of metal in vulnerable areas.
6. By designing to allow replacement of parts; use impingement plates or baffles.
7. By filtering out abrasive contaminants. For steam and gases, trap water droplets.
8. By lowering the temperature.
9. By adding inhibitors.
10. By using cathodic protection.

Cavitation

Cavitation may accompany erosion–corrosion in a worst-possible scenario. It occurs only in high-velocity, turbulent liquids, such as are found with centrifugal pumps, turbines, boat propellers, and the like. A centrifugal pump undergoing cavitation pops and crackles as if nuts and bolts were going through it.

- Five people were standing around a pump listening to the strange noises it was giving off when the pump suddenly disintegrated. No one was killed! In another plant of the same company a compressor exploded from cavitation and killed nine people.

In cavitation the turbulence is so high that in the low-pressure areas the liquid momentarily vaporizes into millions of fine bubbles, which then implode when the pressure increases. Each implosion sends out a shock wave that is focused by other bubbles, firing a microjet of liquid against the metal surface. The microjet strikes the metal with enough force to leave a roughened spot that serves as a nucleation site for more vapor bubbles.

- A boat propeller for outboard motors was made from an experimental 90Al-10Mg alloy. In a test run it developed cavitation damage on the backside (the low-pressure side) within 15 min.

Cavitation appears as a patch of closely spaced pits or a roughened area on the metal. Most of the damage is caused by mechanical action although corrosion may have roughened the surface enough to initiate the cavitation. An example is shown in Figure 6-6. Liquids containing dissolved gases are particularly damaging because vaporization is so easy.

To prevent cavitation

1. Select materials that harden readily by cold work. (Hadfield's manganese steel is 100-years old and still one of the best.)
2. Use resilient coatings and linings.

FIGURE 6-6.   Cavitation appearance on the surface of a sliding bearing. [Reprinted by permission from Kenneth C. Ludema in *Metals Handbook*. 9th ed., Vol. 11, Kathleen Mills, Manager of Editorial Operations. ASM International, Materials Park, OH, 1986, Fig. 9, p. 488.]

3. Prevent entrance of dissolved air if it is not required for passivation.
4. Design to reduce vibration. High-frequency vibration creates low-pressure areas that produce bubbles.
5. Specify a smooth surface finish.
6. Minimize pressure differences in the flow. Prevent pressure from falling below the vapor pressure of the liquid.
7. Inject insoluble gas bubbles into the liquid to cushion the shock waves.
8. Use cathodic protection.

6.2 CORROSIVE WEAR

Corrosion usually develops a film or thick layer of corrosion products on the metal, slowing down further corrosion. When wear is also occurring it removes the protective film and allows corrosion to proceed rapidly.

The wear may be either *abrasion*, in which a hard, rough surface or hard particles under pressure slide across the corroding metal, or *adhesive wear*, in which two smooth surfaces slide over each other with sufficient pressure so that fragments are pulled off one surface and adhere to the other. Frictional heat is produced by the wear, which increases the rate of corrosion attack.

The corrosion first forms the film that is then removed by the wear. The film, being an ionic compound, is generally harder than the metal that it forms on, and is reasonably wear resistant if the film is thin. If the film grows thick, in most cases it is so brittle that wear breaks off this film down to the metal. In a few cases the film is soft and ductile (mainly chlorides, sulfides, phosphates, and salts of fatty acids) so that wear breaks only part of the layer off, keeping both wear and corrosion very low. Metal cutting lubricants often include additives that encourage formation of soft corrosion products.

Corrosive wear on roller and ball bearings produces a reddish-brown coating and very small etch pits over the raceway surface. The oxide corrosion products act as a grinding compound, leaving the balls and ball paths a dull gray color. Damage can be rapid and severe, as Figure 6-7 shows.

- Stray electrical currents passing through machinery bearings that have been penetrated by moisture destroy the bearings extremely rapidly.

Fretting Corrosion

Fretting is a form of corrosive wear in which highly loaded surfaces vibrate or slip only *slightly* in air. Sometimes termed "friction oxidation," fretting is

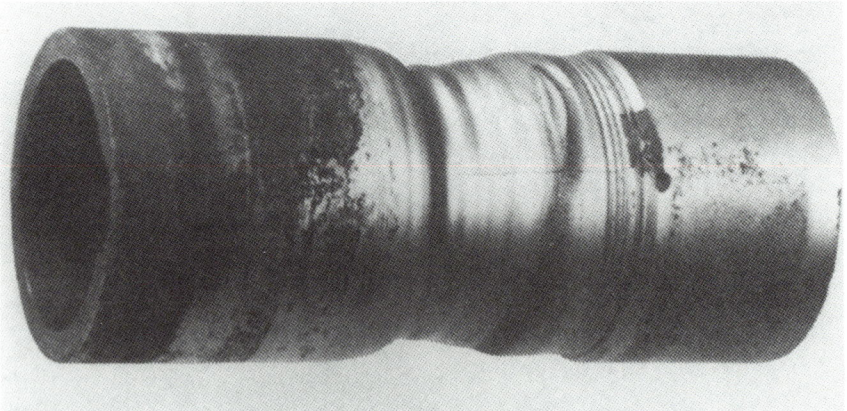

FIGURE 6-7.   Worn 440A stainless steel drive-roller sleeve from belt conveyer carrying anthracite coal in mildly corrosive conditions. [Reprinted by permission from *Metals Handbook*, 8th ed., Vol. 10, Howard E. Boyer, Editor. ASM International, Materials Park, OH, 1975, Fig. 2, p. 135.]

seen in bolted, riveted, keyed, or pinned joints, in splines, couplings, bearings, and press-fitted and shrink-fitted parts. Bolted tie-plates on railroad rails must be tightened frequently. Vibrating electrical contacts may go bad because oxide forms and increases resistance. Fatigue cracks in wire rope are commonly initiated by fretting.

- Automobiles shipped from Detroit to the West Coast by railroad had their engine bearings ruined by fretting before the cars were ever driven.

Ordinary wear increases approximately linearly with amplitude of the sliding motion. Fretting, however, is seldom seen at amplitudes above 25 μm and reaches a maximum at about 7.5 μm.

Fretting discolors the metal surface by the oxide formed: the reddish-brown oxide on steel is often called "cocoa." (The Germans call it "das Blut," i.e., blood.) When streaks of aluminum sheets are shipped the black oxide rub marks on the sheets are termed "traffic marks."

Fretting often initiates fatigue cracks, it destroys close tolerances, and ruins bearing surfaces. The wearing away of a protective coating can initiate galvanic corrosion or crevice corrosion.

The mechanism is now reasonably well understood. The two metal surfaces forced against each other have already been exposed to air and thus have a layer of oxide on their surfaces. When vibration slides the surfaces over each other slightly, the oxide ruptures at asperities (high points) on the surfaces. Bare metal sliding against bare metal will cold weld the asperities on one surface with those on the other. Further sliding breaks the welds. Metal fragments from the break immediately oxidize and the fracture spots on the surface reoxidize. The process repeats itself with every vibration.

Fretting corrosion, in which the environment is a liquid instead of air, has the same basic mechanism as fretting. An example is shown in Figure 6-8. Fretting corrosion occurs, for example, in nuclear reactors at contact points between the fuel element and the pressure tube, with vibration provided by turbulent flow of coolant or pulsing of the pump. A surgical implant fails if a metal plate screwed to bone frets against the surgical screws. Plate-type stainless steel heat exchangers in dairies have undergone fretting corrosion of their thin sheets due to pulsating pump action, contaminating pasteurized milk with unpasteurized milk. Austenitic stainless steel, aluminum, and titanium are particularly susceptible because their passive films can be worn off and the metals are relatively soft.

To prevent fretting and fretting corrosion

1. Lubricate. Greases are commonly used but molybdenum disulfide is much better.

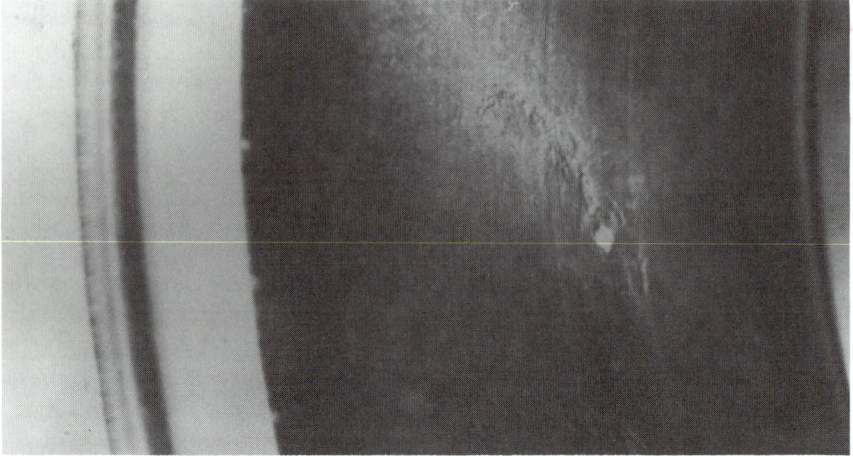

FIGURE 6-8.   Freon-compressor shaft of 4140 steel that failed by fretting corrosion because of a loose bearing. [Reprinted by permission from F. Eric Krueger in *Metals Handbook*, 8th ed., Vol. 10, Howard E. Boyer, Editor. ASM International, Materials Park, OH, 1975, Fig. 3(*c*), p. 158.]

2. Apply soft protective coatings. However, phosphate and sulfide coatings and electroplating, like lubricants, give only short-term protection.
3. Seal out air or seal in a noncorrosive atmosphere. This technique is used in aircraft compasses.
4. Avoid fine surface finish (more asperities contact each other and increase fretting) or put different finishes on the two mating surfaces.
5. Isolate critical parts from vibration.
6. The most-used method is simply to replace bearings and other fretted parts on a regular schedule.

6.3 CORROSION FATIGUE

The cracking of a metal in a corrosive environment may be caused by corrosion fatigue, hydrogen embrittlement, stress-corrosion cracking, or liquid-metal embrittlement. Alternatively, metallurgical factors unrelated to corrosion, such as hot shortness, could be responsible. Corrosion fatigue failures occur in conditions of fluctuating or cyclic stresses that are tensile at least part of the time. Common examples are failures of ship propeller shafts, wire cables used in water or humid air, and pipes cracking from thermal cycling.

The fatigue strengths of metals in air are only about one-half their tensile strengths. Their corrosion fatigue strengths may possibly be no more than 10% of their fatigue strengths. The number of stress cycles that any metal can resist before cracking is always less for corrosion fatigue than for fatigue (shown in Fig. 6-9).

While almost all metals can fail by corrosion fatigue if the cyclic stresses are high, it does very little good to increase the strength of the metal. Good ductility and impact strength help, but the corrosion resistance is the major factor.

- A pump shaft in a chemical plant, made of steel with a fatigue strength of 207 MPa (30,000 psi), failed in 2 months. It was replaced with a stronger steel with twice the fatigue strength. This steel failed in just a little over 2 months. A stainless steel shaft with a fatigue strength of 275 MPa (40,000 psi) was then tried. The result was 10 years of service.

Any corrosive environment can cause corrosion fatigue and the more corrosive it is, the shorter the corrosion fatigue life. Consequently, the remedies for general corrosion also work for corrosion fatigue.

Cracking is transgranular (Pb and Sn are exceptions) with slight or no branching, although many short cracks usually form on the surface. Figure 6-10 shows a typical corrosion fatigue fracture. The wavelike "beach marks" usually seen on ferrous metal fatigue and corrosion fatigue fractures can be detected only at high magnification on some nonferrous metals.

Corrosion fatigue cracks usually initiate from corrosion pits, fretting corrosion, or from a surface roughened by corrosion. In contrast, fatigue, as its name implies, initiates very slowly and only after local slip at the surface creates intrusions that act as stress raisers to start the crack. Corrosion, however, provides stress raisers that greatly speed the slip process for corrosion fatigue.

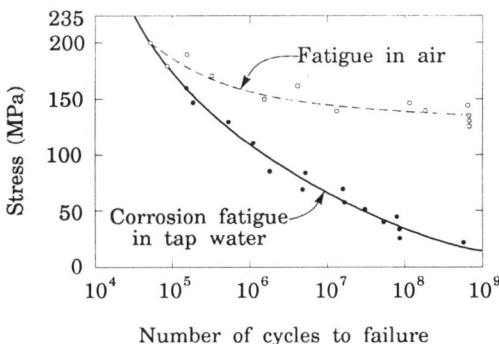

Number of cycles to failure

FIGURE 6-9.   Comparison of fatigue and corrosion fatigue curves for an aluminum alloy. [Reprinted with permission from *Corrosion Basics; An Introduction*, National Association of Corrosion Engineers, 1984, Fig. 5.24, p. 104. Copyright by NACE.]

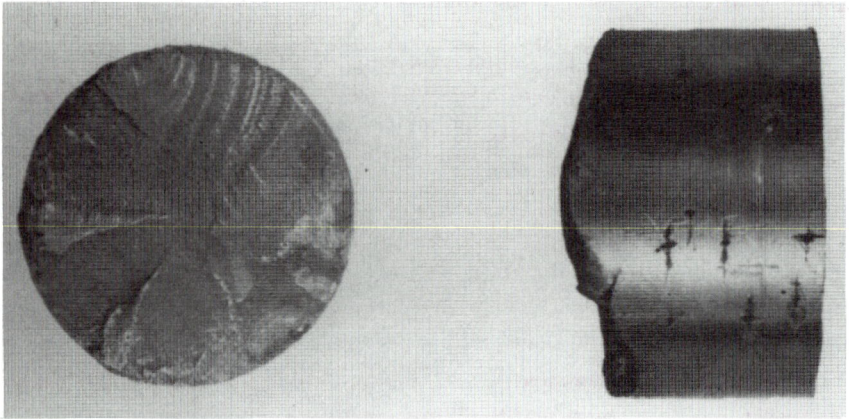

FIGURE 6-10.   Corrosion fatigue. The parallel cracking is characteristic. [From H. R. Copson in *Corrosion Resistance of Metals, and Alloys*, F. L. LaQue and H. R. Copson, Editors. ACS Monograph No. 158, by permission of Reinhold Publishing Corp., New York.] Fig. 1.21, p. 24, 1963.

Corrosion fatigue can be prevented by reducing the stress or reducing the corrosion.

1. Select a material with good ductility and good corrosion resistance.
2. Design to minimize cyclic stresses; reduce the maximum stress, avoid vibration.
3. Specify smooth surface finish. Avoid rough machining, grinding, sheared edges.
4. Put the metal surface in compression: shot peen, nitride, carburize.
5. Avoid heat treatments that decarburize steel surfaces. A soft, decarburized surface fatigues readily.
6. Use butt welds instead of fillet welds. All stress will be concentrated on small fillet welds.
7. Avoid stress raisers: abrupt changes in cross section, sharp corners.
8. Use protective coatings. Zinc and cadmium are often used, but ensure that electroplated coating is not in tension.
9. Add a corrosion inhibitor.
10. Cathodically protect the metal.

These last two techniques will not prevent fatigue, of course, but will prevent corrosion fatigue.

6.4 HYDROGEN DAMAGE

Atomic hydrogen (H) produced by the cathode reaction can diffuse inside metals to embrittle or crack them. In the presence of tensile stresses, complete failure of the metal is often sudden and unexpected.

The source of the hydrogen can be from corrosion in acids, acid pickling and cleaning treatments, electroplating, or $H_2(g)$ atmospheres in high-temperature metallurgical processes (casting, annealing, heat treating, welding, etc.). The cathode process in acid occurs in two stages.

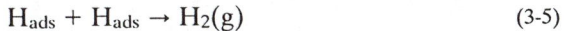

$$H^+ + e^- \rightarrow H_{ads} \tag{3-4}$$

$$H_{ads} + H_{ads} \rightarrow H_2(g) \tag{3-5}$$

Some of the adsorbed atomic H may diffuse into the metal, especially if many surface sites are occupied by "poisons" that interfere with the hydrogen combination step (Reaction 3-5). Such surface poisons include sulfide, cyanide, and arsenate ions. Sulfides and bisulfides can form from corrosion of sulfide impurities in the metal or from an environment such as sour (H_2S-containing) gas.

The hydrogen atom, being very small, diffuses rapidly through the metal lattice interstitially (in the spaces between the metal atoms), and along grain boundaries and dislocation lines. Atomic hydrogen can be removed from the damaging positions by a "bake-out"; for steel, 3 h at 175°C (350°F) is usually sufficient. Probably most of the hydrogen diffuses to traps and becomes harmless, rather than diffusing out of the metal.

- Glass-lined steel vessels must be washed off immediately if any acid is spilled on the *outside* of the vessel. Otherwise, atomic hydrogen diffuses through the vessel wall and spalls off the glass lining because of pressure buildup.

Hydrogen Blistering

Atomic hydrogen diffusing into low-strength steel, copper, or aluminum collects at internal cavities, laminations, and inclusion interfaces where it combines to form gaseous molecular hydrogen (Reaction 3-5). The gas can build up pressures in excess of 300 atm that rupture the metal, forming blisters on the surface. While in high-temperature processes the blisters can exceed 10 cm (4 in.) in diameter, in aqueous corrosion they are commonly a few mm (1/16 in.). The blisters (see Fig. 6-11) often occur in a row, lining up along a slag inclusion, and so are customarily seen with dirty steel. They can be avoided by using low-sulfur steels, not cold rolled, by ultrasonic inspection to detect internal flaws such as laminations, and by corrosion inhibitors.

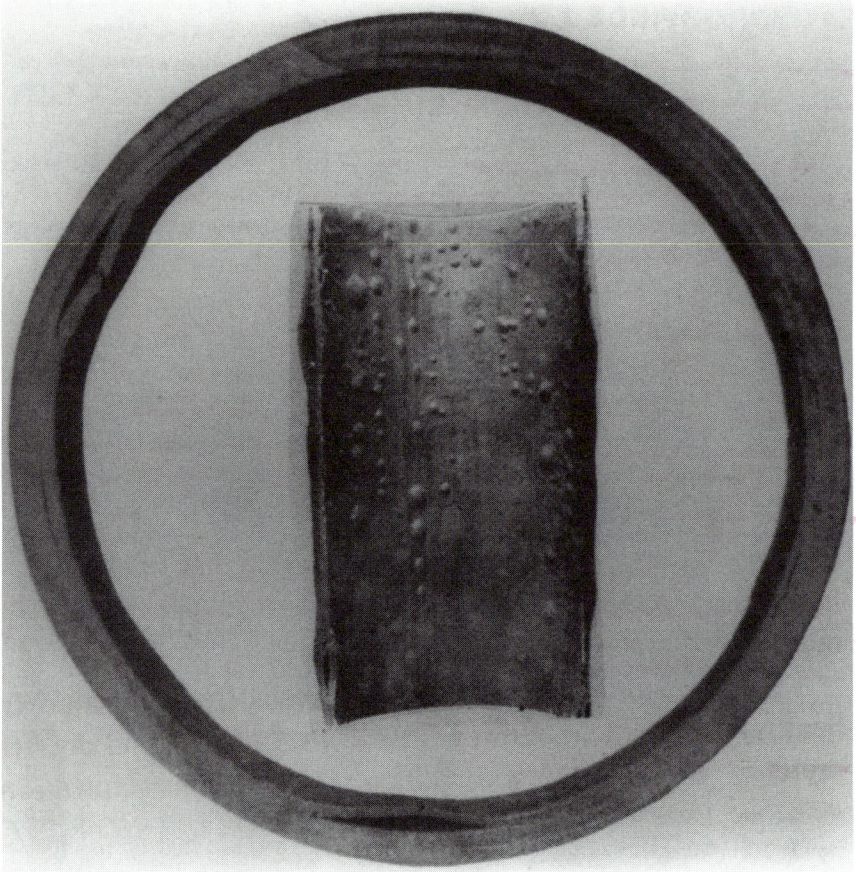

FIGURE 6-11.   Examples of hydrogen blisters in tubing. [Reprinted with permission from *Corrosion Control in Petroleum Production*, National Association of Corrosion Engineers, 1979, Fig. 2.25, p. 18. Copyright © NACE.]

Hydrogen-Induced Cracking

Atomic hydrogen diffusing into a metal can embrittle it. The best-accepted explanation of the embrittlement is a decohesion model in which the hydrogen atoms diffuse ahead of a crack tip and accumulate in regions of high triaxial stress, particularly in grain boundaries and dislocations. The hydrogen alters the bonding between metal atoms, lowering their cohesive forces, producing microcracks ahead of the main crack, and allowing the crack to extend under tensile stresses below the yield strength.

②
Another theory proposes that hydrogen at the crack tip adsorbs on the new crack surfaces as the crack opens, lowering the surface energy and thus the energy required for fracture. The fracture is primarily brittle but is accompanied by some plastic deformation as the crack extends.

A theory of enhanced plastic flow, based on electron-microscopic evidence of plasticity at the crack tip, proposes that atomic hydrogen collects at dislocations at the crack tip and enhances their mobility and the creation of new dislocations. The crack extends by alternate slipping at the tip.

With some metals, notably titanium and zirconium, but not steel, hydrogen reacts to form a brittle hydride phase. With other metals a normally unstable hydride may be stable in the high-stress region ahead of the crack. The crack breaks the brittle hydrides easily and then stops until more hydrides form.

- Hydride formation in zirconium–alloy pressure tubes at the Pickering nuclear reactor near Toronto cost over $1 billion to replace the 720 tubes.

Hydrogen-induced cracking, also called hydrogen stress cracking, is the brittle fracture of a normally ductile alloy caused by a sustained tensile stress and hydrogen interacting with the metal. The cracks are sharp, single cracks that usually form at a fracture stress below the yield strength of the metal. Metals that are often affected are the medium- and high-strength steels, alloy steels, and austenitic and ferritic stainless steels. These metals, however, appear to have a threshold stress below which hydrogen-induced cracking does not occur.

Sulfide Stress Cracking.   The brittle hydrogen-induced cracking of steels in environments containing H_2S and H_2O is commonly termed sulfide stress cracking. The cathode surfaces reduce hydrogen ions and are poisoned by S^{2-} or HS^- ions, allowing the metal to be charged with hydrogen. Cracking normally occurs only if the steel has a Rockwell hardness of HRC 22 or more. The danger of hydrogen-induced cracking or sulfide stress cracking increases with lower temperature [below 80°C (175°F) for steels], which is exactly the opposite of stress-corrosion cracking.

Line-pipe steels often show *stepwise cracking*, which is a combination of hydrogen embrittlement and hydrogen blistering. Internal cracks (small blisters) form along laminations or MnS stringers elongated in the rolling plane and these cracks are linked by transverse cracks. An example is shown in Figure 6-12.

A metal can also be charged with hydrogen by excessive cathodic protection. For example, high-strength, heat-treated steel aircraft bolts coated with zinc have cracked by hydrogen-induced cracking. However, galvanized, high-carbon steel cable strands on suspension bridges hold up well.

To prevent hydrogen damage

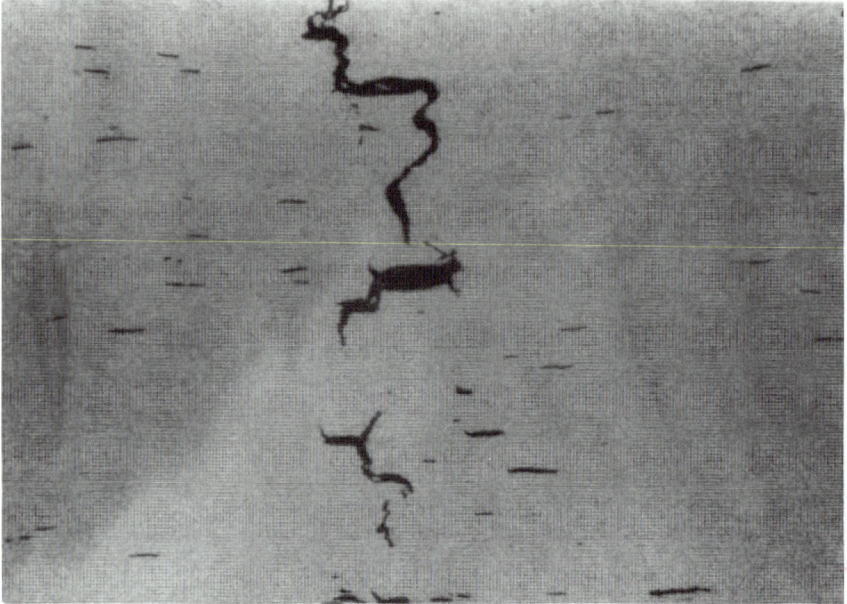

FIGURE 6-12.   Hydrogen stepwise cracking in carbon steel from a catalytic cracker. [Reprinted with permission from R. D. Merrick, "Refinery experiences with cracking in wet H_2S environments," in *Materials Performance*, Vol. 27, No. 1, Jan. 1988, Fig. 6, p. 33. Copyright © by NACE.]

1. Select resistant materials: Monel, Alloy 20, Hastelloy C-276, or very low-strength steels.
2. Use high-strength steels at a maximum hardness of HRC 22.
3. Use low-hydrogen electrodes and weld in dry conditions. Dissociation of water by the electric arc charges the metal with hydrogen.
4. Avoid incorrect acid pickling and plating. Bake out the hydrogen afterwards.
5. Design in compressive stresses.
6. Use corrosion inhibitors but not hydrogen poisons.
7. Control cathodic protection carefully, or better, use anodic protection.
8. Avoid anodic metallic coatings on high-strength steel.

6.5 STRESS-CORROSION CRACKING

Stress-corrosion cracking (SCC) is the fracture of a metal caused by corrosion and a simultaneous sustained tensile stress. As engineering design has required new, stronger alloys and as new technologies have exposed the metals to an ever-wider range of environments, SCC has become the most serious problem facing corrosion engineers today. For example, 30% of all aircraft accidents, major and minor, are caused by SCC. Stress-corrosion cracking is often accompanied by hydrogen embrittlement and is closely related to corrosion fatigue and to liquid-metal embrittlement.

The study of SCC of brass (season cracking) goes back to nineteenth-century British rule in India, where the army noticed that brass cartridges cracked during the monsoon season (see Fig. 6-13). The problem was traced to high humidity plus ammonia vapor from the cavalry stalls combined with the residual stresses left in the metal by necking and crimping the brass to hold the bullet. The problem was solved by reducing the stress and reducing the corrosion: annealing the cases and storing the cartridges in air-tight containers.

- Propane stored in a large tank previously used for liquid ammonia picked up 1 ppm ammonia that cracked all the brass fittings in thousands of homes using the bottled gas. Replacement cost the company $80 million—a century after season cracking was understood.

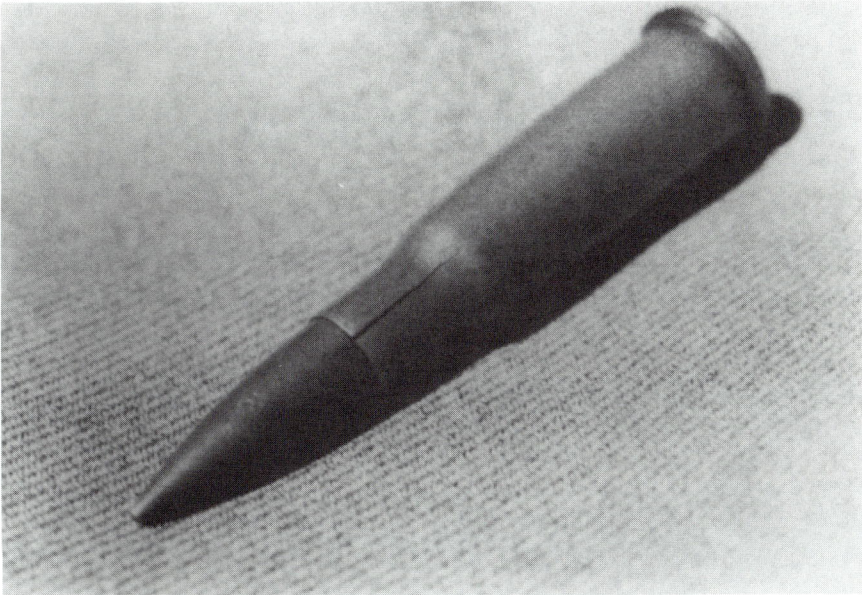

FIGURE 6-13.   Stress-corrosion crack in the neck of a cartridge case from ammonia, high humidity, and high temperature. [Reprinted with permission from Donald J. Wulpi, *Understanding How Components Fail*, ASM International, Materials Park, OH, 1985, Fig. 5, p. 218.]

"Caustic embrittlement" of steel is another term for SCC that goes back to the late nineteenth century. Caustic soda (NaOH) was added in small amounts to boiler water to prevent scaling, but OH^- concentrated under rivet heads and at hot spots, cracking the mild steel. Steam engines were constantly blowing up until the operators quit adding caustic. A modern example is shown in Figure 6-14.

- In 1884 over 10,000 boilers exploded or ruptured in the United States. During the period of 1974–1984 no boilers or pressure vessels failed.

Environments

Virtually all metals crack by SCC in *some* environment, with the stronger metals more susceptible. At one time it was thought that pure metals were immune, and it is true that they seldom if ever crack, but pure metals are infrequently used and rarely have much tensile stress applied to them, as they are not strong.

FIGURE 6-14.   Caustic SCC of a 321 stainless steel expansion bellows in 27-atm (400-psi) steam with entrained, highly alkaline, boiler water, 24 h. [Reprinted with permission from C. P. Dillon, ed., *Forms of Corrosion: Recognition and Prevention*, NACE Handbook 1, National Association of Corrosion Engineers, 1982, Fig. 4.3.4.2, p. 68. Copyright © by NACE.]

Usually each type of alloy cracks in only a few specific environments. Copper alloys do not crack in caustics nor steels in ammonia vapor, for example. Some of the more notorious combinations that do crack are listed in Table 6-1. In general, the environments must cause localized corrosion and, therefore, are not highly corrosive. Very corrosive environments that eat out the cracks faster than they can grow cause uniform attack. Stress-corrosion cracking is much more likely at high temperatures where protective films are weaker. For example, stainless steels in chloride solutions generally crack only above 60°C (140°F), although in contaminated environments (especially at low pH) they may crack at room temperature.

- Titanium alloy parts crack at high temperatures from a single fingerprint on the surface (molten salt).

Stresses

The stress responsible for SCC is the sum of the applied stress, internal stresses in the metal, and fabrication stresses. Internal stresses alone may be sufficient to crack, as they were with the cartridge brass, or fabrication stresses alone, as they sometimes are with welded high-strength steel or

TABLE 6-1   Metal-Environment Combinations Susceptible to Stress-Corrosion Cracking [a]

Alloys	Environments
Carbon steels, moderate strength	Caustic; nitrates; carbonates and bicarbonates; anhydrous liquid NH_3; moist H_2S
Carbon steels, high strength	Natural waters; distilled water; aerated solutions of Cl^-, NO_3^-, SO_4^{2-}, PO_4^{3-}, OH^-; liquid NH_3; many organic compounds
Stainless steels	Chlorides; caustic; polythionic acids; pure water + O_2
Nickel alloys	Hot caustic; molten chlorides; polythionic acids; high-temperature water and steam contaminated with O_2, Pb, Cl^-, F^-, or H_2S
Copper alloys	Ammonia; fumes from HNO_3; SO_2 in air + water vapor; mercury.
Aluminum alloys[b]	Aqueous solutions especially with halogen ions; water; water vapor; N_2O_4; HNO_3; oils; alcohols; CCl_4; mercury
Titanium alloys:	Red fuming HNO_3; dilute HCl or H_2SO_4; methanol and ethanol; chlorinated or brominated hydrocarbons; molten salt; Cl_2; H_2; HCl gas
Zirconium alloys:	Organic liquids with halides; aqueous halide solutions; hot and fused salts; halogen vapors; conc HNO_3
Magnesium alloys	H_2O + O_2; very dilute salt solutions

[a] Only the more common combinations are listed.
[b] Resistances of Al alloys to SCC are detailed in Table 9-4.

force-fit assemblies. The stresses must be tensile or torsional to open up a crack and must be sustained, differentiating SCC from corrosion fatigue. Once a crack has started, the wedging action of solid corrosion products in the crack adds additional stress. (Rust would like to occupy 7 times the volume of the steel that corroded.) The stresses and corrosion must occur simultaneously to extend a stress-corrosion crack.

- A new nuclear reactor with 304 stainless steel tubing throughout but without radioactive fuel, was pressure tested by filling it with tapwater (30 ppm Cl^-) rather than high-purity water that would be used in actual service. When the reactor was heated, every tube cracked. A year later, after repairs, they started to test it again but by mistake filled it with tapwater again! They immediately called a renowned corrosion engineer who advised them to drain it and clean it before it had time to pit. Since the reactor had not been heated the metal had not been stressed and SCC had not started. He was correct.

Cracking

Stress-corrosion cracks initiate in various ways: from notches created by intergranular corrosion, from pitting damage of a passive film, from pits formed by crevice corrosion or erosion corrosion, or from localized attack of slip traces on film-protected surfaces. Figure 6-15 shows SCC that has initiated from a pit.

- In chemical plants, the most common source of SCC is initiation by crevice corrosion under thermal insulation that has gotten wet. An infrared scan can spot the danger by detecting cold spots caused by wet insulation.

The cracks are usually fine, branching cracks, often filled with corrosion products, and running only approximately perpendicular to the applied stress (see Fig. 6-16). The pH inside the cracks is much lower than the bulk solution, since the cracks are crevices with the same reactions as any crevice. Conditions there are favorable for hydrogen evolution even if the bulk solution is basic.

For most metal-environment combinations the cracks are intergranular, unlike corrosion fatigue, but for some combinations, most notably unsensitized stainless steels in chlorides, the cracks are transgranular. A crack may propagate entirely through the metal section or it may stop if the stress concentration has been relaxed or if the crack tip hits an obstacle, such as an unfavorably oriented grain boundary, inclusion, or the like.

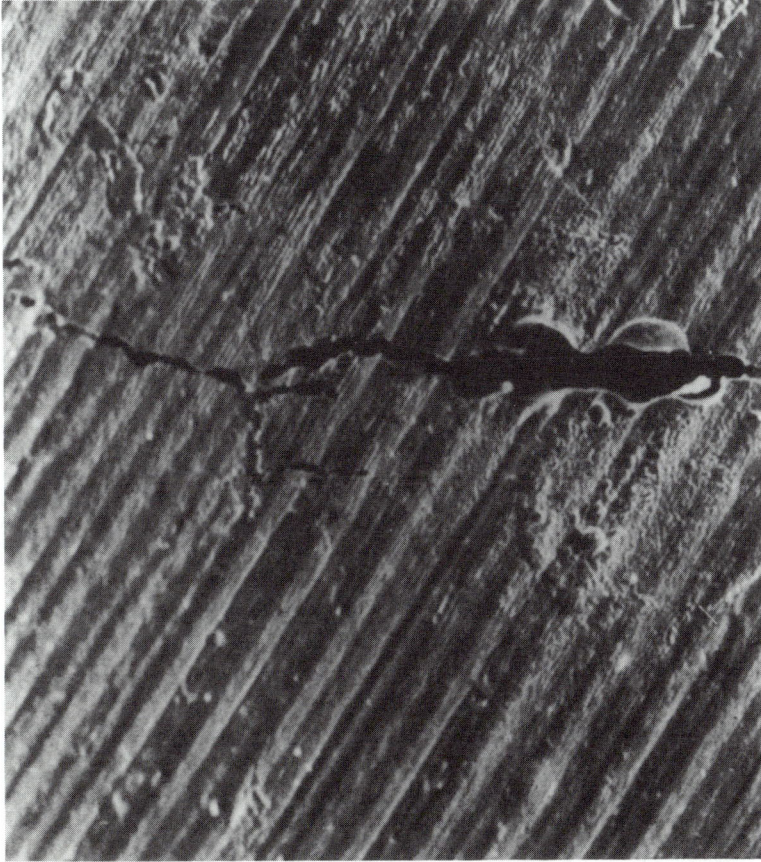

FIGURE 6-15.   Stress-corrosion crack initiating from a corrosion pit in a high-strength turbine steel exposed to oxygenated, demineralized water. [Reprinted with permission from R. H. Jones and R.E. Ricker in *Metals Handbook*, 9th ed., Vol. 13, Joseph R. Davis, Senior Editor. ASM International, Materials Park, OH, 1987, Fig. 6(*a*), p. 149.]

Mechanisms

Stress-corrosion cracking is a general term for several processes, so one or more of the following mechanisms may be operative.

1. Hydrogen embrittlement is a major mechanism of SCC for steels and some other metals. Hydrogen is produced on the crack surfaces that are the cathodes, while the crack tip is the anode. Atomic hydrogen diffuses through the metal ahead of the crack tip and embrittles the metal in that region.

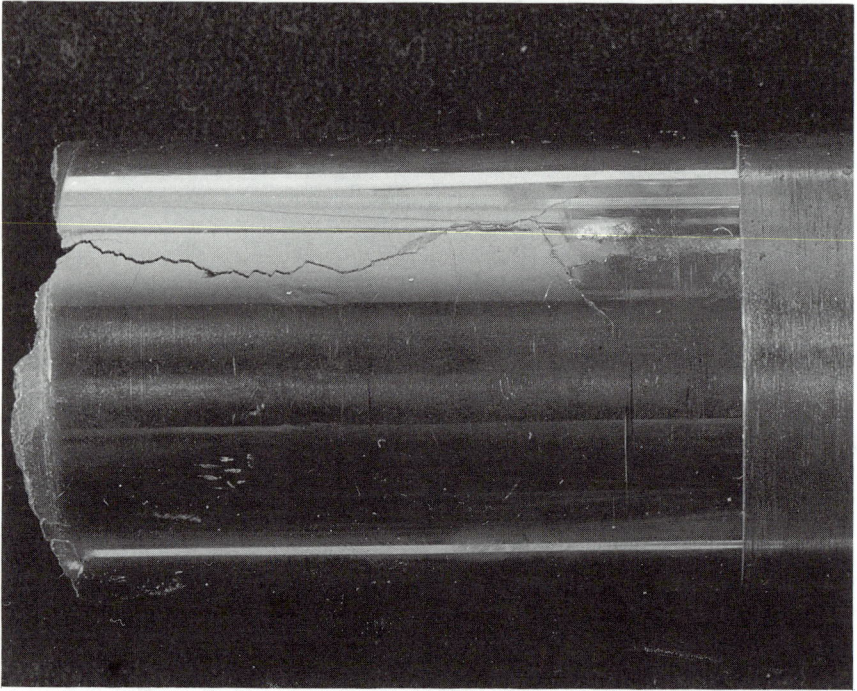

FIGURE 6-16.   Stress-corrosion cracking of a vertical pump shaft sleeve operating in a salt slurry.

2. A film-rupture mechanism is important for passive metals (stainless steel, aluminum, and titanium), for metals that tarnish (brass is an example), and perhaps even for metals with adsorbed ions or adsorption inhibitors. Stress ruptures the film locally, setting up an active–passive cell that causes rapid corrosion at the rupture until the metal repassivates. The corrosion site becomes a stress raiser that ruptures the film again, repeating the cycle.
3. Preexisting weak or corrodible paths, a mechanism proposed 50 years ago, is valid for special cases, such as sensitized stainless steel with its Cr-depleted regions that are both very corrodible and weak. The cracks follow the grain boundaries (see Fig. 6-17). Similarly, aluminum alloys precipitation hardened to near maximum hardness are sensitized and fail by intergranular SCC.
4. An adsorption mechanism in which the damaging species adsorbs on the crack surfaces, lowering the fracture stress, could explain cracking in liquid metals and organic liquids.

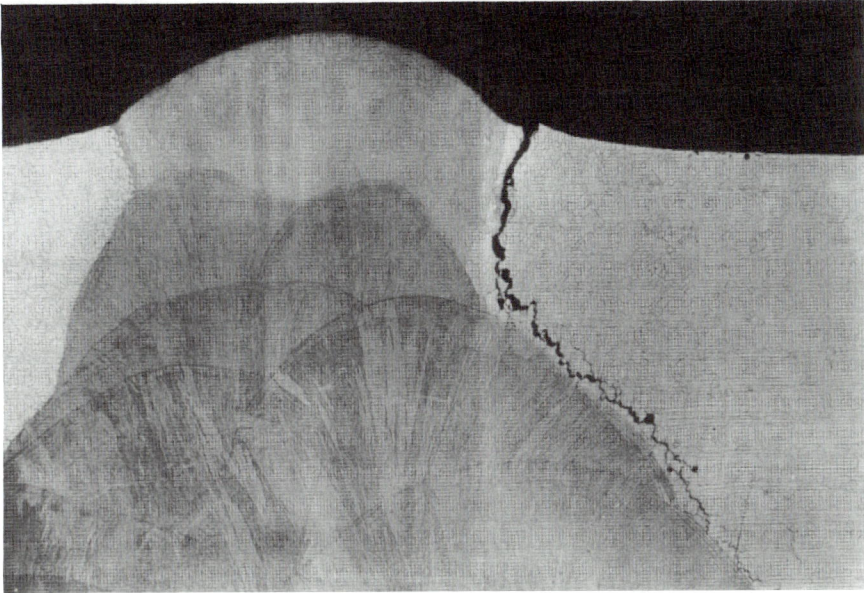

FIGURE 6-17.   Intergranular SCC in the HAZ of a 304 stainless steel weld. Magnification 19×. [Reprinted with permission from Barry M. Gordon and Gerald M. Gordon in *Metals Handbook*, 9th ed., Vol. 13, Joseph R. Davis, Senior Editor. ASM International, Materials Park, OH, 1987, Fig. 6, p. 930.]

Prevention

Prevention or reducing the likelihood of SCC can be accomplished in the following ways.

1. Select a material not known to crack in the specific environment involved; ensure that the microstructure is not susceptible to intergranular attack.
2. Lower internal stresses on the metal and use at the minimum required hardness. A stress relief anneal can double or triple the life of chemical equipment.
3. Design to minimize tensile stresses and stress raisers. Avoid tensile fabricating stresses. Use materials with similar coefficients of expansion.
4. Put in compressive stresses: shot peen, swage, roll in threads.
5. Use as low a temperature as possible.
6. Eliminate the specific species (Cl^-, NH_3, etc., responsible for SCC) from the environment.

7. Prevent concentration of the damaging species: Avoid alternate wetting and drying; remove any crevices.
8. Use protective coatings, such as chemical-resistant paints.
9. Use corrosion inhibitors.
10. Use cathodic protection with great care. Overprotection can cause hydrogen embrittlement.

- Anhydrous liquid ammonia began to be used extensively as a nitrogen fertilizer in the southern United States in the 1950s. Producers quickly became alarmed at the great number (3%) of burst ammonia storage tanks, and for people living in the vicinity of a tank rupture it was a life-threatening disaster. The industry commissioned a study that included a statistical analysis of the failures and laboratory tests to determine the factors responsible.
- The study found that large, high-strength steel tanks were the most likely to crack, that the cause was SCC with dissolved O_2 being the cathode reactant, and that 0.1% H_2O in the ammonia completely inhibited SCC. They recommended (1) all tanks over 3 ft in diameter should be stress relieved, (2) tanks should be thoroughly purged of air, and (3) 0.2% H_2O should be added to the ammonia. The industry accepted *all* the recommendations and completely solved their problem.
- By the 1970s the countries of western Europe had begun using large quantities of ammonia fertilizer but their engineers were much cleverer than the Americans. Obviously, any one of the three remedies would prevent SCC, and while remedies (2) and (3) were easy, remedy (1) was extremely expensive, so they omitted the stress relief. Then their tanks began bursting—first one in Scotland, then a couple in Norway, one in Germany, and so on. They immediately commissioned a study that found (a) the purging was often incomplete, and (b) their SCC had initiated in the vapor regions of the tanks, not the liquid. On cool nights anhydrous ammonia condensed from the vapor onto the tank walls. They now stress relieve their large tanks.
- If you learn from this story that it is stupid to cut corners where a potential disaster is involved, it will be the most important lesson you could learn from this text.

Liquid-Metal Embrittlement

Liquid metals can bring about brittle cracking of many solid metals when tensile stresses, either sustained or fluctuating, are present simultaneously. The problem is a critical one for the nuclear industry where liquid-metal coolants are used, but it also appears frequently when manometers and thermometers containing mercury break around industrial equipment. Even metals that are plated with a low-melting metal, or are soldered or brazed, may be embrittled. The environments responsible (mercury, cesium, gallium, sodium, lead, etc.) are fairly specific. For example, mercury embrittles zinc but not cadmium and also embrittles aluminum but not magnesium. Sodium embrittles copper and aluminum alloys but not nickel (see Table 7-15 for a list of liquid metals known to embrittle major alloys).

• Mercury cannot be shipped by air freight. Why?

Cracking usually appears as a single intergranular crack and is extremely rapid; velocities of 25 cm/s (10 in./s) are often observed. Solid metals near their melting points can also embrittle adjacent metal, with the embrittlement increasing up to the melting point and then continuing at a constant rate. At higher temperatures the embrittlement decreases either because the solid metal becomes more ductile or adsorption of the liquid metal decreases.

The embrittlement is clearly not from electrochemical corrosion. The mechanism is most probably an adsorption-induced reduction in cohesion at the crack tip. Time- and temperature-dependent processes, such as liquid diffusion or corrosion, are not rate controlling.

The usual remedy is to avoid liquid–metal combinations that are known to cause embrittlement. Do not use low-melting metals near their melting point in combination with strong metals. Use electroplating or cladding as a barrier between the solid and liquid.

• A relatively minor fire melted aluminum jacketing protecting thermal insulation on pipes. Molten aluminum contacted nickel alloy, stainless steel, and carbon steel pipes, which then ruptured, releasing hydrocarbons until the entire plant burned at a cost of $100 million. The company has now removed aluminum jacketing from all flammable installations.

STUDY PROBLEMS

6.1   Damage to the blade of a road grader was caused by severe corrosion of the threads of several bolts, although the bolts appeared to be in excellent shape before they gave way. Name and describe the mechanism(s) of the type of corrosion most probably involved.

6.2   A salesperson, to impress a potential customer, demonstrated a new stainless steel shell-and-tube heat exchanger that their company had just installed. The heat exchanger was filled with lake water but they forgot to turn the heat off when they left, so it boiled dry. What sort of corrosion problem could result? Explain.

6.3   Explain the most probable cause of corrosion: An oil well drill pipe has fractured. Numerous deep pits are found inside it.

6.4   Some Al-5%Mg rivets have fractured in a ship superstructure made of aluminum alloy. All metal was painted. Assuming that corrosion played some role in the failures, which of the following suggestions would prevent future failures? Explain. (a) Replace with stainless steel rivets of the same strength, (b) fillet weld rather than rivet, or (c) paint with a zinc-rich paint.

6.5   It took 1 h to transfer 98% H_2SO_4 at 25°C (80°F) through a steel pipeline. A stronger pump was installed that reduced the transfer time to 15 min, but the pipe failed in less than 1 week. Which of the following suggestions could prevent damage to a new steel pipe? Explain. (a) Increase pipe diameter, (b) reduce acid concentration, or (c) align pipe sections carefully.

6.6   A welded steel storage tank containing 5% NaOH at 90°C (195°F) split suddenly after 5-years service. Which of the following changes could prevent another failure? Explain. (a) Keep the tank full, (b) use a higher strength steel, (c) stress relieve all welds, or (d) lower the pH to 7.

6.7   A cast steel, steam condensate line has a large hole through an elbow. Horizontal and vertical sections of the steel line are unaffected. Which of the following changes could prevent another failure? Explain. (a) Increase velocity to prevent condensation, (b) make the elbow with a larger radius of curvature, (c) make the elbow from a more corrosion-resistant metal, or (d) make the elbow from a stronger metal.

6.8   A chlorinator at a water pumping station contains liquid Cl_2 in a stainless steel line with brass fittings. Technicians found that spraying a 15% ammonia solution over the connections is an excellent way of detecting leaks. (A dense, white smoke of NH_4Cl is produced.) Comment.

6.9   A heat exchanger has erosion–corrosion inside its tubes. Which of the following corrective actions could be used to prevent further damage? Explain. (a) Decrease the tube diameter, (b) extend tubes beyond tube sheets at the inlet end, (c) strengthen the tubes by heat treatment, or (d) apply a resilient rubber coating.

6.10 In what ways could SCC differ from corrosion followed by stress overload? Explain differences in (a) metal appearance after fracture, (b) the corrosive environment, and (c) possible differences in type of stress.

6.11 While brass valves are commonly used underground in water distribution systems, the valves in one area began cracking. The metal was found to be red-brown in spots instead of the original yellow. Name the type of corrosion and explain its mechanism.

6.12 How could you tell whether a failure is corrosion fatigue or SCC?

6.13 A pump impeller handling a corrosive liquid shows a small patch of closely spaced pits. What is your diagnosis and what is your cure for the disease?

6.14 Steam power turbine blades have been found to have "streaky-looking pits." Which of the following suggestions would be desirable? Explain. (a) Install a water trap, (b) lower the operating temperature, (c) zinc coat the blades for galvanic protection, or (d) stress relieve welds.

6.15 To avoid any chance of SCC a fabricator stress relieved an entire 316 stainless steel reactor at 650°C (1260°F) followed by a 10-h furnace cool. Should he do anything else to ensure safe operation? Explain.

6.16 The corrosion resistance of steel is improved by raising the pH to the range of 8–11. One plant regularly adds LiOH or NaOH for this purpose *unless* the system experiences wetting and drying. Why the exception?

6.17 A series of 10-year-old stainless steel heat exchangers, ordinarily cooled with recirculating river water, was cooled with untreated river water for 48 h during an emergency. A few weeks later five of the coolers failed by extensive cracking. Discuss the causes.

6.18 A hydrogen cooler was given a hydrostatic pressure test and the test water containing sediment was allowed to remain in it for an extended period of time. The cooler tubes, made of admiralty brass, were then found to be cracked. Explain.

6.19 A small, deep-drawn oxygen cylinder made of Muntz metal (60Cu-40Zn) cracks down one side. Discuss possible causes.

6.20 An expansion joint for a steam-heated reboiler was welded from 321 stainless steel followed by a complete stress relief at 620°C (1150°F) and a 10-h furnace cool. Intergranular cracking occurred in 6 weeks. Massive carbide precipitates were found in the microstructure. What errors in materials selection, welding, heat treatment, or operation could have been responsible?

6.21 Replacement parts for aircraft, consisting of an aluminum coupling swaged onto a magnesium alloy rod, were ruined when the aluminum cracked. Stress-corrosion cracking was suggested because the magnesium was badly corroded, but the parts had never been put into service—no stress was ever applied. Discuss.

Memos

6.22 Memo to: Corrosion Engineer
From: Chet Bailey, Production Supervisor
We have two cast steel pipes connected together and only one of them is pitting. Supposedly, they are identical but upstream from the joint everything looks fine. The other half is badly pitted for several inches

and looks spongy. The pipe carries a CO_2–H_2O solution at $40 \pm 10°C$, 55 atm and 4 m/s. We just happened to find it when we took it apart trying to trace a water hammer problem. You know it all, so tell us what's wrong and what to do about it.

6.23 Memo to: Corrosion Engineer

From: Chet Bailey, Production Supervisor

We have a vertical cooler lined with 330 brass that some moron claimed was corrosion resistant. The cooler contains HCl vapor and a little air at temperatures up to 150°C. The brass doesn't seem to be pitted or etched but it's cracked at every sharp bend, letting the gas through to the steel shell. And it's probably news to you, but HCl eats the hell out of steel. We're going to have to shut down for repairs or relining before the whole cooler comes down. Can we braze the cracks? Do we have to replace it? With what? We need an answer now. ~solder w/med w/high melt point

6.24 Memo to: Corrosion Engineer

From: Chet Bailey, Production Supervisor

We've had to shut down the entire west mill because of one burst line. This pipe was a good quality 18–8 stainless steel, cold formed and welded, 5 years old. Temperatures were moderate—not over 100°C. In fact, the break was at the coolest part of the line, which was all thermally insulated with a good polyethylene wrap around the whole thing. The line was carrying 140-MPa virgin naphtha stock with 24 vol% H_2 and 0.21 vol% H_2S, and that's flammable in case you didn't know.

The fractured pipe has a lot of hairline cracks but no rust. There was some white powder on the surface but it was found to be just $AlCl_3$ from the asbestos insulation. When we rebuild, what can we do different?

7

Corrosion in Major Environments

No metal is suitable under all conditions for any one environment, but the corrosion engineer will immediately think of certain particularly compatible environment–metal combinations that will work in most circumstances. Anyone involved in materials selection for corrosive environments should remember the 11 environment–metal combinations in Table 7-1.

7.1 NATURAL ENVIRONMENTS

The Atmosphere

Atmospheres are classified as rural, industrial, or marine in an effort to describe the different types of corrosion a little more accurately, but this classification is sadly inadequate. The wide variations in rainfall, humidity, temperature fluctuations, wind, and pollutants prevent classification schemes from giving more than a mere indication of corrosion rates.

Two variables deserve particular mention.

1. The type and amount of pollutants carried in the air. Most important of these is SO_2 gas, commonly produced by industries burning high-sulfur coal. With moisture and oxygen, SO_2 produces dilute sulfurous and sulfuric acids, which are extremely corrosive to almost all metals, except the compatible combination with lead. Solid pollutants, even dust, tend to be hygroscopic and hold moisture against the metal surface.

2. The fraction of the time that the metal remains wet. While rain plays a part, it may wash off contaminants from the metal surface and actually reduce

131

TABLE 7-1   Compatible Combinations

Environment	Metal	Comments
The atmosphere		
low stresses	Aluminum	May be anodized
high stresses	Galvanized steel	Zinc hot-dipped
Water	Copper, Cu alloys	Freshwater or seawater
Soil	Steel	Plus cathodic protection
Sulfuric acid		
dilute	Lead	Below 70% acid
concentrated	Steel	Above 70% acid
Hydrochloric acid	Hastelloy B	Developed for HCl
Nitric acid	Stainless steels	
Alkalies	Nickel, Ni alloys	
Food	Tin	Coating on steel
Strong oxidizers	Titanium	Even at high temperature

corrosion. Condensation of moisture, from changes in humidity and temperature, is more important because it puts a thin, air-saturated film on the metal. In field tests, time of wetting is often measured by a galvanic probe with flat electrodes separated by a thin insulator strip.

Atmospheric corrosion normally starts off at a very high rate and then slows down to an approximately constant rate after a few years as a solid layer of corrosion products develop on the surface. Test results must include the test duration to have any meaning.

Aluminum.   Aluminum and the atmosphere are a particularly compatible combination with a corrosion rate that settles down to a constant value in about 2 years. In rural atmosphere aluminum may remain bright for years but when pollutants are present it turns a dull gray with mild surface roughening. Galvalume is the trade name for a 55Al-44Zn coating on steel that combines the inherent corrosion resistance of aluminum with the sacrificial protection offered by zinc.

Zinc.   Zinc's excellent resistance to atmospheric corrosion is due to an insoluble basic carbonate film that develops to protect the surface. However, if SO_2 or other acidic pollutants are present in industrial environments the corrosion rate may be 10 times as great as in rural atmospheres. Galvanized zinc coatings on steel combine corrosion resistance with high strength to make an outstanding composite material. Although zinc is anodic to steel and will sacrificially protect the steel, zinc corrodes more slowly than steel. In a seacoast environment zinc typically corrodes only one twenty fifth the rate of steel and, more importantly, zinc's corrosion rate decreases with time while steel's rate increases. The life of a zinc coating is directly proportional to its thickness.

Weathering Steels. Some high-strength, low-alloy (HSLA) steels have been developed especially for atmospheric environments where they form a thick, dense rust layer that protects them from further rusting (Fig. 7-1). Commercial trade names include Cor-Ten, Mayari R, and Stelcoloy. The steels have 0.2% C and approximately 1% Mn for toughness along with about 0.4% each of Cu, Ni, Si, and Cr to develop the dense, protective scale and additional strength. During the first year of exposure they rust like any ordinary steel but within 2 years rusting virtually ceases and their color changes to a deep brown, which is attractive enough that they are usually left unpainted. Weathering steels are used extensively for bridges and highway guardrails because of their combination of moderate cost, high strength, and low maintenance. They are not satisfactory in marine or some industrial atmospheres. An example of a weathering steel bridge is shown in Figure 4-1.

Table 7-2 lists atmospheric corrosion rates for common metals. Keep in mind that the rates given are indicative of service but the data are not reproducible. The weather and the pollutants will never be quite the same again.

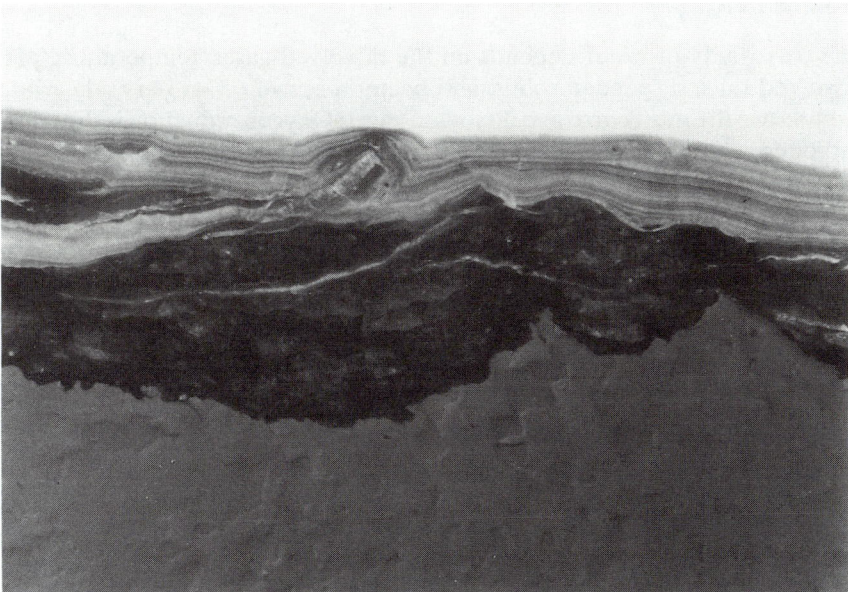

FIGURE 7-1.   Protective scale developed on Stelcoloy weathering steel after 10 years of exposure. Photographed Nov. 1966. [Courtesy of Stelco Inc.]

TABLE 7-2   Corrosion Rate of Metals in the Atmosphere

Metal	Years	Average Corrosion Rate (μm/y) Rural	Industrial	Marine
Carbon steel	3.5	14.6	24.0	414.
Carbon steel	7.5	10.1	13.9	
Carbon steel	15.5	7.6	8.7	
Gray cast iron	6–12		61	102
304 Stainless Steel	5–15		0.1	<0.03
Nickel	20	0.2	2.6	0.2
Aluminum	10–30	0.1	2.5	3.8
Galvalume (Al-Zn)	13	0.4	0.3	0.6
Copper, ETP	20	0.4	1.4	1.3
Cartridge brass	20	0.5	3.1	0.2
Lead, chemical	20	0.3	0.4	0.5
Tin	20	0.5	1.7	2.9
Magnesium AZ31B	3	13.	25.	27.
Zinc	20	1.1	6.3	1.5

Sources: Aluminum with Food and Chemicals, 1964; Craig, 1989; Davis, 1987; De Renzo, 1985; Shreir, 1976; Slunder and Boyd, 1983.

Freshwater

The corrosivity of water depends on the dissolved gases, temperature, pH, dissolved salts, suspended solids, and bacteria present. Oxygen and carbon dioxide are the most corrosive dissolved gases: Oxygen reduction is the usual cathode reaction but CO_2 dissolves in water to form H_2CO_3, which ionizes slightly to form H^+ and bicarbonate HCO_3^- ions. While CO_2 is more soluble than O_2 it is much less corrosive because the carbonic acid ionizes only very slightly.

Chlorine, added in small amounts to water to kill bacteria, dissolves to form HClO and HCl acids in low concentrations that can damage copper and copper alloys. Ammonia also attacks copper and can cause SCC of its alloys. If copper is corroded by ammonia it usually is a decomposition product from nitrogen compounds added to water to prevent corrosion of steel, but occasionally a natural water is found to be high in nitrogen.

Increasing temperature reduces the solubilities of dissolved gases but increases diffusion rates, reaction rates, and ionic mobility, which increases conductivity of the water. Consequently, increasing temperature may reduce or increase corrosion depending on which factor is rate controlling. A differential temperature cell can be set up with copper if the hot metal is 65°C (150°F) or more above the cold end. Copper corrodes at the hot end and plates out at the cold end. Corrosion of copper alloys in steam is shown in Table 9-5.

Below about pH 8, steel forms loosely adherent oxides, but at higher pH the oxide layer is extremely strong and adherent. High pH also precipitates $Ca(OH)_2$ and lime ($CaCO_3$), which provides an additional barrier to O_2.

Naturally soft water, and softened water, are very corrosive to metals partly because dissolved salts increase water conductivity, but hard water (containing Ca^{2+}, Mg^{2+}, and HCO_3^- ions) can precipitate a protective scale. Chloride, sulfate, manganese, sulfide, phosphate, and nitrate ions all damage passive films. For steel, the corrosivity of water is proportional to the chloride content up to about 6000 ppm.

Copper and copper alloys make a compatible combination with freshwater but steel is also often used where corrosion is less important than cost. Tin-coated metals are commonly used to handle distilled water. Table 7-3 shows typical corrosion values for many common metals in various waters.

Seawater

Seawater has an average salt content of 3.5%, which consists of approximately 77.8% NaCl, 10.9% $MgCl_2$, 4.7% $MgSO_4$, 3.6% $CaSO_4$, 2.5% K_2SO_4,

TABLE 7-3 Typical Corrosion Rates of Common Metals in Freshwater

Metal	Water	Corrosion Rate (mm/y)
Carbon steel	Clean river water	0.06–0.09
Carbon steel	Polluted river water	0.06–0.15
Carbon steel	Natural hard water	0.15
410 Stainless steel	Tap water	Unattacked
304 Stainless steel	Tap water, 20°C(68°F)	<0.0025
304 Stainless steel	Tap water, 70°C (158°F)	<0.0025, small pits
Nickel	Distilled water, 25°C (77°F)	<0.0025
Nickel	Domestic hot water	<0.005
Aluminum	Tap water	Satisfactory
Copper	Potable water	0.005–0.025
Copper	Very soft water	0.0025–0.125
Brass	Unpolluted	0.0025–0.025
Chemical lead	Tap water, 25°C (77°F)	0.009
Chemical lead	Aerated lake water	0.114
Magnesium	Pure water, 100°C (212°F)	Up to 25
Mg-Zn alloys	Pure water, 100°C (212°F)	0.25–0.005
Tin	Distilled water, 20°C (68°F)	0
Zinc	Hard water	0.0025–0.005
Zinc	Soft river water	0.020
Zinc	Soft tap water	0.005–0.010

Sources: Craig, 1989; Davis, 1987; Slunder and Boyd, 1983.

0.2% $MgBr_2$, and 0.3% $CaCO_3$ plus small amounts of a dozen other ions, with a pH buffered at 8.1–8.3. In the polar regions and near large rivers the salinity is appreciably less. Seawater has a conductivity about 400 times greater than freshwater and is biologically active, tending to coat metal surfaces with biofouling films that may shield the metal from the water and reduce corrosion or may set up crevice corrosion.

The high chloride content pits most passivating metals at low velocities but as velocities increase above 1 m/s (3 ft/s) pitting decreases. Titanium and Hastelloy C do not pit at all.

Compatible combinations with seawater are copper and copper alloys, such as admiralty brass (70Cu-29Zn-1Sn), aluminum bronze, and silicon bronze. Table 9-5 shows the corrosion resistance of some major copper alloys in saltwater.

Corrosion rates in seawater vary greatly due to temperature, depth, oxygen content of the water, formation of biofilms, and velocity, but some typical rates are given in Table 7-4. Conditions are ambient temperatures and near-surface immersion unless otherwise specified. Corrosion resistance of the principal plastics is listed in Table 9-6.

- In answering a Memo, one student wrote, "If seawater is cheap and plentiful you may want to consider this approach to solve your problem."

TABLE 7-4   Corrosion Rates of Common Metals in Seawater

Metal	Conditions	Corrosion Rate (mm/y)
Carbon steel	1.5 years	0.120
Carbon steel	4.5 years	0.085
Carbon steel	8.5 years	0.070
Carbon steel	At 1700 m	0.020
Cast iron		0.210
410 Stainless steel	All depths	1.270 pitting
304 Stainless steel	4 years	0.0002 pitting
Nickel		0.025
Hastelloy C		0
Monel 400	At 704 m	0.033 pitting
Aluminum	4 years	0.0017 pitting
Copper	4 years	0.018
Admiralty brass		0.008
Silicon bronze		0.004
Lead		0.015
Magnesium		Unsuitable
Zinc	At surface	0.015

Sources: Craig, 1989; Shreir, 1976.

Soil

Soils are classified according to their texture: the proportions of sand, silt, and clay that compose them. Clay has the finest particle size, under 0.002 mm in diameter, silt runs 0.002–0.02 mm, fine sand 0.02–0.20 mm, and coarse sand 0.20–2 mm. Soil also will generally have 1–5% organic matter, although peats can have as much as 95%. In a good soil, around 50% of the total volume will be pore space between the mineral particles, filled with moisture and air. The moisture, usually only a thin surface film on the mineral particles, holds dissolved salts and gases. Air in the pores contains 10–20 times as much CO_2 as atmospheric air.

While a sandy (light) soil drains well and tends to be noncorrosive, peat and wet clay (heavy soils) are the worst. Unaerated soils combined with a high sulfate content and neutral pH are a perfect breeding ground for sulfate-reducing bacteria that cause particularly severe attack.

Loam is a friable mixture of sand, silt, and clay. To be friable (crumbly), a loam will have a clay content of only 10–25% of the total mineral content. A typical loam might run 40% sand, 40% silt, and 20% clay. Mucks contain 15–50% organic matter and peats are more than 50% organic.

The single most-important property of a soil in determining its corrosivity is its electrical resistivity. Table 7-5 indicates the corrosivities usually associated with various resistivity ranges.

The resistivity ρ is a property of the electrical conductor (the soil in this case) where

$$R = \rho \ell / A \qquad (7\text{-}1)$$

in which R is the resistance in ohms (Ω), ℓ is the length of the conduction path, and A the cross-sectional area of the conductor.

Oxygen concentration cells are established quickly in underground corrosion, with O_2 high near the earth's surface but almost zero below the water table. Soil is inhomogeneous: Bits of clay hold moisture and create a crevice

TABLE 7-5   Relationship Between Soil Resistivity and Corrosivity

Resistivity Range	Corrosion
<1000 $\Omega \cdot$ cm	Very severe
1000–5000	Severe to moderate
5000–15000	Moderate to light
>15000	Light

where they contact metal, whereas sandy regions become cathode areas with high O_2 concentrations. Organic matter holds moisture and decomposes into a variety of corrosive chemicals, and microbial corrosion is always a possibility.

In soils the metal most often used is steel because of its combination of cost, strength, and moderate corrosion. If a life of more than a few years is needed, the steel is coated and cathodically protected. The range of results of long-term (10–14 years) corrosion tests conducted by the National Bureau of Standards is given in Table 7-6.

- The natural gas distribution systems in the United States annually average nearly one gas leak per mile of buried pipe, requiring a maintenance cost of $800 million and growing.

7.2 ORGANIC ENVIRONMENTS

Organic Acids

Formic Acid.   Formic acid, HCOOH, is used for dyeing and finishing textiles and paper, for insecticides, and production of other chemicals. Car-

TABLE 7-6   Corrosion of Common Metals in Soils

Soil/Metal	Corrosion Rate (mm/y)	Maximum Pit Depth (mm)
Clay loam, good aeration		
Steel	0.012	1.72
Gray cast iron	<0.014	3.93
Aluminum[a]	0.001	0.53
Copper, tough pitch	0.001	0.15
Yellow brass	0.001	<0.15
Chemical lead	0.002	0.45
Zinc	0.007	0.43
Galvanized steel	0.002	<0.15
Peat, high S^{2-}, poor aeration		
Steel	0.064	2.98
Gray cast iron	0.063	4.33
Aluminum[b]	Destroyed	1.6+
Copper, tough pitch	0.035	1.16
Yellow brass	0.065	2.63
Chemical lead	0.002	0.83
Zinc	0.110	2.53
Galvanized steel	0.067	2.22

[a]Fine sandy loam; fair aeration.
[b]Muck; very poor aeration.
Source: Romanoff, 1957.

bon steels are attacked rapidly. Type 304 stainless is usually used for storing the acid but Alloy 20 stainless is necessary for high temperatures. The Hastelloy C nickel alloys are excellent, as is titanium, but in anhydrous formic acid, titanium corrodes catastrophically (see Table 7-7).

Acetic Acid.   Acetic acid, CH_3COOH, (vinegar acid) is widely used in production of plastics, dyes, insecticides, and other chemicals. Glacial acetic is the term for the 99.8% pure acid. Aluminum is commonly used as container material for storage and shipment but it is rapidly attacked at high temperatures. Carbon steel is unsuitable for any concentration, even at room temperature, as shown in Table 7-8, so Type 304 stainless steel is the lowest grade ferrous metal used. Aeration is particularly bad for nickel and copper alloys. Titanium is satisfactory at all concentrations up to the boiling point but high-strength titanium alloys are susceptible to SCC.

Citric Acid.   Citric acid, $HOOCCH_2C(OH)(COOH)CH_2COOH$, derived from lemon, lime, and pineapple juice, is used primarily as a flavoring, so consequently any contamination by corrosion products cannot be tolerated. Type 304 stainless steel is satisfactory at moderate temperatures but 316 is recommended for all concentrations up to the boiling point. Table 7-9 shows typical corrosion rates.

Fatty Acids.   Organic acids derived from natural fats and oils, mostly with 16 or 18 carbon atoms arranged in a straight chain with a single acid group are termed fatty acids. They are used in soaps, detergents, lubricants, and

TABLE 7-7   Corrosion of Common Metals in Formic Acid

Metal	Acid Concentration (%)	Temperature (°C)	Aeration	Time	Rate (mm/y)
304 Stainless steel	90	Boiling	None	48 h	12.7
Nickel	90	Room			0.1
Nickel	90	100 (212°F)			0.45
Hastelloy C	90	Boiling	Yes	48 h	0.18
Aluminum	5	Room		84 d	0.147
Aluminum	5	100 (212°F)			31.6
Monel 400	50	Boiling	None		0.025
Copper	5	Boiling		96 h	0.02
Copper	90	Boiling		96 h	0.22
Lead	85	60 (140°F)			0.032
Zinc	4.6	100 (212°F)	None	4 h	30.3
Galvanized steel	2.5	Room	In vapors	30 d	0.119

Sources: Behrens, 1987–1989; Craig, 1989; Slunder and Boyd, 1983.

TABLE 7-8   Corrosion of Common Metals in Acetic Acid

Metal	Acid Concentration (%)	Temperature (°C)	Aeration	Time (d)	Rate (mm/y)
Steel	5	Room			1.14
304 Stainless steel	50	60 (140°F)			0.0003
304 Stainless steel	Concentrated	60 (140°F)			0.081
Alloy 20	98	128 (262°F)			0.05
Nickel	5	116 (240°F)	None		0.007
Nickel	50	Boiling			0.12
Monel 400	50	Boiling			0.05
Aluminum	5	25 (77°F)		84 d	0.042
Aluminum	5	Boiling			12.4
Aluminum	Glacial	Boiling			0.17
Copper	6	25 (77°F)	Yes		0.67
Copper	100	100 (212°F)	None		0.1–0.4
Copper	Concentrated	100 (212°F)	Yes		2.5–8.0
70–30 Brass	1.1	60 (140°F)	Yes		0.1–0.4
Tin	100	Boiling	None		4.267

Sources: Behrens, 1987–1989; Britton, 1952; Craig, 1989.

TABLE 7-9   Corrosion Rates of Metals in Citric Acid Solutions

Metal	Acid Concentration (%)	Temperature (°C)	Aeration	Time (d)	Rate (mm/y)
304 Stainless steel	10	99–102 (210–215°F)			0.210
304 Stainless steel	58	54 (130°F)		5	0.217 pits
316 Stainless steel	10	99–102 (210–215°F)			0.013
316 Stainless steel	58	54 (130°F)			0.021 pits
Aluminum	2	100 (212°F)			1.5
Tin	6.4	25 (77°F)	None	7	0.052

Sources: Aluminum with Food and Chemicals, 1964; Britton, 1952; Craig, 1989; Polar, 1961.

rubber products. Corrosion products often catalyze their decomposition, turning them rancid. Table 7-10 lists a few corrosion rates.

Petroleum

Refined petroleum is not corrosive to metals because electrochemical corrosion is not possible in the absence of an electrolyte. However, crude oil commonly contains moisture, sulfur compounds, and carbon dioxide plus dissolved salts.

For moderate temperatures carbon steel is the metal most commonly used to handle petroleum. However, aluminum's corrosion rate is lower than steel's so that it can be used for pipelines and distillation towers. With very corrosive crudes at high temperatures, stainless steels or even nickel alloys are required (see Table 7-11).

Human Body

Medical uses for metal within the human body include surgical implants to replace bones (see Fig. 7-2), and also for reconstructive and cosmetic surgery, heart valve parts, and such. Porous, sintered metals are used where ingrowth of flesh or bone is needed.

TABLE 7-10   Corrosion Rates of Metals in Fatty Acids

Metal	Acid	Temperature (°C)	Aeration	Time (d)	Rate (mm/y)
304 Stainless steel	Cottonseed	275 (525°F)	Moderate	50.5	0.967 pits
316 Stainless steel	Mixed, crude	100 (212°F)	Strong	43	0.124 crevice corrosion
316 Stainless steel	Tallow	245 (475°F)		147	0.003
Nickel	Tallow	245 (475°F)		147	0.4
Inconel	Tallow	245 (475°F)		147	0.008
Aluminum	Stearic	<270 (<520°F)			Resistant
Aluminum	Oleic	To bp			Resistant
Aluminum	Linoleic, Linolenic	To bp			Resistant
Copper	Tallow	220 (425°F)		84	0.068
Tin	Stearic	330–340 (625–645°F)			Corroded

Sources: Aluminum with Food and Chemicals, 1964; Behrens, 1987–1989, Britton, 1952, Polar, 1961.

TABLE 7-11   Corrosion Rates of Metals in Crude Oil

Metal	Crude	Temperature (°C)	Time (h)	Rate(mm/y)
Carbon steel	Refinery crude, 0.6%S	260 (500°F)		0.093
Carbon steel	Refinery crude, 0.6%S	400 (750°F)		1.56
Cast steel	Petroleum vapor, 12 atm	345 (650°F)	1000 h	33.9
304 Stainless steel	Refinery crude, 0.6%S	260 (500°F)		0.002
304 Stainless steel	Refinery crude, 0.6%S	400 (750°F)		0.015
304 Stainless steel	Petroleum vapor, 12 atm	345 (650°F)	1000 h	0.023
316 Stainless steel	S-contg. oils	350 (660°F)		<0.5
Hastelloy C	With brine, 1.4%S	63 (145°F)		Excellent
Aluminum		250–500 (500–900°F)		Satisfactory
Copper, ETP				Good
70–30 Brass	S-contg. oils	150 (300°F)		>1
Titanium		Room		0.007–0.589

Sources: Craig, 1989; Davis, 1987; De Renzo, 1985.

Corrosion must be at a very low rate with corrosion products that do not poison or damage tissue, cause pain or blood clotting, or interfere with tissue growth or body reactions. Some people are extremely sensitive to metal ions or develop a sensitivity. For example, stainless steel sometimes causes rashes or pain because of the nickel ions produced in corrosion.

• About 9% of women and 1% of men are allergic to nickel.

Implant metals must passivate in the high-chloride fluids of the body and so are restricted to just a few alloy systems. Cobalt–chromium alloys are the most used: Haynes Stellite 21, a Co-Cr-Mo-C alloy, and Haynes Stellite 25, a Co-Cr-W-Ni alloy. Titanium and titanium alloys are becoming increasingly popular although pure titanium has low strength and can only be used in reconstructive surgery. Type 316L stainless steel is used primarily for temporary implants, but its corrosion resistance is improved considerably by electropolishing, which removes surface defects that could initiate pitting.

All types of corrosion, including galvanic corrosion, attack implants. Pitting and crevice corrosion are the most common problems but fretting corrosion of plate and screw prosthetic devices also occurs. Corrosion fatigue, even at a walking frequency, is particularly serious for the Co-Cr alloys but seldom is a problem for titanium alloys.

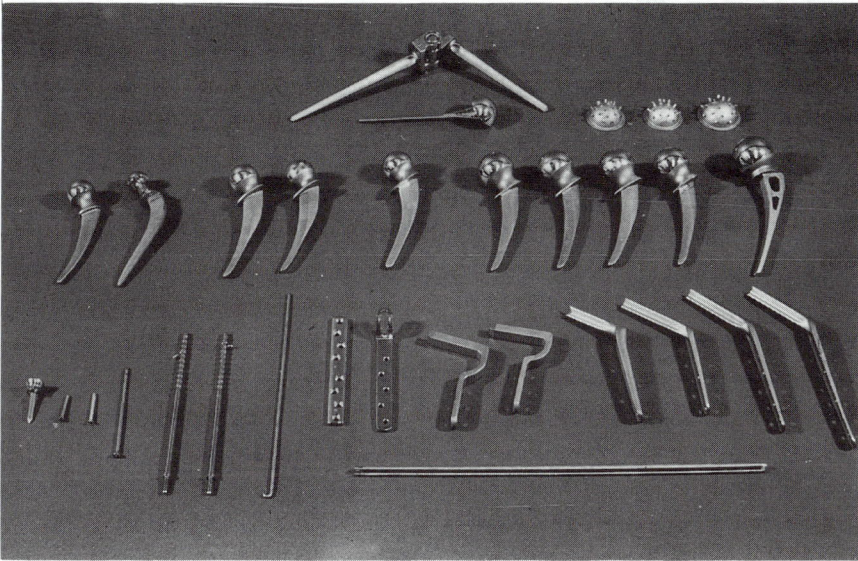

FIGURE 7-2.   Examples of medical implants removed because of infection, pain from corrosion, or corrosion fatigue.

- A magnesium plate screwed to a fractured bone with stainless steel screws disappeared even before the fracture healed, much to the amazement of the doctors who examined the x-rays.

Dental alloys are subjected to a much wider variety of corrosives than medical implants but they are also much easier to replace. Fillings are commonly Ag-Sn-Cu-Zn alloy amalgams while plates, bridges, crowns, and so on, which involve fusion to porcelain, are made of a wide variety of alloys based on gold, palladium, nickel, or cobalt. Orthodontic wires are commonly stainless steel, Co-Cr-Ni alloys, Ni-Ti alloys, or β titanium.

In the presence of high aeration, high chloride, pH varying from 8 to 4.5 or less, and sometimes even sulfide ions, corrosion conditions are severe. However, saliva is usually well buffered (maintaining constant pH) and proteins, and the like, act as corrosion inhibitors. The chloride present makes pitting and crevice corrosion possible, while even galvanic corrosion occasionally occurs between two adjacent metals. In that case the electric current that flows through the gums causes intense pain. Also, bacterial corrosion can occur from SRB or acid-producing bacteria.

Sewage and Waste

A U.S. government study in 1986 estimated that the nation would have to spend $136.5 billion for waste-water treatment by the year 2000. Few metals can withstand the chlorides, sulfates, hydrogen sulfide, and bacteria present in municipal sewage or the extreme fluctuations in composition of industrial waste water. Combine these chemicals with high-temperature processes for oxidation and incineration, which are becoming frequently necessary, and the list of satisfactory materials becomes quite limited.

Austenitic stainless steels (304, 316, 316L, etc.) can withstand chloride concentrations up to 400 mg/L (or ppm) in municipal sewage sludge for oxidation processes at temperatures of 175–315° (350–600°F) without pitting or cracking. At higher chloride contents, even over 5000 mg/L, titanium performs with no problem. Hastelloy C-276 and C-22 also resist these high-chloride, high-temperature environments well.

With less severe environments, lower chlorides and lower temperatures, other nickel alloys such as Inconel 625 and Hastelloys B and G have served well. For sewage handling and treatment at ambient temperatures, good resistance is reported for aluminum alloys, copper, and copper alloys except high-zinc brasses, which dezincify.

One failure analysis of type 409 stainless steel in rotating-biological-contactor units handling raw waste water with up to 3000 mg/L of chloride showed that cracks always initiated at pits under bacterial colonies that grew at welds and HAZs. In another study of chloride-containing mist with H_2S at waste-water treatment facilities, galvanized steel rusted and pitted badly but aluminum performed well.

7.3 MINERAL ACIDS

Sulfuric Acid

Sulfuric acid, generally considered the most important industrial chemical, is used in making fertilizers, chemicals, paints, in petroleum refining, and so on. The concentrated acid has a concentration of 95–98%.

Steel is the most common material for storage and transport of H_2SO_4 concentrations of 70% or higher near room temperature. At 50°C (122°F) the corrosion rate of steel reaches 0.5 mm/y for the concentrated acid and more than 1 mm/y for concentrations less than 90%. Fig. 7-3 and table shows pure metals and alloys suitable for H_2SO_4 solutions.

Chemical lead (containing 0.01% Ag) corrodes at less than 0.1 mm/y at 100°C (212°F) for all H_2SO_4 concentrations up to 70%, which explains why lead and dilute sulfuric acid are listed as a compatible combination (Table 7-1).

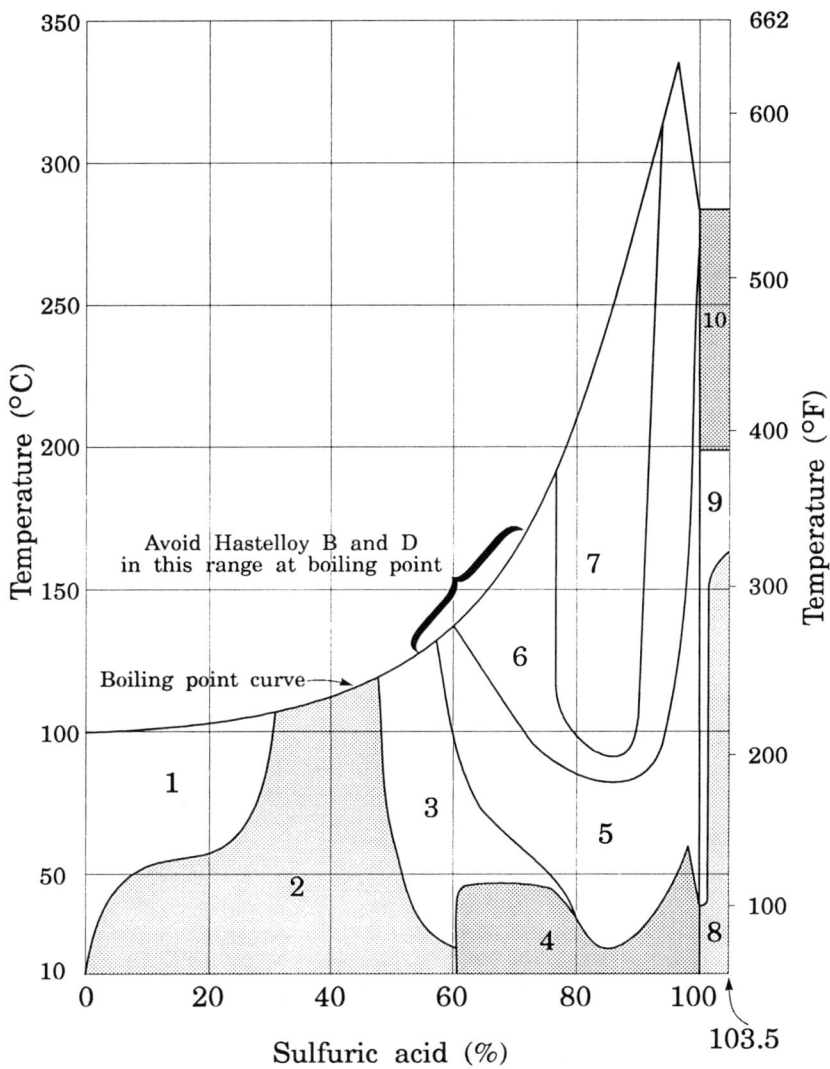

FIGURE 7-3. Materials with less than 0.5-mm/y corrosion in sulfuric acid (see Table for details) (Reprinted with permission from G. A. Nelson, *Corrosion Data Survey*, Shell Development Co., 1960.)

Materials with less than 0.5-mm/y Corrosion in Sulfuric Acid (see Fig. 7-3).

Zone 1

Alloy 20	Impervious graphite	Platinum
10% Al bronze[a]	Lead	Rubber[b]
Copper[a]	Molybdenum	Silver
Glass	Monel[a]	316 Stainless[c] steel
Gold	Nionel	Tantalum
Hastelloy B and D	Ni-Resist	Tungsten
Haveg 43 phenolic	Platinum	Worthite stainless steel
Illium G		Zirconium

Zone 2

Alloy 20[b]	Lead	Silicon iron
10% Al bronze[a]	Molybdenum	Silver
Copper[a]	Monel[a]	316 Stainless[f] steel
Glass	Nionel	Tantalum
Gold	Ni-Resist[e]	Tungsten
Hastelloy B and D	Platinum	Worthite[d]
Haveg 43	Rubber[b]	Zirconium
Impervious graphite		

Zone 3

Alloy 20[b]	Lead	Silicon iron
Glass	Molybdenum	Tantalum
Gold	Monel[a]	Worthite[d]
Hastelloy B and D	Platinum	Zirconium
Impervious graphite		

Zone 4

Alloy 20	Lead[g]	Steel
Glass	Ni-Resist	Tantalum
Gold	Platinum	Worthite
Hastelloy B and D	Silicon iron	Zirconium[b]
Impervious graphite[g]	316 Stainless steel[h]	

Zone 5

Alloy 20[d]	Impervious graphite[b,g]	Silicon iron
Glass	Lead[b,g]	Tantalum
Gold	Platinum	Worthite[d]
Hastelloy B and D		

Zone 6

Glass	Hastelloy B and D[i]	Silicon iron
Gold	Platinum	Tantalum

Materials with less than 0.5-mm/y Corrosion in Sulfuric Acid. *(continued)*

	Zone 7	
Glass	Platinum	Tantalum
Gold	Silicon iron	

	Zone 8	
Alloy 20	Hastelloy C	Steel
Glass	Platinum	Worthite
Gold	304 Stainless steel	

	Zone 9	
Alloy 20	Gold	304 Stainless steel
Glass	Platinum	Worthite

	Zone 10	
Glass	Gold	Platinum

Source: Treseder, R. S., Ed., *Alloys for sulfuric acid service,* from Corrosion Data Survey, NACE, 1974, as reprinted in NACE Corrosion Engineers Reference Book, 1980, pp. 78, 79. Reprinted with permission from the National Association of Corrosion Engineers. Copyright © NACE.

[a]Air free.
[b]Up to 75°C (170°F).
[c]Up to 10%, aerated.
[d]Up to 65°C (150°F).
[e]Up to 20%, 25°C (80°F).
[f]Up to 25%, 25°C (80°F), aerated.
[g]Up to 96%.
[h]Above 80%.
[i]0.5–1.1 mm/y.

Hydrochloric Acid

Hydrochloric acid is actually a solution of hydrogen chloride gas in water so that the concentrated acid contains only 36.5–38 wt% HCl. The acid is used in acidizing oil wells, as a chemical intermediate, for ore reduction, and in a wide variety of other industrial processes.

Hydrochloric acid presents a formidable challenge for materials selection because it is the most difficult common acid to handle. Since it is a reducing acid with a high concentration of chloride ions, metals will not passivate in it. Figure 7-4 (see table 7.11B) shows materials suitable for HCl as a function of temperature and concentration. Plastics such as polyethylene, polypropylene, and fiberglass reinforced plastics are often used for HCl. Table 9-6 shows the resistance of plastics to mineral acids.

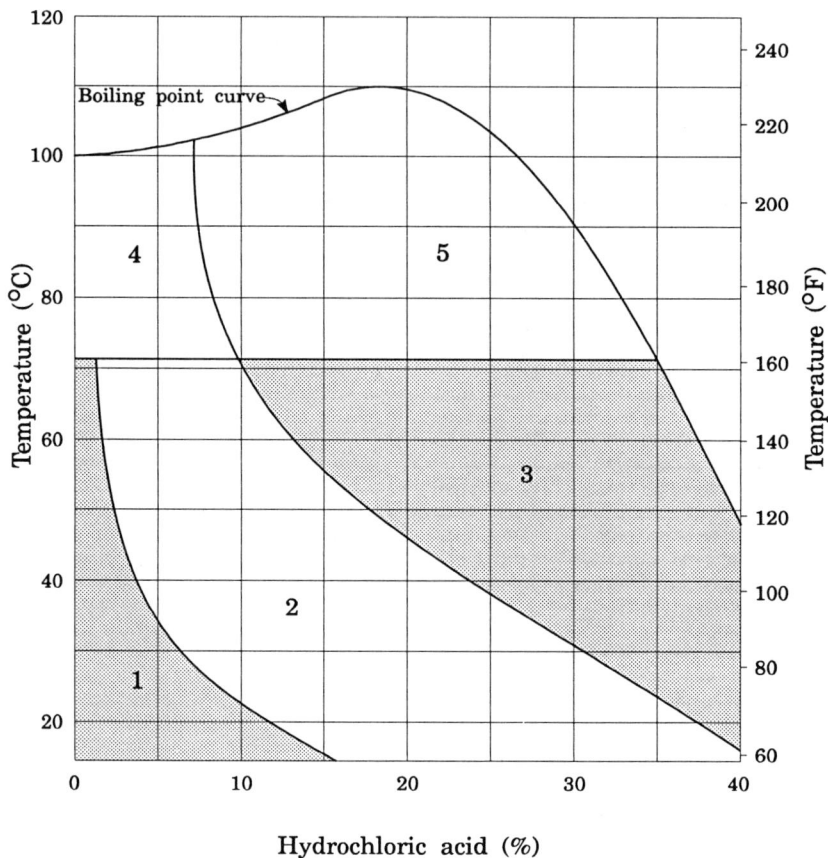

FIGURE 7-4. Materials with less than 0.5-mm/y corrosion in hydrochloric acid (see Table for details). Reprinted with permission from G. A. Nelson, *Corrosion Data Survey*, Shell Development Co., 1960).

Materials with less than 0.5-mm/y Corrosion in Hydrochloric Acid (see Fig. 7-4).

Zone 1		
Chlorimet 2	Molybdenum	Si bronze[a]
Copper[a]	Monel[a]	Silver
Durichlor[b]	Nickel[a]	Tantalum
Glass	Platinum	Titanium[c]
Hastelloy B	Rubber	Tungsten
Haveg	Saran	Zirconium

Zone 2		
Chlorimet 2	Impervious graphite	Si bronze[a]
Durichlor[b]	Molybdenum	Silver
Glass	Platinum	Tantalum
Hastelloy B	Rubber	Zirconium
Haveg	Saran	

Zone 3		
Chlorimet 2	Impervious graphite	Saran
Durichlor[b]	Molybdenum	Silver
Glass	Platinum	Tantalum
Hastelloy B[d]	Rubber	Zirconium
Haveg		

Zone 4		
Chlorimet 2	Impervious graphite	Tantalum
Durichlor[b]	Platinum	Tungsten
Glass	Silver	Zirconium
Hastelloy B[d]		

Zone 5		
Chlorimet 2	Impervious graphite	Tantalum
Glass	Platinum	Zirconium
Hastelloy B[d]	Silver	

Source: Treseder, R. S., Ed., *Alloys for hydrochloric acid service*, from Corrosion Data Survey, NACE, 1974, as reprinted in NACE Corrosion Engineers Reference Book, 1980, pp. 80, 81. Reprinted with permission from the National Association of Corrosion Engineers. Copyright NACE.
[a]Air free.
[b]$FeCl_3$ free
[c]Up to 10%, 25°C (80°F).
[d]Chlorine free.

Nitric Acid

Nitric acid, a powerful oxidizer, is usually handled with stainless steels or 14.5% silicon cast iron. Titanium is also suitable as Figure 7-5 and table shows.

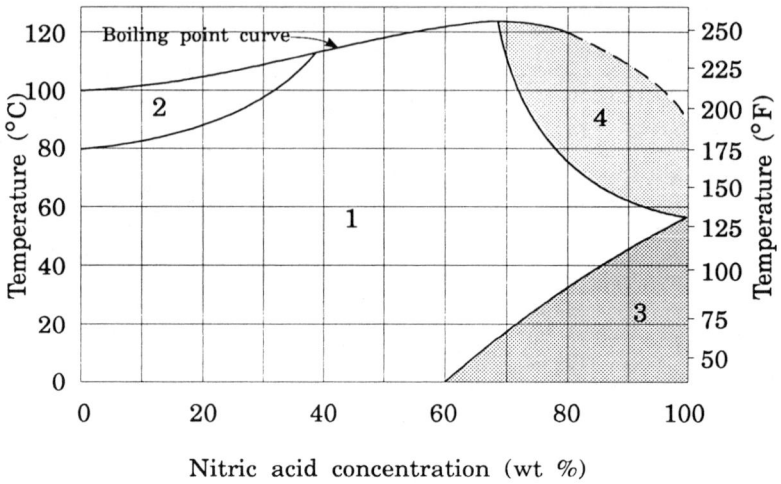

FIGURE 7-5. Materials with less than 0.5-mm/y corrosion in nitric acid (see Table for details).

Materials with less than 0.5-mm/y Corrosion in Nitric Acid (see Fig. 7-5).

Zone 1

Alloy 20	Incoloy 825	316 Stainless steel
20Cb3	Platinum	Tantalum
FEP	PTFE	Titanium
Glass	Si cast iron	Zirconium
Gold	304 Stainless steel	

Zone 2

Alloy 20	FEP	304 Stainless steel
20Cb3	Incoloy 825	316 Stainless steel
Glass	Platinum	Tantalum
Gold	PTFE	Zirconium

Zone 3

Alloy 20	FEP	304 Stainless steel
Aluminum	Platinum	316 Stainless steel
20Cb3	PTFE	Titanium
Glass	Si cast iron	Zirconium
Gold		

150

Materials with less than 0.5-mm/y Corrosion in Nitric Acid. *(continued)*

Zone 4		
FEP	Platinum	Tantalum
Glass	Si cast iron	Zirconium
Gold		

Sources: De Renzo, 1985; Davis, 1987; Fontana, 1987; Dillon, 1986a; Dillon, 1986b.

Nitric acid is used in production of fertilizers, dyes, drugs, explosives, and so on. The ordinary concentrated acid is approximately 70%, while concentrations above 85% are called "fuming nitric acid" because they give off red or colorless gases of nitrogen oxides. Fuming nitric acid can have a pyrophoric reaction with titanium if the water content is less than 1.5% and NO_2 is greater than 2.5%. Red fuming nitric acid may crack titanium by SCC.

Phosphoric Acid

The concentrated acid of H_3PO_4 is an 85% syrup, usually handled at the boiling point with type 316 or Alloy 20 stainless steel. Lead can be used up to 200°C (400°F) for concentrations up to 80% acid and is widely used in phosphoric acid manufacture. Corrosion resistance of various metals is given in Figure 7-6 (see table 7.11D). Phosphoric acid is used primarily for fertilizer

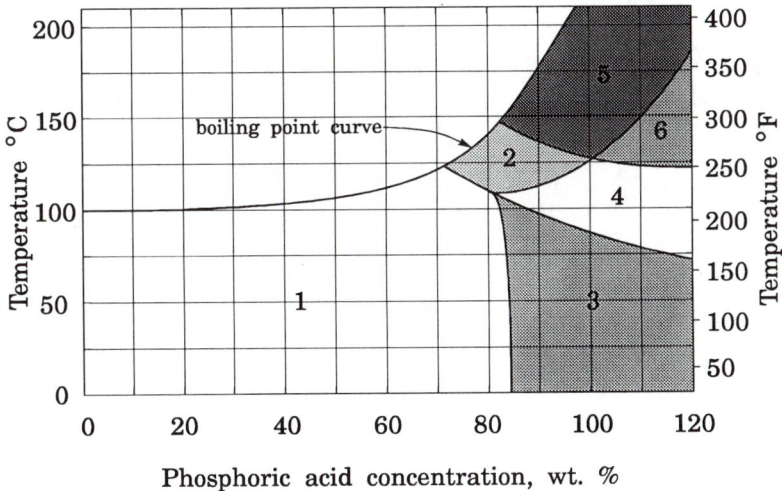

FIGURE 7-6.   Materials with less than 0.5-mm/y corrosion in fluoride-free phosphoric acid (see Table for details).

Materials with less than 0.5-mm/y Corrosion in Fluoride-Free Phosphoric Acid (see Fig. 7-6).

Zone 1		
Carbon	Hastelloy G-3	Silver
FEP	Inconel 625	316 Stainless steel
Fiberglass resin	Lead	317 Stainless steel
Gold	Platinum	Tantalum
Graphite	PTFE	Vinyl ester
Hastelloy C	Si cast iron	Zirconium
Hastelloy C-4		

Zone 2		
Alloy 20	Hastelloy G-3	Si cast iron
Borosilicate glass	Incoloy 825	Silver
Gold	Platinum	Tantalum
Hastelloy C-4		

Zone 3		
Borosilicate glass	Hastelloy C	316 Stainless steel
20Cb3	Platinum	317 Stainless steel
Fiberglass resin	Si cast iron	Tantalum
Gold	Silver	Vinyl ester

Zone 4		
Alloy 20	Hastelloy G-30	Silver
Borosilicate glass	Incoloy 825	316 Stainless steel
Gold	Platinum	317 Stainless steel
Hastelloy C-4	Si cast iron	Tantalum

Zone 5		
Borosilicate glass	Platinum	Silver
Gold	Si cast iron	Tantalum
Hastelloy G-3		

Zone 6		
Gold	Si cast iron	317 Stainless steel
Hastelloy G-3	316 Stainless steel	Tantalum
Platinum		

Sources: Craig, 1989; De Renzo, 1985; Davis, 1987; Dillon, 1986a; Dillon, 1986b; Graver, 1985; Schweitzer, 1986.

production but also for pickling metals, for corrosion inhibitors, in soft drinks and pharmaceuticals, and a wide variety of minor applications.

The industrial wet-process acid, made by reacting phosphate rock with sulfuric acid, contains F^- ions and H_2SO_4, which makes it much more corrosive than the pure acid. High-silicon cast iron is rated good to excellent in phosphoric acid, and tantalum has superior resistance, but both are unacceptable if F^- is present.

7.4 COMMON INORGANICS

Alkalies

The term "alkali" refers to the hydroxides and carbonates of the alkali metals (Li, Na, K, etc.) and the ammonium NH_4^+ ion. In more general terms, an alkali is any strong base that produces OH^- ions in water. Those alkalies that are important in industry and present particular corrosion problems for metals are KOH, NaOH, and NH_4OH.

Nickel and nickel alloys make compatible combinations with solutions of caustics (strong, corrosive bases) and are used extensively in caustic production. Figure 7-7 shows the corrosion rates of pure nickel in sodium hydroxide concentrations at temperatures up to the boiling point of the solutions. Figure 7-8 is a similar diagram for Types 304 and 316 stainless steels. Note that stainless steels may suffer SCC at high temperatures, unlike the nickel alloys. Iron and carbon steels can handle caustic up to about 70% and 80°C (175°F), also with the danger of SCC. Table 7-12 lists resistance of many pure metals and alloys in a wide variety of alkali solutions.

In alkali solutions of up to 40% concentration at temperatures up to about 50°C (120°F), lead corrodes at a satisfactory rate. Magnesium is suitable for any concentration of alkali at room temperature. Tantalum, the most corrosion resistant of all metals, cannot withstand strong alkalies even at room temperature. The resistance of plastics to alkalies is shown in Table 9-6.

Concrete

Reinforcing steel in concrete would seem to be in a near-ideal environment: alkaline conditions with a heavy coating to shield the metal from moisture. But sometimes the steel corrodes and the volume of rust exerts pressure on the concrete, cracking or spalling it, which allows more air and water to reach the steel.

The problem is most acute on highway bridges (see Fig. 7-9) but is also seen in parking garages, on marine structures at or near the waterline, and process industry plants. The main culprit is chloride. For bridges and garages

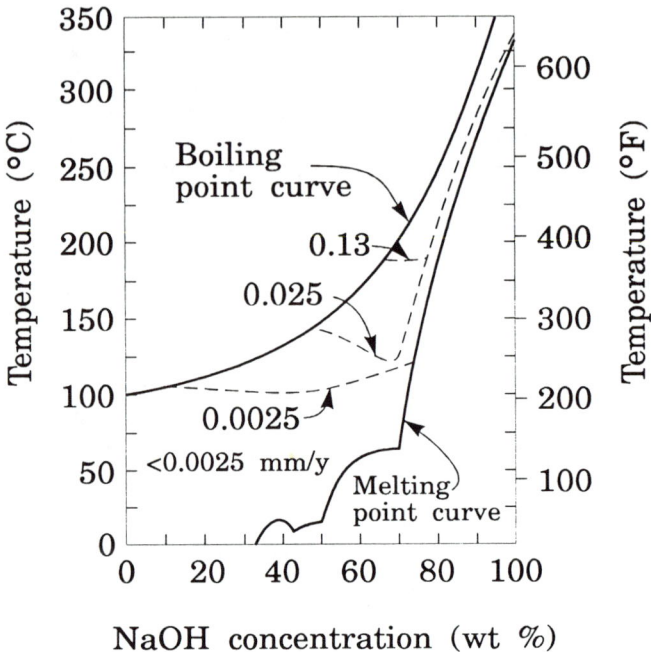

FIGURE 7-7. Corrosion of Nickel 200 and 201 in NaOH solutions. [From *Corrosion Engineering Bulletin 2*, Courtesy of Inco Alloys International.]

the source is deicing salt used in the northern half of the United States and in southern Canada. As moisture penetrates pores in the concrete, chloride ions and O_2 are also carried along to develop acid conditions next to the steel reinforcing bars (rebar) by the usual crevice corrosion mechanism. Chloride penetration is not uniform so local anodes and cathodes are created. Since rust has approximately seven times the volume of the steel corroded, the swelling breaks up the concrete. Because of rebar corrosion the U.S. Federal Highway Administration (FHWA) has rated 243,000 of their bridges as deficient. Somewhere in the nation every 2 days a bridge collapses.

Prevention of rebar corrosion is being tried in several ways.

1. Improvement in concrete quality: reducing the water/cement ratio, adding latex emulsions, and so on, to slow chloride penetration.
2. Barrier systems added to concrete surface: dense concrete overlays and waterproof sealers.
3. Inhibitors added to cement.
4. New deicing salts: without chloride or with inhibitors.

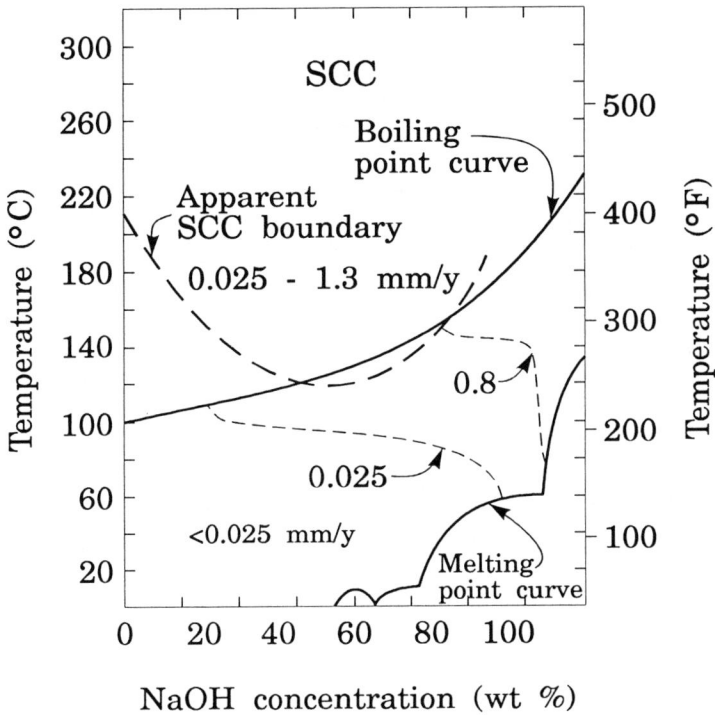

FIGURE 7-8. Corrosion of 304 and 316 stainless steels in NaOH solutions. [C. W. Funk and G. B. Barton, "Caustic Stress Corrosion Cracking," Paper 54, presented at *Corrosion/77*, National Association of Corrosion Engineers, 1977. Copyright © by NACE.]

5. Organic coatings on rebar; epoxy coatings have proved to be excellent.
6. Galvanized rebar; zinc corrodes but the steel is given long life.
7. Impressed-current cathodic protection; for existing structures this is the only proven technique, according to the FHWA. Even prestressed and poststressed bridge decks are being tested despite the danger of over-protection and hydrogen embrittlement.

Fused Salts

Molten salts have been used for many years to heat treat metals in constant-temperature baths, but more recently molten salt eutectics have become considered as coolants for nuclear power reactors. Below about 800°C (1500°F) nitrates and nitrites can be used, although halides are generally required for higher temperatures.

TABLE 7-12   Metals That Are Fully Resistant to Alkali Solutions

NH₄OH

Metals: 20Cb-3, Hastelloy B, Hastelloy C, Incoloy 800, Incoloy 825, Inconel 600, nickel 200, Types 304, 316, 430 stainless, titanium, zirconium

Ca(OH)₂

Conditions: 10–20%, boiling
Metals: Tantalum, 20Cb-3, copper, Cu-Ni alloys, Hastelloy B, Hastelloy C, Incoloy 800, Incoloy 825, Inconel 600, Monel 400, nickel 200, Si bronze, Types 304, 316, 430 stainless steel

Conditions: 50%, boiling
Metals: 20Cb-3, copper, Cu-Ni alloys, Hastelloy B, Hastelloy C, Incoloy 825, Si bronze, titanium

KOH

Conditions: 5%, room temperature
Metals: Tantalum, 20Cb-3, Cu-Ni alloys, Hastelloy B, Hastelloy C, Incoloy 800, Incoloy 825, Monel 400, nickel 200, Types 304, 316, 430 stainless, titanium, zirconium

Conditions: 27%, boiling
Metals: 20Cb-3, Cu-Ni alloys, Hastelloy B, Hastelloy C, Incoloy 800, Incoloy 825, Monel 400, nickel 200, Types 304, 316 stainless steel, titanium, zirconium

Conditions: 50% boiling
Metals: 20Cb-3, Hastelloy B, Hastelloy C, Incoloy 800, Incoloy 825, Monel 400, nickel 200, Type 316 stainless steel, titanium

NaOH

Conditions: 10%, room temp.
Metals: 20Cb-3, Cu-Ni alloys, Hastelloy B, Hastelloy C, Incoloy 800, Incoloy 825, Inconel 600, Monel 400, nickel 200, Types 304, 316, 430 stainless steel, titanium, zirconium

Adapted from De Renzo, 1985.

Corrosion attack can be caused by the salt or its impurities in several ways.

1. By dissolving the metal or one metal from an alloy.
2. By selective attack of an alloy phase, grain boundary, or the like.
3. By pitting.
4. By electrochemical reaction—very rapid at the high temperatures in the high-conductivity electrolytes.

FIGURE 7-9.   Rebar exposed by corrosion in support pillar of bridge. [By permission of Harco Technologies Corporation.]

Corrosion in nuclear reactors is often followed by redepositing the metal in a cooler section of the system.

Any of several corrosion mechanisms may be involved.

1. Salt often fluxes off protective films.
2. Salt may contain ions more noble than the metal.

$$Fe + 2AgCl \rightarrow 2Ag + FeCl_2$$

3. In NaCl, molten sodium may distill off so that even noble metals are attacked.

$$Ag + NaCl \rightarrow AgCl + Na\uparrow$$

4. If the salt has oxygen-containing anions that can be reduced to form oxide ions, the O^{2-} can attack the metal.

Nitrates and Nitrites.   Up to 500°C (930°F) nitrates and nitrites can be contained in steel because it forms a passive Fe_3O_4 film that can be strengthened considerably by chromium additions to the steel. Nickel or stainless steels are used at higher temperatures. Tables 7-13 and 7-14 show corrosion rates of various metals in nitrate salt mixtures.

Halides.   The halides (fluorides, chlorides, bromides, and iodides) dissolve protective passive films, so stainless steels cannot be used at high temperatures. Inconel 600 has been often used but chlorides can dealloy it intergranularly, removing the chromium and leaving a porous metal region at the grain boundaries. Chlorides also corrode carbon steels rapidly by preferentially attacking the carbides. In general, the greater the iron and chromium in an alloy, the greater the corrosion. If chromium corrodes, chromium dendritic crystals then form in the cold zone of a cooling system. Even noble metals tend to dissolve in molten alkali chlorides but Hastelloy N (Ni-6.5Mo-6.9Cr-4.5Fe) appears the most promising.

Other Common Salts.   In molten cyanides used to surface-harden steels, the nickel alloys or ferritic stainless steels with over 20% chromium can best withstand the cyaniding temperatures of 760–870°C (1400–1600°F). In sulfates, any of the high-temperature alloys high in chromium passivate well. Molten hydroxides, however, selectively oxidize the chromium, so that nickel

TABLE 7-13   Corrosion Rates in Eutectic Molten Salts at 250–400°C (480–750°F)

Salt Mixture (mol%)	Carbon Steel (mm/y)	304 Stainless steel (mm/y)
$86.3NaNO_3$, $8.4NaCl$, $5.3Na_2SO_4$	0.015	0.001
$94KNO_3$, $6KCl$	0.023	0.008
$58 LiCl$, $42KCl$	0.063	0.020

Adapted from Davis, 1987.

TABLE 7-14   Corrosion Rates of Metals in Eutectic Molten Salts at High Temperatures

Metal	$54KNO_3 +$ $46NaNO_3$ 675°C (1250°F)	$27BaCl_2 + 40KCl +$ $33NaCl$ 845°C (1550°F)	$70LiF + 23BeF_2 +$ $5ZrF_4 + 1ThF_4 +$ $1UF_4$, 663°C (1225°F)
Inconel 600	0.3 mm/y	0.2 mm/y	
Hastelloy N	0.33		0.002 mm/y
Hastelloy S	0.4	0.01	
Inconel 625	0.74		
Hastelloy X	1.04	0.080	
Incoloy 800	1.07		
316 Stainless steel	2.1		
304 Stainless steel	2.67	0.16	0.028
Nickel 200	8.18		
Maraging steel			0.014

Adapted from Davis, 1987.

is more resistant. Austenitic stainless steels are used for molten carbonates up to 500°C (930°F) while up to 700°C (1290°F) alloys with at least 50% Cr are suitable, as is aluminum-coated steel. Above 700°C no metal can withstand molten carbonates.

Liquid Metals

Molten metals create corrosion problems in metallurgical operations such as refining, casting, and so on, but far more important today is their use as coolants in nuclear power reactors. Because liquid metals are not ionized and, therefore, not electrolytes, electrochemical corrosion does not occur. Instead, corrosion damage occurs in the following ways.

1. By dissolving the solid metal or one component from an alloy.
2. By selective attack of an alloy phase or grain boundary.
3. By alloying or forming intermetallic compounds with the solid metal at its surface, or along grain boundaries, or as subsurface precipitates.
4. By liquid-metal embrittlement.

After the corrosion takes place, mass transfer and redeposition in cooler areas is a frequent problem.

A list of liquid metals known to cause liquid-metal embrittlement with a variety of important metals and alloys is given in Table 7-15.

Table 7-16 shows the approximate ranges of corrosion rates in liquid metals at their melting points and at 600°C (°F).

TABLE 7-15   Liquid-Metal Embrittlement

Embrittled Metals	Liquid Metals
Steels	Bi, Cd, Cu, Ga, Hg, In, Li, Pb, Pb-Sn solders, Sb, Sn, Te, Zn
Aluminum	Ga, Hg, In, Li, Na, Pb-Sn solders, Sn-Zn alloys
Copper	Bi, Cd, Ga, Hg, In, Li, Na, Pb, Sb, Tl
Gold, silver	Ga, Hg
Magnesium	Zn
Nickel	Cd dissolved in Cs
Tantalum	Hg
Titanium alloys	Cd, Hg, Zn
Zinc	Ga, Hg, In, Pb-Sn solder
Zirconium	Ca, Cd, Sr, Zn

Adapted from Davis, 1987.

TABLE 7-16. Resistance of Metals to Molten Metals[a]

		Al	Bi	Bi-Pb	Cd	Ga	Hg	In	K,Na	Li	Mg	Pb	Sb	Sn	Tl	Zn
FERROUS METALS mp°C		660°	271°	125°	321°	30°	-39°b	157°	63°/98°	181°	650°	328°	631°	232°	304°	420°
Low carbon steel	600°C		G	L	G	P	G	P	L	P	G	G		P		P
	m.p.	P	G	G	G	P	G	P	G	G	G	G	P	L		P
Gray cast iron	600°C	P			G		G		L		G	G		P		P
	m.p.		G		P	P	G	P	G	P	G	G	P	L		P
Ferritic stainless	600°C	P	P	L	P	P	G		G	G	G	G		P	G	P
	m.p.		P		P	L	P	G	G			P	P	P	G	P
Austenitic stainless	600°C	P	P	L	P	P	G	G	G	G	P	P	P	P	G	
	m.p.		P		P	L	G	G	G	G	G	G		L	G	
NONFERROUS METALS																
Aluminum	600°C	P	P	P	P	P	P	P	P	P	P	P	P	P		P
	m.p.	G	G	G	P	P	P	P		P	P	G	P	P	G	P
Beryllium	600°C	G	G	G			G				G	G		G		
	m.p.	G	G	G	G	P	G	P	G			G				L
Chromium	600°C	G	G		G	P	G	P	G		G	G		G		
	m.p.	L	G		G	P	P	P	G			P	P	G		P
Copper	600°C	P	P		P	P	P	P	G	P	G	P	P	P	P	P
	m.p.	L	L		P	P	P	P	G	P		L		P	P	P
Be copper	600°C	P	P		P	P	G	P	G	P	P	P	P	P	P	L
	m.p.	P	P		P	P	P		P	P		P		P	P	
Brass, tin bronze	600°C		P			P	G		P	P	P					
	m.p.		P			P	P			P						

160

Material		Mo		Ni		Hast. A,B,C		Ni-Cr		Ni-Cu		Nb		Pb, Sn		Pt,Au,Ag		Ta		Ti		W		Zr	
		600°C	m.p.	600°C	m.p.	600°C	m.p.	600°C	m.p.	600°C	m.p.	600°C	m.p.	600°C	m.p.	600°C	m.p.	600°C	m.p.	600°C	m.p.	600°C	m.p.	600°C	m.p.
		G	G	P	P	P		G		P	P	L	G	P	P	P	P	G	G	P		L	G	G	G
		G	G	P	P		P	G	G	P				P		P	P	G	P	G	P		G	P	P
		P	P	P	P	P		P	G	P	P		G	P	P	P		G	P	L	L		G	P	P
		P				P		P	P	P				P	P	P		G	P	L	L		P	P	P
																		G	G			P	P	G	G

aG is Good: < 0.025 mm/y; L is Limited: 0.025–0.25 mm/y; and P is Poor: > 0.25 mm/y.

bRatings for Hg at "mp" are for 25°C (80°F).

Sources: Miller, 1952; Davis, 1987; De Renzo, 1985; Graver, 1985; Schweitzer, 1986.

Molybdenum — 600°C, m.p.
Nickel — 600°C, m.p.
Hastelloys A,B,C — 600°C, m.p.
Ni-Cr alloys — 600°C, m.p.
Ni-Cu alloys — 600°C, m.p.
Niobium — 600°C, m.p.
Lead, tin — 600°C, m.p.
Pt, Au, Ag — 600°C, m.p.
Tantalum — 600°C, m.p.
Titanium — 600°C, m.p.
Tungsten — 600°C, m.p.
Zirconium — 600°C, m.p.

Liquid sodium and potassium oxidize rapidly in air, producing dissolved oxides that are extremely corrosive. Inert atmospheres are often necessary to exclude air.

Mercury is contained by ferrous metals with virtually no corrosion in stagnant, isothermal conditions but attack is severe in a thermal gradient or liquid flow situation unless a few milligrams per litre of magnesium, sodium, or titanium inhibitor is added to the mercury.

Molten aluminum attacks all known metals and alloys.

STUDY PROBLEMS

7.1 If a coating is anodic to the base steel, does this mean that it should corrode more slowly than steel? If the coating is cathodic to the base steel, should it corrode more slowly than steel? Explain.

7.2 Estimate the useable life of a steel beam 3/8-in. thick with 5 mil galvanizing when immersed in seawater. Strength requires 0.200 in. thickness of steel plus a 50% safety factor. Assume uniform corrosion.

7.3 Estimate the total corrosion of steel immersed in seawater for 6 years.

7.4 Estimate the total corrosion of steel in industrial air after 5 years.

7.5 Estimate the total corrosion of steel in industrial air after 2 years.

7.6 Can you suggest a reason why corrosion of steel in river water might be less than in hard water (Table 7-3)?

7.7 The corrosion rate of magnesium in freshwater is given in Table 7-3. How could magnesium be used advantageously in freshwater?

7.8 Stainless steel may pit in hot tap water (Table 7-3). Is oxygen solubility an important factor here? Explain.

7.9 A soil box 4.0 × 3.1 × 20.7 cm is packed with soil. The resistance between the pins inserted in the soil 12.4 cm apart is measured as 1200 Ω. What is the soil resistivity? Estimate the corrosivity of the soil.

7.10 A soil box 4.0 × 3.1 × 20.7 cm is filled with seawater. A current of 10 mA is passed through the water. The potential difference between pins spaced 12.4 cm apart in the water is measured as 245 mV. What is the resistivity of the seawater?

7.11 A soil box 4.0 × 3.1 × 20.7 cm is packed with a soil sample. The resistance between pins inserted in the soil 12.4 cm apart is measured as 17.5 kΩ. What is the relationship between resistance and resistivity in this standard-size soil box?

7.12 Explain why molten metals, which conduct electric current very well, are not considered to be electrolytes. Define "electrolyte." Do molten salts meet your definition?

7.13 Sketch a graph of corrosion versus time for rebar in concrete. Explain its shape.

7.14 Construct a graph for lead in sulfuric acid showing temperatures and concentrations where lead corrodes at less than 0.5 mm/y.

7.15 Can high-silicon cast iron be used in nitric acid? Explain.

7.16 A Type 304 stainless steel reactor contains 20% NaOH at its normal boiling point. Estimate the expected corrosion rate. Is SCC possible?

7.17 A nickel-lined, mild steel reactor contains 20% NaOH at its normal boiling point. Estimate the expected corrosion rate. Is SCC possible?

7.18 Can steel be used to contain mercury at room temperature?

7.19 Can steel be used to contain molten cadmium at 500°C (930°F)?

7.20 Can tantalum be used to contain mercury at 500°C (930°F)?

7.21 Why do federal laws prohibit shipment of mercury by air freight?

8

Corrosion Measurement and Failure Analysis

·Corrosion testing involves evaluation of materials, coatings, or inhibitors that allegedly will be satisfactory for a certain corrosive environment. Tests may be made on new products or just to see if the products being delivered meet their specifications.

A corrosion test must yield results that satisfy one criterion: Are they valid? Does the test truly indicate how this product will perform in service? To be valid the test has to be reasonably reproducible, but reproducibility does not necessarily make it valid. In some cases the test does not have to give the same corrosion rate that will show up in service, but only has to rank products in the proper order. Cost or quick results are often wrongly emphasized as most important, although certainly these must both be considered in designing the test.

8.1 TYPES OF TESTS

Tests use either the real corrosive environment or a simulated environment. Hopefully, this environment will perform the same way but will be more reproducible, cheaper, or faster.

- A synthetic seawater is available commercially for corrosion testing. It is considerably more reproducible than the natural environment but also considerably more expensive.

Tests may use the real service conditions for the material, they may simulate service, or in many cases they omit service conditions. Service conditions are the fluid velocities, the turbulence, the applied stresses, the thermal gradients and heat transfer conditions, the crevices—all the factors other than the material and environment themselves that in some circumstances might significantly influence corrosion.

Laboratory Tests.   In the lab, tests can be run in the real environment but more often are conducted in a simulated environment for better reproducibility. Small test specimens, commonly called "coupons," are placed in the environment with service conditions simulated only if they appear to be important.

Pilot Plant Tests.   Testing in a pilot plant is an excellent way of selecting materials of construction for the full-scale plant. The environment is real (approximately) with all the variations that can occur during shutdowns and when things just go wrong. Coupons can be exposed with simulated service or no service, but parts of the pilot plant itself can be built of candidate materials that will be in service conditions very closely simulating real service. Unfortunately, the trend these days is to omit the pilot plant stage in order to save about a year of time, but often this omission has later proved to be an extremely costly error in judgment.

* One plant costing in the hundreds of millions of dollars was built after brief pilot plant tests in an old garage without any tests on materials. The plant opened with great fanfare and ran almost 1 week. It then ran sporadically for the next 2 years while it was rebuilt.

Field and Plant Tests.   Field tests typically involve large numbers of coupons placed in the real environment (exposed to sea air, buried, put in a process tank, etc.) but usually without service conditions, such as stresses or heat transfer. Coupons are often fastened to a rack, but electrically insulated from the rack and from each other to avoid galvanic effects. Paints and other coatings are most often tested in this way as shown in Figure 8-1.

Figure 8-2 shows a specimen rack mounted under an automobile. Several automobiles were used to average out differences in driving patterns, roads, garaging, and cleaning.

Coupons are also used to monitor corrosion rates inside pipes and vessels that cannot be easily inspected. A coupon of the same metal as a pipeline is inserted in the line and removed from time to time for inspection and measurement. It is assumed that a coupon of the same metal and in the same environment will corrode at the same rate as the pipe. Coupons are usually not welded because rapid cooling rates of small specimens give metallurgical structures different than those for large pipes.

Service Tests.   Actual service tests involve testing metals, coatings, or inhibitors in real operation. The product that holds up best will be the one eventually chosen. A railroad can order 10,000 spikes from four manufacturers, try them, and see how they actually perform. A pipe carrying corrosive liquid can be constructed of several short test sections of different materials,

FIGURE 8-1.   INCO test facility for atmospheric corrosion at Kure Beach, NC. [From *Corrosion in Action*, The International Nickel Company of Canada, Ltd., Toronto, 1955, Fig. 70, p. 28. Courtesy of Inco Alloys International.] General supervision by F. L. LaQue

insulated from each other, of course. These tests certainly give the most convincing results but testing in this way may be very costly or even dangerous. Service tests should be made only after lab tests have narrowed the choices down to a few good candidates.

- One young engineer, completely frustrated at trying to keep three stainless steel pumps working with a very corrosive mixed acid, decided to have three new pumps built out of three different alloys and compare their performance, even though those pumps cost about $250,000 apiece.

8.2 LABORATORY TESTS

Lab tests can have one important advantage over all other types of tests: better reproducibility. An investigator who runs a series of tests could compare his/her results with tests run 5 years ago in Berlin or Bangladesh. For

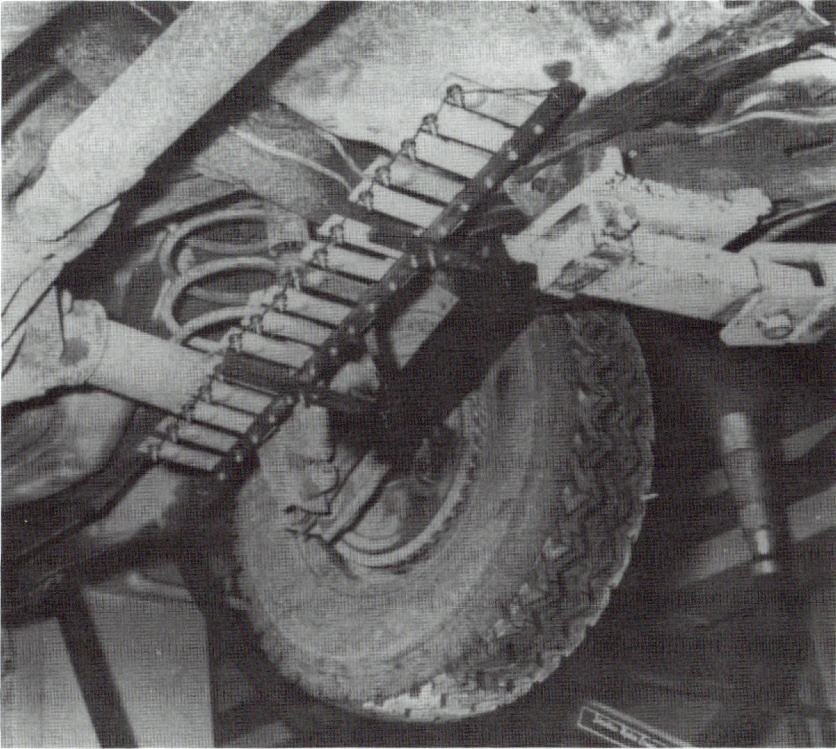

FIGURE 8-2.   Test rack of specimens bolted beneath automobile to test resistance to road salt, slush, and moisture. Two 13 × 3-cm specimens are bolted together with a 0.15-cm gap between them to investigate crevice corrosion. [R. McDonald and R. R. Ramsingh, "Eighteen months of under-body automotive materials testing," in *Materials Performance*, Vol. 24, No. 4, April 1985, Fig. 2, p. 49. Copyright © by NACE.]

this reason the environment is usually simulated rather than being the varying, real environment. Service conditions are often omitted where it is assumed that they are not important, although they can be simulated as the examples in Table 8-1 indicate.

Lab tests are good for screening out entirely unsuitable products quickly. The environment is sometimes made more aggressive than the real-life situation, but the results can be completely misleading if a crucial factor is omitted. Accelerated tests are risky. Very few accelerated tests are completely successful in ranking materials and virtually none has been well correlated to real corrosion rates.

At the very least the volume of solution should be 100 L/m² of surface

TABLE 8-1   Simulation of Service Conditions

Service	Simulation Examples
Fluid velocity	Rotating disk holding test coupons
Cavitation	Coupon fastened to ultrasonic transducer
Crevices	Rubber band wrapped around coupon
Tensile stress	U-bend test specimen
Heat transfer	Coupon fastened to soldering iron
Vapor, water line	Partial immersion

exposed in order that corrosion products do not contaminate the solution too quickly. Also, coupons of different materials must be tested separately if the corrosion products of one could affect the results of another. Planned-interval tests (discussed later) can be used to monitor any changes in the corrosivity of the environment.

In selecting a lab test to simulate real conditions the investigator would be well advised to consider suitable standard tests prepared by ASTM and NACE. A brief summary of some of the ASTM tests is given in Table 8-2.

• Monel fasteners were being ordered for a new plant at a cost of $125,000. Alternate immersion tests in the lab showed corrosion rates 120 times those of published data for continuous immersion. The order was switched to another alloy in time to save the cost of the fasteners and many hours of expensive down-time for replacement and clean-up.

Planned-Interval Tests

Corrosion involves both the reaction of the environment and the reaction of the metal. Planned-interval tests (ASTM G31) show two things: how the corrosivity (reactivity) of the environment is changing with time and how the corrodibility of the metal is changing with time. Solution corrosivity may change because corrosion product concentration increases, because the concentration of the original corrosive species is being depleted, or because microorganisms are growing or dying in the solution. Metal corrodibility usually decreases with time because corrosion products tend to form a protective layer, but corrodibility may increase if corrosion greatly increases surface roughness or a protective coating becomes damaged.

A planned-interval test consists of the following steps:

1. Put a set of specimens into the environment and take them out at intervals.
2. Occasionally, put in a new specimen during the same test intervals to

TABLE 8-2   Selected ASTM Corrosion Tests

Type of Test	Standard	Environment	Specimen
Lab immersion	G31	Solutions	All metals
Coupons	G4	Plant	All metals
Coupon	G52	Seawater	All metals
Weight-loss	D2688	Water	All metals
Visual rusting	D610	Any	Painted steel
Anodic polarization	G5	$1\ N\ H_2SO_4$	Any metal
Cyclic polarization	G61	Aqueous	Fe,Ni,Co alloys
Electrical resistance	D2776A	Water	Any metal
Linear polarization	G59	$1\ N\ H_2SO_4$	Any metal
Linear polarization	D2776B	Water	Any metal
Corrosion potential	G69	$NaCl + H_2O_2$	Al alloys
Electrolytic corrosion	B627	Various	Cr plating
Electrical potential	C876	Concrete	Steel rebar
Intergranular corrosion	A262	Acids	Austenitic stainless
Intergranular corrosion	G28	H_2SO_4	Ni-Cr alloys
Intergranular corrosion	G67	HNO_3	Al-Mg alloys
Exfoliation	G34,G66	Salt soln	Al alloys
Pitting	G46	Any	Any metal
Crevice, pitting	G48	$FeCl_3$ soln	Stainless
Filiform corrosion	D2803	Humid	Organic coated
Stress-corrosion cracking	G38	Any	C-ring
Stress-corrosion cracking	G30	Any	U-bend
Stress-corrosion cracking	G39	Any	Bent beam
Stress-corrosion cracking	G49	Any	Tension
Stress-corrosion cracking	G58	Any	Welded
Stress-corrosion cracking	G44	NaCl soln	Al, ferrous, etc.
Stress-corrosion cracking	G36	Boiling $MgCl_2$	Stainless
Stress-corrosion cracking	G37	$CuSO_4$, $(NH_4)_2SO_4$	Brasses
Atmospheric	G50	Weather	All metals
Atmospheric	B117	Salt spray	Metals; coatings
Atmospheric, accelerated	D1654	Various	Coated metals
Oxidation	G54	Air	All metals

compare with the original results. This procedure checks to see if solution corrosivity is changing.

3. Compare corrosion of old specimens (specimens tested for a long time) with new (short-term) specimens exposed during the same time period. This procedure shows how metal corrodibility changes with time.

A very brief example will illustrate planned interval testing. Three specimens are put into the environment at the beginning of the test (Day 0) and removed at intervals.

Sample	Day In	Day Out	Wt Loss (g/m^2)
A	0	1	4
B	0	14	56
C	0	15	60
D	14	15	5

A new sample (D) is put in on Day 14. Its corrosion of 5 g/m^2 in the 2-week-old solution compared with the 1-day test (A) in the original fresh solution (4 g/m^2) shows that the solution has become more corrosive with time. To find how metal corrodibility is changing with time, compare the new test specimen (D) with the old ones (B and C) in the same 2-week-old solution. A 5 g/m^2 loss is greater than the weight change for the old specimens: (60–56 g/m^2) = 4 g/m^2, so the metal corrodibility has decreased in 2 weeks.

Monitoring solution corrosivity shows when it is time to change the solution, but even if the solution has altered, the metal corrodibility can be determined.

*Electrochemical Tests

Electrochemical tests produce results more quickly than any other laboratory test. An entire polarization curve showing both anodic and cathodic behavior of a metal in an environment with a wide range of oxidation and reduction potentials requires only minutes to a few hours.

• Some old-time corrosion engineers are leery of these new-fangled electro-chemical tests, arguing that tests run in a few minutes cannot tell what will happen when a thick layer of corrosion products builds up on a metal. That is true, but electrochemical tests are still invaluable.

The most-used method is the potentiostatic or potentiodynamic polarization scan (ASTM G5), which uses a test cell like the one shown in Figure 8-3, a reference electrode connected to the cell via the salt bridge probe, and an electronic device called a potentiostat. The specimen to be tested is the working electrode while the auxiliary electrode is usually platinum or graphite.

The potentiostat monitors the potential of the specimen by comparing it with the reference electrode. If the potential is not the desired value set on the instrument, then the potentiostat alters the cell current flowing between working and auxiliary electrodes to bring the potential to the desired value. The potential and resulting current are recorded, the potential is increased slightly, and the process is repeated. The potential can be increased either stepwise (potentiostatic) or continuously (potentiodynamic), and the result is an anodic polarization curve, such as the sketch in Figure 8-4a or b.

The polarization curves show how the solution potential affects the cor-

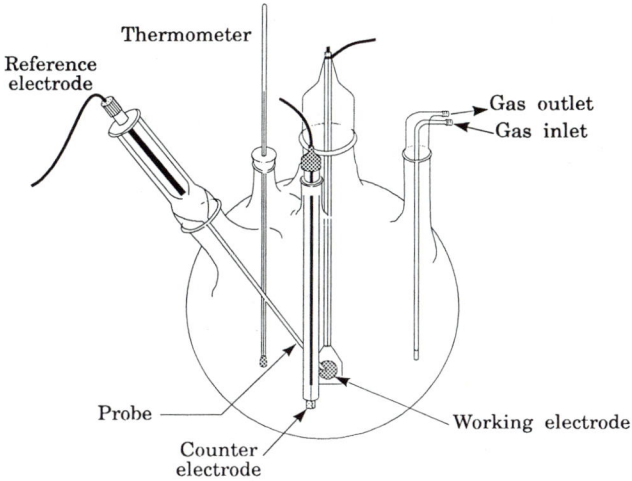

FIGURE 8-3.   Polarization cell.

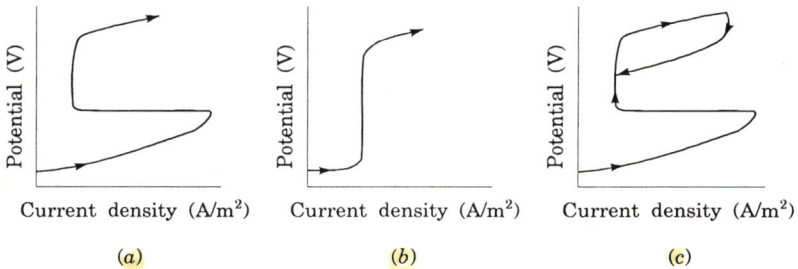

FIGURE 8-4.   (a) Anodic polarization curve for an active–passive metal. (b) Anodic polarization curve for a metal passive at its rest potential. (c) Cyclic potentiodynamic polarization curve.

rosion rate, as calculated from the current density by Faraday's law (Eq. 2-21). They show how well the metal passivates and even if aeration of the solution will passivate: If the critical current density for passivation (the peak current density) is no more than about 1 A/m² the metal will passivate in an air-saturated solution.

Cyclic polarization (ASTM G61) adds one more complication to the polarization curve but gives valuable information about susceptibility to crevice corrosion of active–passive metals. The anodic polarization curve is run up to the transpassive region to some fixed current density and then the potential is decreased stepwise or continuously as before. The pits that have developed in the transpassive region in many cases do not immediately repassivate when the potential is lowered into the passive region but create a hysteresis loop, such as the one shown in Figure 8-4c. The area of this loop or the potential where the reverse curve crosses the forward curve is used as an approximate and relative measure of susceptibility to crevice corrosion.

Mechanical Tests

A wide variety of test methods and samples is used to measure susceptibility of metals to corrosion fatigue, hydrogen embrittlement, SCC, and sulfide stress cracking. A few of the more common tests are described.

The cantilever beam test is a constant-load type used to test for hydrogen embrittlement and SCC. A diagram of the apparatus is illustrated in Figure 8-5. The specimen may be notched, precracked, or smooth. The environment may be in a container surrounding the specimen as in the figure or it may be fed by a wick to the specimen surface. A series of tests is run at different loads to find how load affects time to failure.

The slow strain rate technique (SSRT, also called CERT, constant extension rate technique) used for SCC is simply a tensile test with the specimen surrounded by the corrosive environment but with the test run at an extremely slow strain rate (10^{-6}/s for steel) to give time for the corrodent to act. Consequently, the test may last several days. The stress–strain curve obtained is compared with a test of the same metal in air to evaluate the extent of susceptibility to SCC.

Fracture mechanics tests are used to study corrosion fatigue, hydrogen embrittlement, and SCC. A double cantilever beam specimen, precracked by fatigue, is commonly used. See Figure 8-6 for one type. A constant or cyclic load is applied and the crack velocity monitored. The SCC specimens usually

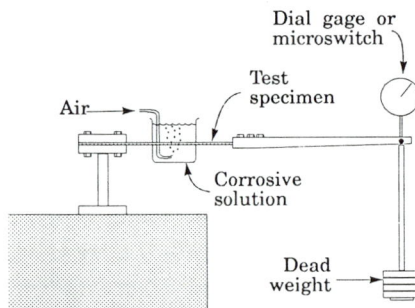

FIGURE 8-5.   Cantilever beam test apparatus.

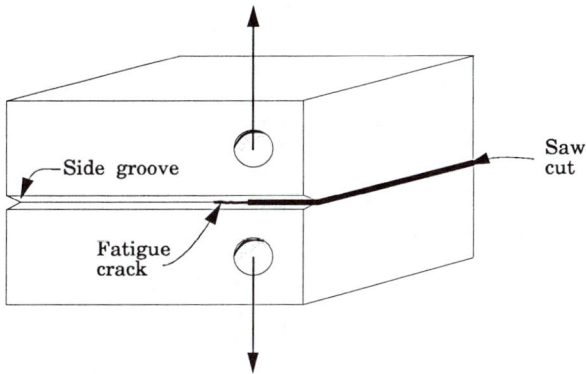

FIGURE 8-6.   Fracture mechanics test specimen.

have the side walls grooved to prevent the crack from branching and to keep the crack front straight.

Sulfide stress cracking is most often tested in a constant load apparatus, such as that shown in Figure 8-7, according to NACE Test Method TM01-77. A calibrated proving ring holds the specimen in tension so that cracking can be monitored by changes in deflection of the ring. The environment is an acidified 5% NaCl solution saturated with H_2S.

8.3 TEST SPECIMENS

Specimens should be in the same form as the metal when it is in service; that is, if the metal is to be cast, rolled, drawn, heat treated, or welded, test specimens should be processed the same way. Testing welded samples is particularly important because of the complex metallurgy produced by the weld.

Small specimens have the advantage that they can be weighed and measured more precisely than larger specimens. The disadvantage of small, flat specimens is that they are mostly edges that corrode much faster than surfaces. The edges should be rounded or preferably masked with electroplater's tape. Large samples detect pitting much better because only one pit on a square metre of surface indicates a serious problem. The surface area of the specimen measured before testing is used in all calculations of corrosion rate.

Surface finish affects corrosion rates, so coupons should be finished to match the equipment finish. The most common surface finish is polishing with 120 grit abrasive paper, which leaves a smooth but not highly polished

FIGURE 8-7.   NACE test proof ring for sulfide stress cracking. [Courtesy of CorTest Inc.]

surface. The samples must then be degreased with a solvent, such as acetone, before testing.

Tests should even be made on metals that have already been tried and are known to be unsatisfactory. This procedure is a check on the test method to see if it gives the same results as actual service does.

Replicate specimens are strongly advised—the more the better, as duplicate weight-loss coupons will usually give no better than ±10% reproducibility. To decide whether the difference between two sets of test results is significant or not, Rossini's criterion is often used. For results $A \pm a$ and $B \pm b$, where A and B are the average values for the two sets, and a and b are their standard deviations, if

$$|A - B| \geq 2(a + b) \qquad (8\text{-}1)$$

to test whether difference b/w two sets of results is significant

there is a 90% confidence limit that A is significantly better than B. For example, if corrosion tests of replicate samples of stainless steel from two different suppliers give averages of 0.0341 ± 0.0008 and 0.0356 ± 0.0008 mm/y, is the first significantly better than the second? $|A - B| = (0.0356 - 0.0341$ mm/y$) = 0.0015$ mm/y, which is less than $2(a + b) = 2(0.0008 + 0.0008$ mm/y$) = 0.0032$ mm/y. Thus, no significant difference is found.

- One old-time corrosion man said he never ran duplicate (2) specimens in tests. If they come out the same, why bother? If they come out different, which one is right? You will have to go back and run the tests all over again. He always ran at least triplicate specimens or just one.

As simple as it may seem, one of the most troublesome problems in corrosion testing is just keeping track of which sample is which. Careful records must be made of every detail of metallurgical history and sample preparation, and the samples must be permanently identified. A common method of identification is to stamp a number in one corner of the coupon, while another method is to cut notches in one edge. This procedure must be done unambiguously, so that the code is not confused if the sample is turned over.

- Some people object to stamped numbers because they sometimes cause SCC. But that is great! You now have very important additional information.

After testing is completed the coupons must be cleaned of all corrosion products and dried before they are weighed and inspected visually or microscopically for pitting. Corrosion products are removed by brushing with a stiff-bristled brush, by dipping in inhibited acids, or electrolytically by cathodic reduction. Details of cleaning methods, cleaning solutions, and assessment of corrosion damage are given in ASTM Standard G1.

Coupon tests have several disadvantages. They cannot detect rapid changes in corrosivity and they often do not reveal corrosion problems that may take a long time to initiate, such as pitting, erosion–corrosion, and SCC. It must be remembered that corrosion rates of coupons are really only *proportional* to corrosion rates of the equipment, not equal, because of the differences in mass, the difference in the ratio of solution volume to metal surface area, the difference in turbulence around the coupon, and other such factors.

8.4 ELECTRONIC PROBES

Weight-loss tests of coupons are considered the time-proven, standard way to measure corrosion rates, but some of their disadvantages have been pointed out. Industry now uses electronic methods extensively to monitor corrosion in critical areas of their equipment. The probes are used to troubleshoot corrosion problems, to protect equipment from unexpected changes in conditions, and even for process control in some cases, although pH electrodes and conductivity sensors respond faster and so are used more often.

Galvanic Probes

These probes are used in situations where corrosion is not supposed to be happening. They consist of two electrodes, usually steel and brass, which will set up a galvanic corrosion cell if an electrolyte is present, thus allowing current to flow to sound an alarm or turn on equipment to correct the problem (Fig. 8-8.) For example, current flow may indicate the presence of oxygen that has leaked into a closed system, or the failure of a corrosion inhibitor. While the galvanic probe does not measure the corrosion rate of the equipment, it does indicate that a corrosion situation exists, and can even be used to turn on a sump pump or take some other corrective action.

- You would not use a galvanic probe in flue gas. Why not? (Remember that it must set up an electrochemical cell.)

Electrical Resistance Probes

The probe consists of a wire loop (Fig. 8-9) made of the same metal as the equipment it is designed to protect. As the wire corrodes its cross-sectional area is reduced and its electrical resistance increases proportionally. A record of the probe's resistance as it increases with time usually shows approximately a straight line while any change in slope indicates a change in the corrosion rate of the equipment. The results usually correlate well with weight-loss coupons and, in fact, the probe is sometimes referred to as an "automated weight-loss technique" because it is simply using an electrical method for measuring the metal loss.

  This probe is the most common type used everywhere in industry and will work in any environment (flue gases, oil, etc.) except in good electrical conductors (liquid metals and molten salts) or if conductive films form (magnetite, iron sulfides, copper deposits, etc.). Corrosion losses are averaged by this technique so it will not detect brief surges in corrosion (response time about 5 min) and usually does not detect pitting. Some probes have a flat strip of metal as the loop, instead of the wire loop, in order to increase their sensitivity to pitting but the surface area is still so small it is unlikely to pit. The electrical resistance changes markedly with temperature so if the temperature of the system fluctuates more than a few degrees the electrical

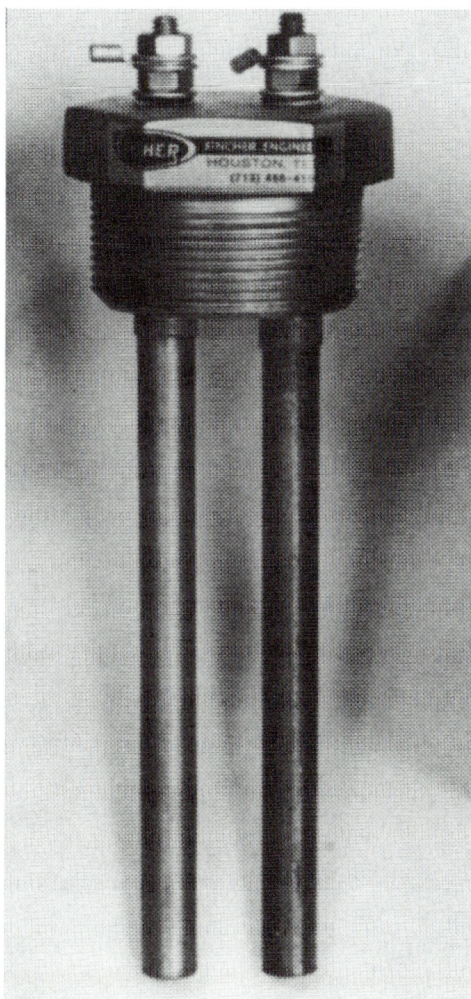

FIGURE 8-8. Galvanic probe. [Reprinted with permission from *Corrosion Control in Petroleum Production*, TPC Publication No. 5, NACE Handbook 1, National Association of Corrosion Engineers, 1982, Fig. 3.7, p. 35. From Fincher Engineering. Copyright © by NACE.]

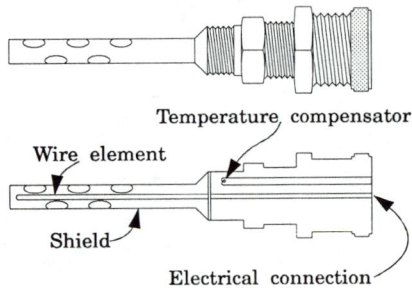

FIGURE 8-9. Electrical resistance probe showing internal wiring. [Copyright © ASTM. Reprinted with permission.]

resistance probe must have an internal noncorroding sensor as a standard for comparison. The perforated shield in Figure 8-9 serves only to protect the wire from damage. One of several commercially available electrical resistance probes has the trade name of "Corrosometer," pronounced corr-o-SOM-eter.

- Electrical resistance probes are now being tested in U.S. military aircraft in inaccessible crevices where salt, moisture, and dirt may accumulate. They are all wired to a central data logger.

Linear Polarization Probes

These are also called LPR or "polarization resistance" probes, an unfortunate term that creates confusion with the electrical resistance probes—with which they have nothing in common. One commercial linear polarization probe has the trade name "Corrater."

The principle of linear polarization is based on the observation that in a corrosion cell, if the electrodes have a very small dc potential difference between them, within about ±10 mV of the corrosion potential, the measurable current that flows between the two electrodes is approximately linearly related to their potential difference (hence, the term "linear polarization"). More importantly, the applied current density i_{app} is directly proportional to the corrosion rate through the equation

$$\frac{\Delta E}{i_{app}} = \frac{B}{i_{corr}} \tag{8-2}$$

in which ΔE is the potential difference between the two electrodes (a fixed value of 10 or 20 mV), B is a constant, and i_{corr} is the corrosion current density that is directly proportional to corrosion rate by Faraday's law (Eq. 2-21).

The constant B can be evaluated for each metal–environment corrosion system by calibrating the linear polarization measurements from results of weight-loss coupons, but more often B is just approximated, since it is about 0.02 V for most corrosion systems.

The probe has two identical electrodes made of the metal whose corrosion rate is to be measured. The electrodes are connected through an external circuit to a battery and ammeter, which apply the potential ΔE and measure the resulting current density i_{app}, respectively. The corrosion rate found from this ammeter reading is the *instantaneous* value, not a long-time average, so the probe is ideal for detecting any sudden changes in the system and can respond immediately.

The linear polarization probe can only be used in an electrolyte. Some probes have a third electrode as a reference electrode, which is necessary if the electrolyte conductivity is low, such as for distilled water. The probe is much more sensitive than weight-loss coupons if the corrosion rate is very low, and of course coupons might take months to develop any measurable weight loss while the linear polarization probe responds instantly. Various types of linear polarization probes are shown in Figure 8-10.

- Extensive tests were conducted a few years ago to establish how much better the three-electrode probes were than two-electrode probes. The two-electrode type actually came out better (although not significantly).

Since the linear polarization probe measures the current flow through an electrolyte, anything that interferes with the current flow, such as an oil film on the probe or fouling, gives spurious results. The probe gives no information on localized attack, for example pitting or SCC.

*Electrochemical Theory of Linear Polarization

The Stern–Geary equation, upon which the linear polarization is based, is

$$\frac{\Delta E}{i_{app}} = \frac{b_a b_c}{2.3\,(b_a + b_c)\,i_{corr}} \tag{8-3}$$

where b_a and b_c are the anodic and cathodic Tafel slopes, respectively. The anodic and cathodic linear polarization curves are illustrated in Figure 8-11 with the straight-line region having a slope of $\Delta E/i_{app}$. Since volts divided by amperes equals ohms of resistance, the name "polarization resistance" has become popular for this technique. It can be shown that the polarization is only precisely linear when anodic and cathodic Tafel slopes are equal, which is very unlikely, but that does not invalidate the usefulness of the linear polarization technique for measuring instantaneous corrosion rates.

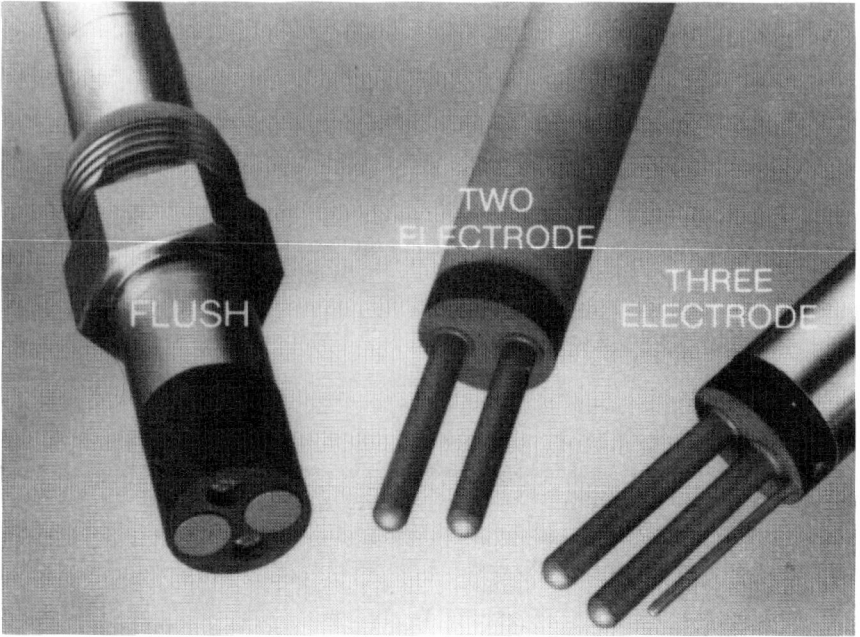

FIGURE 8-10.   Linear polarization probes. [Corrater® LPR probes, products of Rohrback Cosasco Systems, Inc.]

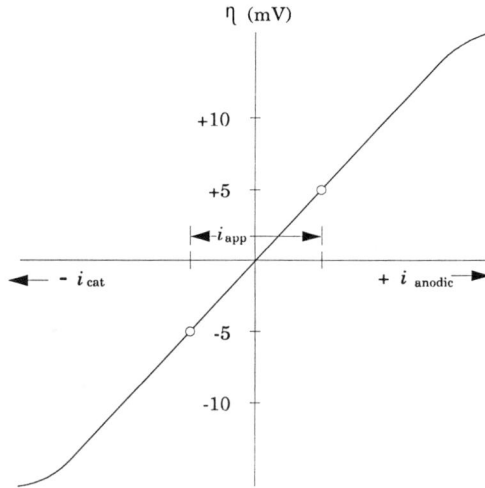

FIGURE 8-11.   Linear region of polarization.

180

8.5 FAILURE ANALYSIS

Investigation of a corrosion problem may be in answer to the recurrent question, "Why did it fail?" or it may be made before failure when the question is, "What is the problem here?" These questions must be answered along with another question that is seldom asked but is implied and is really the important question to answer: "How can this be prevented from happening again" ?

- A doctor who can tell you the name of your disease but has no cure to offer is not very helpful.

Often times the investigator feels that is also his/her role to fix the blame for the failure, but that is seldom the case. That decision is usually best left up to the investigating commission, arbitrator, coroner's jury, and so on.

- The blades in a large industrial fan disintegrated within months after installation. The plant sued the fan manufacturer who in turn sued the foundry that made the blades. An independent lab analyzed the alloy and fixed the blame on the foundry for using an alloy known to be susceptible to SCC. Further investigation showed that cracks had initiated in the fan hubs due to high tensile fabrication stresses. The fan manufacturer had incorrectly attached the blades and had also *specified* the alloy.

Before examination can begin, agreement must be reached between all concerned parties as to whether destructive or only nondestructive methods can be used. The investigation should start with an on-site examination if at all possible. In many cases the cause and remedy will be obvious so that no further study is necessary, but more complicated problems may require a thorough laboratory study or even an extensive testing program if the seriousness of the problem warrants and if money and time are available.

Preliminary Examination

The failure and the metal adjacent to it should be examined visually and by hand-held magnifier. A magnification of 10–20× is usually found best for corrosion work. Note all identifying marks on the equipment—they may reveal specifications as well as the manufacturer. A sketch of the failure can often show more than a photograph, especially if notations of special features are made on the sketch. Record dimensions on the sketch.

Observe the operation of nearby equipment to determine any relationship to the failure. Take written notes on all details. Try to account for every detail observed, every scratch, every discoloration. For major failures, immediately take photographs, preferably in color, as permanent evidence before obtaining samples. Failed parts get scrapped or disappear very quickly.

- I once had a steel tank the size of a small truck disappear.

Gather a history of the failed equipment and the environment: drawings, specifications, daily log sheets, inspection reports, but do not believe everything implicitly. Human beings often make mistakes in recording data or sometimes fake records out of laziness or to cover up their errors.

Sampling

Samples of the environment should be taken for chemical analysis, particularly for analysis of trace impurities that may alter corrosion rates appreciably from handbook values. In some cases samples of the metal itself cannot be taken for legal reasons or because the equipment is still operational. In that case perhaps only scrapings of corrosion products can be sampled but try to avoid mixing dirt and grease, coatings, and mill scale with the corrosion products. Where samples of metal can be taken, metal from the undamaged area, as well as damaged metal, should be obtained.

A common way to sample metal is to cut out the damaged area with a cutting torch, but care must be taken to cut out a sample large enough that heating from the torch does not alter the metallurgical structure, coatings, or corrosion products in the damaged area. Even sawing may generate excessive heat or contaminate the metal with coolant containing lubricants and corrosion inhibitors.

Immediately label each sample and identify the precise location from which it was removed. Marking the samples and photographing before cutting can often be the best method of identification.

Once samples have been taken they must be protected from further environmental changes and mechanical damage before they get to the laboratory. Plastic vials and sealable plastic bags are often used as containers.

- An unprotected metal sample that has been bouncing around in the back of a pickup truck for a week before being delivered to the lab contains more misleading damage than true evidence. (Yes, it happens.)

Protect the samples from handling by bystanders: People seem to have an irresistible urge to mate two fracture surfaces to see if they will fit together again. This procedure invariably scratches the surfaces and may destroy vital evidence of the fracture.

Laboratory Examination

Some samples may require cleaning before examination. A solvent in an ultrasonic bath will remove most deposits, but brushing with a soft-bristled brush may be necessary. Washing with water is usually avoided to prevent further corrosion. If corrosion products are to be removed from the sample or part of it, inhibited pickling solutions, such as those listed in ASTM Specification G1, are necessary. Where corrosion products or coatings are to be preserved in place, impregnating the sample with a cold-setting resin works well.

As basic as it may seem, it is best to start with a verification that the material involved is actually the material specified, although chemical analyses showing minor deviations from specifications are not often significant. Any experienced investigator can recount numerous horror stories where mistakes were made in alloys, welding rods, or coatings. For nondestructive testing, chemical identification kits for rapid spot tests of alloys are commercially available.

- A large 316 stainless steel reaction vessel with a hemispherical bottom was fabricated from four curved sections and welded together on site. One section began pitting almost immediately and was found to be 304 stainless steel due to some inexplicable error.

Nondestructive examination of cracks can be made by Magnafluxing magnetic metals or by dye penetrants. Internal flaws are revealed by ultrasonic tests and radiography.

Metal microstructures are best examined by cross-sectioning the sample and examining it at magnifications of 100–600× with a metallurgical microscope. Surface examination requires the indispensable stereoscopic microscope with magnifications of 5–50×, which reveals much more to the eyes than can be shown on a two-dimensional photograph with its very limited depth of field. The scanning electron microscope (SEM), however, has an almost unlimited depth of field at high magnifications and so is particularly helpful in photographing crack surfaces.

The hardness test, simplest of all mechanical tests, yields a great deal of information about the strength and ductility of a metal and in most cases is nondestructive. Portable testers can measure hardnesses on the original equipment without the need for sampling. Other mechanical tests often performed are tensile, fatigue, and impact tests. These require the machining of small specimens from samples of the failed metal, but take note that small samples generally show better mechanical properties than the full-scale equipment. The probability of metal having flaws of a critical size increases with the amount of metal being tested.

STUDY PROBLEMS

8.1 From the following four tests, how are the liquid corrosiveness and the metal corrodibility changing with time? Explain how you came to your conclusions.

Interval (d)	Weight Loss (g/m^2)
0–5	9
0–10	17
0–15	24
10–15	7

8.2 Lab tests showed 304 stainless steel to have a corrosion rate of 0.013 mm/y in a heavy asphalt at 400°C (750°F). When 304 was used for the flash section of a vacuum distillation tower with the same asphalt at 400°C (750°F) the tower had to be shut down after 22 days because of severe pitting. What sort of tests could have been made to prevent this error?

8.3 From the following data, graph (a) solution corrosivity as a function of time, and (b) metal corrodibility versus time.

Day In–Out	Penetration (µm)	Day In–Out	Penetration (µm)
0–1	1.3	0–11	15.7
0–3	4.1	0–13	18.3
0–5	6.9	0–15	21.1
5–6	1.8	15–16	2.0
0–6	8.4	0–16	22.4
0–8	11.4	0–18	24.9
0–10	14.2	0–20	27.2
10–11	2.0	20–21	2.0
		0–21	28.4

8.4 From the following four tests, how are the liquid corrosiveness and the metal corrodibility changing with time? Explain how you came to your conclusions.

Interval (d)	Weight Loss (mg)
0–5	0.20
0–10	0.30
0–15	0.40
10–15	0.10

8.5 You need a better paint for your plant equipment. Built of carbon steel and exposed to the weather and industrial air, it has badly corroded, especially by pitting and crevice attack. Describe the sort of field test you would set up.

8.6 In laboratory corrosion tests, why are samples usually leaned against the side of a beaker instead of being placed flat on the bottom?

8.7 You are asked to perform corrosion tests on a variety of coatings (electroplated, anodized, etc.) of various thicknesses to be considered for the aluminum shaft of a new type of mechanical pencil. Summarize your research proposal concerning the type of test and the test environment.

8.8 The following data are from a planned-interval laboratory test of corrosion coupons in a simulated industrial solution. Give your reasons for your conclusions as to (a) how the solution corrosivity changes with time, and (b) how the metal corrodibility changes with time.

Days (In–Out)	Wt Loss (mg)	Days (In–Out)	Wt Loss (mg)
0–1	17	0-5	83
0–2	35	0-6	95
0–3	52	5-6	11
2–3	18	0-7	105
0–4	69		

8.9 The data below are from the first planned-interval lab test of an experimental nickel alloy in a synthetic combustion condensate. (a) How often should the environment be renewed? Explain. (b) It was predicted by a technician that the corrosion products on this alloy would not be protective. Do the data bear this out? Explain.

Day In–Out	Wt Loss (µg)
0–1	42
0–2	83
1–2	43
0–3	124
2–3	37
0–4	166
3–4	28
0–5	206
4–5	21

8.10 Compare an electrical resistance probe with a polarization resistance probe by listing one important advantage that each has over the other.

8.11 The linear polarization method is widely used in industry to monitor corrosion. (a) What is measured to determine the corrosion? (b) In what sort of corrosive conditions is the method unsuitable?

8.12 The results of planned-interval tests are given below. (a) How is the solution corrosivity changing with time? (b) How is the metal corrodibility changing with time? State mathematical equalities or inequalities that prove your conclusions.

Days	Wt Loss (g)	Days	Wt Loss (g)
0–1	7.0	0–6	32.0
0–2	12.5	0–7	36.5
0–3	17.5	7–8	4.5
3–4	5.0	0–8	41.0
0–4	22.5		

8.13 Planned-interval tests in a tank of seawater gave the following results:

Sample No.	1	2	3	4	5	6	7	8	9
Day In/Out	1/3	1/5	1/7	1/9	1/11	1/13	1/15	7/9	13/15
Wt. Loss (mg)	5	10	16	22	29	35	42	7	9

How are solution corrosivity and metal corrodibility changing with time? Explain.

8.14 Lab tests of coupons in beakers of deaerated distilled water showed very low corrosion rates in short-term tests but higher corrosion rates in longer term tests. List at least two possible reasons.

8.15 Corrosion damaged 304 stainless steel heat exchanger tubing carrying 24% acetic acid and an inhibitor at 20–85°C (68–185°F). Laboratory tests were made as follows: a disk specimen of each alternative material was cut from rolled rod, weighed, submerged in a beaker of the plant solution at 85°C (185°F) for 2 weeks, removed, and reweighed. The material with the least weight loss was selected. Offer some criticism.

8.16 An HCl gas scrubber made of reinforced phenolic has failed after 2 years of service. Several alternative materials have been suggested. How would you go about choosing one?

8.17 Describe the arrangements for testing that you would make to evaluate the effectiveness of different inhibitors in preventing pitting of aluminum in an industrial solution.

8.18 Chet Bailey asks one of his engineers to run some corrosion tests to choose a suitable material for a new pipeline for seawater. The engineer chooses eight materials that might be suitable and immerses them 30 days in a drum of seawater. He then selects the material with the least weight loss. Comment on this problem.

8.19 Boiling seawater is said to be less corrosive than hot seawater [e.g. 65°C (150°F)]. Why?

8.20 Suppose you were assigned the problem of evaluating several different materials being considered for a new high-pressure reaction vessel. Describe briefly the laboratory tests you would make.

Memos

8.21 Memo to: Corrosion Engineer
From: Chet Bailey, Production Supervisor
We have a tank in the Processing Section that we've been having trouble with ever since it was installed. It holds 4% sulfuric acid with large amounts of sodium chloride at room temperature and is under violent agitation. The stainless steel is pitted badly and corrosion is disastrous at every crevice.
I won't listen to any more of your brilliant ideas on solving the problem. I have obtained samples of practically every material on the market and I want them tested in my tank under our actual operating conditions—not in the lab where you read a dial on a gadget that's supposed to give some magic answer. Let me know what you need, how much time you're going to take, how much space you're going to take up in the tank, and everything else that's going to interfere with our production.

8.22 Memo to: Corrosion Engineer
From: Chet Bailey, Production Supervisor
We're having continual trouble with pitting of cold water pipes and the manager suggested we put you to work for a change. We want some sort of short-term lab tests on copper piping of different grades and manufacturers so that we can order the best one for the new building. For once earn your pay and give us some answers and fast. Give me a short outline of just what you intend to do and I'll send you all the samples you need.

8.23 Memo to: Corrosion Engineer
From: Chet Bailey, Production Supervisor
We're having continuing corrosion problems with anything buried in the ground around here: pipes, tanks, posts, and so on. The plant is built on a landfill that seems to chew up all metals. We want you to design some sort of tests to solve this problem. We want definite answers with real numbers that will clearly show which metals are the best, not weak excuses for why it can't be done. Give me a short outline of just what you intend to do and I'll send you all the samples you need.

9

Materials Selection

Corrosion problems can be solved in the following ways:

1. Select a material that is resistant to the corrosive environment.
2. Give metal a protective coating.
3. Change the service conditions, such as temperature, pressure, or velocity.
4. Change the environment chemistry, such as pH, concentration, aeration, or impurities.
5. Add a corrosion inhibitor.
6. Shift the electrical potential of the metal by cathodic or anodic protection.
7. Modify the design of the equipment or system.
8. Let it corrode and replace it (often a viable alternative!).

The methods listed above are the accepted ways of dealing with a corrosion problem, but not all of them apply in a given situation. In particular, the corrosion engineer often cannot change the service conditions or environment chemistry. These may be as unalterable as the ocean, or nearly as unalterable: an industrial process that is running fairly smoothly where any change will be fanatically opposed by the production people.

Most corrosion problems originate with either improper design or improper materials selection. However, a good choice of material can overcome severe environmental conditions and even some deficiencies in design.

Once the engineer has determined that there is no danger of a catastrophe, deciding which way to combat corrosion usually comes down to the economics of the situation. The most commonly used method of analyzing the alternatives is by the technique described in NACE Standard RP-02-72 (Direct Calculation of Economic Appraisals of Corrosion Control Measures,

1972), in which the discounted cash flow method is explained with numerous examples. This standardized approach also allows calculation of present worth after taxes and the equivalent uniform annual cost. Another clear explanation of the equivalent annual cost method as applied to cathodic protection has been published recently (Webster, 1992).

To give some idea of the relative costs of materials, at the present time the approximate price per foot of a 0.61-m (24-in) diameter pipe is $30 for carbon steel, $60 for filament-wound polyester, $260 for 316 stainless steel, $600 for Monel 400, and $640 for Hastelloy C-276. However, relative costs vary considerably for different shapes and sizes.

9.1 STAINLESS STEELS

Stainless steels are usually the first choice for a "probably corrosive" environment with unknown properties, because these alloys are resistant to a wide range of oxidizers, but they cannot withstand strong reducing solutions, such as hydrochloric acid. Stainless steels *can* be corroded, despite their name.

A stainless steel is defined as a ferrous alloy containing enough chromium to passivate in some environment. At least 10% Cr (but usually ~12%), is required to form a passive film (mainly Cr_2O_3). The carbon content is typically quite low in stainless steels, so the terms "carbon steels" and "steels" refer to *non*stainless steels, since stainless steels have nothing in common with ordinary steels in terms of corrodibility.

- One renowned corrosion engineer continually griped that his tax money had been spent on a stainless steel moon buggy when a carbon steel vehicle would have served just as well. However, corrosion might have been severe in the humid environment of Cape Canaveral, Florida before the launch, and someday the moon buggy will be brought back to earth for its historical importance.

The stainless steels are classified into five general groups according to their metallurgical structures, with the choice of which one to use depending not only on corrosion resistance but also on required strength and cost. Table 9-1 lists the wrought stainless steels, their compositions, and their UNS (Unified Numbering System), numbers that will be used more in the future than they are at present. Wrought alloys are formed by rolling, drawing, forging, and the like, as distinguished from cast metals.

Martensitic Stainless Steels

Martensitic stainless steel can be hardened by air cooling and tempering, like many alloy carbon steels. The corrosion resistance of martensitic stainless

TABLE 9-1　Compositions of Wrought Stainless Steels

Type	UNS	Cr(%)	Ni(%)	C(%)	Other Elements (%)	Typical Uses
			Martensitic Types			
403	S40300	11.5–13.0		0.15 max		Turbine valves
410	S41000	11.5–13.5		0.15 max		Furnace parts
414	S41400	11.5–13.5	1.25-2.5	0.15 max		Petroleum towers
416	S41600	12.0–14.0		0.15 max	0.6Mo max	Screws, bolts
420	S42000	12.0–14.0		0.15 min		Cutlery
422	S42200	11.5–13.5	0.5–1.0	0.20–0.25	1Mo, 1W, 0.2V	Generator blades
431	S43100	15.0–17.0	1.25–2.5	0.20 max		Aircraft parts
440A	S44002	16.0–18.0		0.60–0.75	0.75Mo	Cutters
440B	S44003	16.0–18.0		0.75–0.95	0.75Mo	Valve seats
440C	S44004	16.0–18.0		0.95–1.20	0.75Mo	Springs
			Ferritic Types			
405	S40500	11.5–14.5		0.08 max	0.2Al	Boiler tubing
409	S40900	10.5–11.75	0.5 max.	0.08 max	0.7Ti	Mufflers
429	S42900	14.0–16.0		0.12 max		
430	S43000	16.0–18.0		0.12 max		Architecture
434	S43400	16.0–18.0		0.12 max	1Mo	Auto trim
436	S43600	16.0–18.0		0.12 max	1Mo, 0.6Nb	
439	S43035	17.0–19.0	0.5 max	0.07 max	0.1Al, 1Ti	Welded equipment
442	S44200	18.0–23.0		0.20 max		Furnace parts
444	S44400	17.5–19.5	1.0 max	0.025 max	2Mo, 0.025N, 0.7(Ti + Nb)	
446	S44600	23.0–27.0		0.20 max	0.25N	Burner nozzles
			Duplex (Ferritic–Austenitic Type)			
329	S32900	23.0–28.0	2.5–5.0	0.20 max	1–2Mo	
			Precipition-Hardening Types			
PH13–8Mo	S13800	12.25–13.25	7.5–8.5	0.05 max	2Mo, 1Al, 0.01N max	Pressure vessels
15–5PH	S15500	14.0–15.5	3.5–5.5	0.07 max	3Cu, 0.3Nb	Pressure vessels
17–4PH	S17400	15.5–17.5	3.0–5.0	0.07 max	4Cu, 0.3Nb	Missiles
17–7PH	S17700	16.0–18.0	6.5–7.75	0.09 max	1Al	Aircraft
			Austenitic Types			
201	S20100	16.0–18.0	3.5–5.5	0.15 max	6Mn, 0.25N max	Auto trim
202	S20200	17.0–19.0	4.0–6.0	0.15 max	8Mn, 0.25N max	Food vats

TABLE 9-1   Compositions of Wrought Stainless Steels *(Continued)*

Type	UNS	Cr(%)	Ni(%)	C(%)	Other Elements (%)	Typical Uses
			Austenitic Types			
205	S20500	16.5–18.0	1.0–1.75	0.12–0.25	15Mn, 0.4N	
301	S30100	16.0–18.0	6.0–8.0	0.15 max		Railway cars
302	S30200	17.0–19.0	8.0–10.0	0.15 max		Architectural
303	S30300	17.0–19.0	8.0–10.0	0.15 max	0.5Mo, 0.15S min	Nuts, bolts
304	S30400	18.0–20.0	8.0–10.5	0.08 max		Laundry kettles
304L	S30403	18.0–20.0	8.0–10.5	0.03 max		Dairy tanks
305	S30500	17.0–19.0	10.5–13.0	0.12 max		Spun kettles
308	S30800	19.0–21.0	10.0–12.0	0.08 max		304 welds
309	S30900	22.0–24.0	12.0–15.0	0.20 max		Chemical processing
310	S31000	24.0–26.0	19.0–22.0	0.25 max		Furnace parts
314	S31400	23.0–26.0	19.0–22.0	0.25 max	2Si	
316	S31600	16.0–18.0	10.0–14.0	0.08 max	2.0–3.0Mo	Pipes, tanks
316L	S31600	16.0–18.0	10.0–14.0	0.03 max	2.0–3.0Mo	Welded piping
317	S31700	18.0–20.0	11.0–15.0	0.08 max	3.0–4.0Mo	Heat exchangers
321	S32100	17.0–19.0	9.0–12.0	0.08 max	0.4Ti	Pharmaceuticals
330	N08330	17.0–20.0	34.0–37.0	0.08 max	1Si	
347	S34700	17.0–19.0	9.0–13.0	0.08 max	1Nb	Kitchen sinks
348	S34800	17.0–19.0	9.0–13.0	0.08 max	0.1Co, 1Nb, 0.1Ta max	Nuclear
384	S38400	15.0–17.0	17.0–19.0	0.08 max		Bolts

steel is good in the atmosphere but only fair in chemical solutions, although it is better in the hardened state than annealed. The tensile strengths are typically 500–1000 MPa (70,000–140,000 psi) with a few reaching 2700 MPa (390,000 psi). While hardening makes these steels somewhat brittle, they are used for applications where high strength must be combined with moderate corrosion resistance.

The basic martensitic alloy is Type 410 with modifications made to it for special purposes: more Cr and Ni added for better corrosion resistance, C increased for high strength or decreased for better toughness, P and Si increased to improve machinability, Se added to give better machined surfaces, or high-temperature strength and toughness improved with Mo, V, and W.

Ferritic Stainless Steels

The body-centered cubic structure of ferritic stainless steels cannot be hardened by heat treatment and hardens only moderately with cold work. They are generally straight chromium types with 11–27% Cr and low C content. Their corrosion resistance is good with almost total immunity to stress-corrosion cracking (SCC). Although they are not strong, having tensile strengths in the 400–650-MPa (60,000–90,000-psi) range, they do have good ductility and formability, and are excellent at high temperatures if high strength is not required.

Type 430 ferritic stainless steel is the basic alloy that is modified with alloy additions for specific applications: Cr increased to improve scaling resistance or decreased to improve weldability; Al added to prevent martensite formation; P, S, and Si added for machinability; Mo added to resist pitting; or Ti and Nb to prevent sensitization.

"Super-ferritic" stainless alloys, recently developed, are Cr-Mo steels with extremely low C and N interstitial impurity levels. Their corrosion resistance is excellent, even equivalent to the austenitic stainless alloys in many environments and they are superior in resistance to SCC. They have 50% higher thermal conductivities than the austenitics, but still only one-half that of steel. The commercial alloys, not having standard identification numbers yet, are described by the Cr–Mo contents, such as 18–2, 26–1, and 29–4–2, which is 29%Cr-4%Mo-2%Ni.

Austenitic Stainless Steels

These alloys are the most used and generally the most corrosion resistant of the stainless steels. In addition to their Cr, they contain Ni, which strengthens the passive film and increases resistance to strong acids, while also stabilizing the face-centered cubic crystal structure. These stainless steels have excellent corrosion resistance and very good formability. Austenitic stainless steels cannot be hardened by heat treatment but the common grades work harden extremely well. They are nonmagnetic, unlike the other groups of stainless steels, although cold working can produce enough ferrite to make them very slightly magnetic.

- Testing with a magnet is quite useful in preventing mix-ups in stainless steel alloys but you should not assume that a magnetic stainless steel necessarily has inferior corrosion resistance—the super-ferritics being a case in point.

The 200-series alloys have low Ni contents with high Mn to keep them austenitic. These alloys are a bit cheaper than the comparable 300-series alloys but with just as good corrosion resistance, and are slightly stronger. The tensile strengths of the 200- and 300-series austenitics commonly lie in the 490–860-MPa (70,000–125,000-psi) range although some are as high as 1170 MPa (170,000 psi).

Type 302 was the basic 18–8 (Cr-Ni) alloy which was improved by various alloy modifications: Cr was increased for better corrosion and heat resistance, Ni increased to lower work hardening and resist carburization, Mo added for pitting resistance, C decreased or Nb and Ti added to prevent sensitization, N added for strength, S, P, and Se for machinability, or Si increased for the best heat resistance.

Duplex Stainless Steels

The duplex stainless alloys, with approximately equal amounts of ferrite and austenite, have recently been developed to get the excellent corrosion resistance of the austenitics while improving SCC resistance. The basic alloy is 28Cr-6Ni but N, Mo, Cu, Si, or W may be added to control the ferrite–austenite structure balance or to improve corrosion resistance. The ferrite in the structure raises tensile strengths to the 680–900-MPa (100,000–130,000-psi) range, with yield strengths over twice those of wholly austenitic alloys, even though the alloys are not heat treatable. Some of the newer duplex alloys have nitrogen additions that give excellent pitting resistance and help prevent alloy segregation between the ferrite and austenite phases. Cracks from SCC that can propagate easily through the austenitic fcc phase are stopped by the bcc ferrite, as Figure 9-1 shows.

Precipitation-Hardenable Stainless Steels

They can be heat treated to tensile strengths in the 900–1100-MPa (130,000–160,000-psi) range with one alloy going to 2140 MPa (310,000 psi). Corrosion resistance in most cases is only fair, so these alloys are used where high strength and hardness are essential in a mild or only moderately corrosive environment. The precipitation-hardenable stainless steels have little Ni but contain other alloying elements that will form fine precipitates of an intermetallic compound during aging or tempering.

Cast Stainless Steels

The cast alloys have a different system of identification numbers than their wrought counterparts, somewhat different composition ranges, and often different crystalline microstructures. Table 9-2 compares some cast stainless steels with their approximately equivalent wrought types.

The corrosion resistance of the cast alloys is much the same as the wrought alloys except that the cast alloys are more resistant to SCC because they usually have a small amount of ferrite in their microstructure along with high silicon contents required for good fluidity in casting (Fig. 9-1).

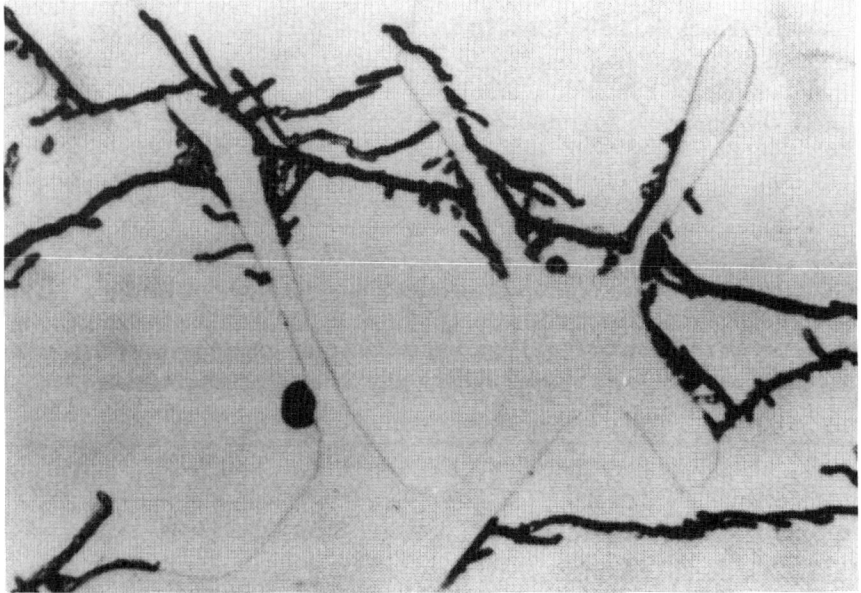

FIGURE 9-1. The SCC cracks blocked by ferrite pools in a cast stainless steel. [Reprinted with permission from Raymond W. Monroe and Steven J. Pawelin *Metals Handbook*, 9th ed., Vol. 13, Joseph R. Davis, Senior Editor. ASM International, Materials Park, OH, 1987, Fig. 21, p. 582.]

TABLE 9-2   Cast Stainless Steels

ACI Type[a]	Similar Wrought Type	Most Common Cast Microstructure
CA-15	410	Martensite
CA-40	420	Martensite
CB-30	431, 442	Ferrite and carbides
CC-50	446	Ferrite and carbides
CF-3	304L	Ferrite in austenite
CF-3M	316L	Ferrite in austenite
CF-8	304	Ferrite in austenite
CF-8C	347	Ferrite in austenite
CF-8M	316	Ferrite in austenite
CF-12M	316	Ferrite in austenite or austenite
CF-16F	303	Austenite
CF-20	302	Austenite
CG-8M	317	Ferrite in austenite
CH-20	309	Austenite
CK-20	310	Austenite

[a]The numbers in the ACI designations indicate maximum carbon content in hundredths of a percent.

194

9.2 NICKEL AND NICKEL ALLOYS

Nickel

Commercially pure nickel 200 has high corrosion resistance, especially to alkalies, combined with mechanical properties similar to mild steel, and good weldability. Nickel and nickel alloys are widely used in the food industry and are frequently selected for service in chlorine, hydrogen chloride, and chlorinated hydrocarbons. They are very resistant to high-temperature air and to SCC.

However, nickel alone cannot withstand strong oxidizing solutions, such as nitric acid, ferric chloride, or aerated ammonium hydroxide, and is not resistant to seawater. Thus some nickel alloys have been especially formulated to overcome these deficiencies. Nickel and its alloys are disasterouly attacked by high-temperature sulfur-containing gases. Table 9-3 gives the compositions of nickels and nickel alloys commonly used for corrosive environments.

Nickel-Copper Alloys

The Monels are various modifications of the basic 70Ni-30Cu alloy. Monel 400 has corrosion resistance similar to nickel but is stronger and cheaper. The

TABLE 9-3   Compositions of Corrosion-Resistant Nickel Alloys

Alloy	UNS No.	Ni	Cr	Mo	Cu	Fe	Other
Nickel 200	N02200	99.5					0.08C max
Nickel 201	N02201	99.5					0.01C max
Monel 400	N04400	66.5			31.5	2.5max	2Mn max
Monel K-500	N05500	66.5			29.5		2.7Al, 0.6Ti
Inconel 600	N06600	76	15.5			8	1Mn max
Inconel 625	N06625	61	21	9		5max	3.6Nb
Incoloy 800	N08800	32.5	21			46	1.5Mn max
Incoloy 825	N08825	42	21.5	3	2.2	30	1Ti
Hastelloy B-2	N10665	69	1	28		2	1Co, 1Mn
Hastelloy C-4	N06455	65	16	15.5		3	2Co, 1Mn
Hastelloy C-22	N06022	56	22	13		3	2.5Co, 3W
Hastelloy C-276	N10276	57	15.5	16		5.5	2.5Co, 4W, 1Mn
Hastelloy G-3	N06985	44	22	7		19.5	5Co,2Cu,1.5W, 1Si,1Mn
Hastelloy N	N10003	71	7	17		5max	
Illium 98		55	28	8.5	5.5		1Mn, 1Si
Illium B		50	28	8.5	5.5		3.5Si, 1Mn
Illium G		56	22.5	6.5	6.5	5	1Mn, 1Si
20		29	20	2.5	4	42	1.5Si max
20Cb-3	N08020	34	20	2.5	3.5	38	0.5Nb, 1Mn

Monels resist seawater well and can handle caustics without SCC. They are the metals usually selected for hydrogen fluoride (HF) service, although corrosion rates are rather high if the acid is aerated, and SCC has been reported for cold-worked Monels in HF so stress relief is advisable.

Nickel–Molybdenum Alloys

Hastelloy B and Chlorimet 2 are trade names for alloys that are particularly resistant to hydrochloric acid at all concentrations and all temperatures up to boiling, as well as other nonoxidizing acids. These alloys are not resistant to oxidizing environments, not even aerated hydrochloric acid. Compositions have evolved from the basic 70Ni-30Mo alloy.

Nickel–Chromium–Iron Alloys

Several of these alloys are technically stainless steels but are grouped with the nickel alloys because of their high nickel contents, high corrosion resistance (and high price). These alloys have better corrosion resistance than the 300-series austenitic stainless steels; the high nickel content resists reducing environments and the chromium resists oxidizing ones.

Alloy 20, sold under the trade names of Carpenter 20, Durimet 20, and so on, with a long history of satisfactory service in a variety of environments, has been improved as 20Cb-3 by even higher nickel and stabilization with niobium. Incoloy 800 shows excellent resistance to high-temperature air because of its high iron content. Incoloy 825, stabilized with titanium and containing more nickel than Alloy 20Cb-3 shows similar or slightly better resistance in some environments.

Inconel 600 is particularly resistant to exhaust gases containing a mixture of air, CO_2, and steam, as well as environments of caustic and high-temperature chlorine. Hastelloy G-3 is very resistant to hot sulfuric acid, phosphoric acids, mixed acids, hydrofluoric acid, and, of course, alkalies.

Nickel–Chromium–Molybdenum Alloys

This group, which includes Inconel 625, the various Hastelloy C alloys, and the Illium alloys, contains molybdenum to improve their resistance to pitting. Inconel 625 is commonly used to handle hot oxidizing and carburizing gases and hot liquids where stainless steels might be attacked intergranularly or suffer SCC. Hastelloy C is extremely resistant to strong oxidizers at moderate temperatures and to hot contaminated acids and solvents. Hastelloy C-22 shows superior resistance to both strong oxidizers and reducing conditions without pitting or SCC.

9.3 OTHER METALS AND ALLOYS

Carbon Steel

Steels are not corrosion-resistant metals. They are the most used metals only because of their excellent mechanical properties, weldability, and cost. In corrosive environments steel usually is coated, inhibitors are added, cathodic protection is applied, or the steel is just replaced when rusted out.

Unprotected carbon steel, however, is used extensively to store and handle alkaline solutions, such as anhydrous NH_3, NH_4OH, and $NaOH$, although SCC can occur in anhydrous NH_3 at welds that have not been stress relieved or in hot solutions of caustic soda ($NaOH$). Steel is compatible with H_2SO_4 at 70% concentration or greater near room temperature and is widely used in H_2SO_4 production plants. The alloyed "weathering steels," described in Section 7.1, show good resistance to atmospheric corrosion because of their very dense, protective rust layer.

- Accidental substitution of a short length of plain carbon steel pipe for stronger 5Cr-0.5Mo steel in a refinery caused a fire and a $700 million lawsuit.

Cast Iron

Cast irons contain 2–4% carbon as graphite or iron carbide along with 0.5–2.5% silicon to make them flow easily during casting. Ordinary gray iron and white iron are extremely brittle, but they can be converted into ductile iron or malleable iron by alloy addition or heat treatment. In all cases they have two-phase microstructures that set up microscopic galvanic cells with corrosion rates similar to unalloyed steels, but they are the cheapest of all engineering metals and can give long-term service in thick sections. Gray cast iron is susceptible to gradual graphitic corrosion in soils and stagnant or slow-flowing solutions.

High-silicon cast irons (> 14% silicon) are extremely corrosion resistant to nearly everything except hydrofluoric acid, sulfites, and caustics. The high hardness of these cast irons also gives them excellent resistance to erosion. They are relatively inexpensive, being marketed under trade names, such as Duriron and Durichlor. These cast irons are so corrosion resistant that they are used for anodes for impressed-current cathodic protection, but they are extremely brittle and practically unweldable.

High-nickel austenitic cast irons, such as Ni-Resist, contain 14–32% nickel, which makes them fairly resistant to concentrated H_2SO_4, H_3PO_4, HCl, and organic acids, while being almost immune to alkalies.

Aluminum and Magnesium Alloys

Aluminum. Aluminum is a very reactive metal in the standard emf series (Table 2-1); it immediately reacts with air to form a passive film consisting of two layers: an inner, compact, amorphous oxide and an outer, thicker, more permeable hydrated oxide. Aluminum is naturally compatible with the atmosphere and withstands many solutions well if the pH lies between about 4 and 9, as the Pourbaix diagram (Fig. 2-11) shows. Strong acids and moderately strong bases destroy aluminum's passive film. Chloride ions are particularly damaging because they attack the film only at weak spots and pit the aluminum. Many chlorinated organic solvents and alcohols can attack aluminum alloys disastrously, sometimes explosively.

The 1000 series of wrought aluminum alloys is essentially pure aluminum with different amounts of impurities. The corrosion resistance of all these alloys is good although second-phase impurity precipitates can weaken the passive film locally and allow the corrosive to attack the underlying metal. The 2000 series consists of Al-Cu alloys that are strengthened by precipitation of an aluminum–copper intermetallic compound. These two-phase alloys are strong but not corrosion resistant, as Table 9-4 shows. The 4000, 6000, and 7000 series are also two-phase alloys when strengthened and all are

TABLE 9-4   Corrosion Resistance of Typical Aluminum Alloys

Alloy	Chemical Analysis[a](%)	Temper	Condition	Corrosion[b] General	SCC	Tensile Strength [c](MPa)
1060	99.6 Al min	0	Annealed	A	A	69
1060	99.6 Al min	H18	Cold worked	A	A	130
2014	4.5 Cu	T6	Pptn.[e] hardened	D	C	485
3003	1.2Mn	0	Annealed	A	A	110
3003	1.2Mn	H18	Cold worked	A	A	290
4032	12.5 Si	T6	Pptn. hardened	C	B	380
5050	1.2 Mg	H3	Cold worked, stabilized	A	A	220
5052	2.5 Mg	H38	Cold worked, stabilized	A	A	290
6061	1Mg, 0.6Si	T6	Pptn. hardened	B	A	310
7178	7Zn, 3Mg, 2Cu	T6	Pptn. hardened	C[d]	B	607

Source: Adapted from Davis, 1987.
[a]Chemical analyses are incomplete. Only major elements are listed.
[b]Corrosion tests with NaCl solution in immersion or spray. Ratings are relative: A being the most corrosion resistant.
[c]To get pounds per square inch (psi), multiply megapascals (MPa) by 145.
[d]In thick sections the rating would be E.
[e]pptn. = precipitation.

readily corroded, but not as badly as the 2000 series where the second phase contains relatively noble copper.

Wrought alloys of the 3000 series, containing manganese, have excellent corrosion resistance. Some manganese-containing precipitates are present but they have almost the same potential as the aluminum solid solution. These alloys are widely used in food processing equipment and chemical plants.

The 5000 series of aluminum–magnesium alloys are essentially one phase with general corrosion resistance as high as commercially pure aluminum and better resistance to chlorides and alkaline solutions. However, alloys that contain more than 3% Mg may form an almost-continuous intergranular precipitate of Al_8Mg_5 that makes them susceptible to exfoliation and SCC.

Magnesium Alloys. Magnesium performs satisfactorily in rural atmospheres by forming a gray film, and is reasonably serviceable in industrial and marine atmospheres provided the design does not trap moisture to set up a galvanic cell. Seawater and chlorides in general are particularly severe on magnesium alloys. Magnesium alloys cannot withstand acids (except conc HF and pure H_2CrO_4!) but they are much more resistant than aluminum to dilute alkalies.

Galvanic corrosion is especially dangerous for magnesium alloys because they are anodic to all other engineering metals. Careful design is necessary to electrically insulate the magnesium or to separate it sufficiently for the distance effect to stifle the galvanic corrosion cell. Stress-corrosion cracking susceptibility generally increases with aluminum content, with most Al-free alloys being nonsusceptible.

Copper and Copper Alloys

Copper and copper alloys make compatible combinations with freshwater and seawater. They are also very resistant to the atmosphere and to non-oxidizing acids. In acids, reducing H^+ to $H_2(g)$ is not possible because copper is more noble than hydrogen and, therefore, would be more easily reduced than hydrogen. However, if the acid contains dissolved O_2 or is an oxidizing acid, such as HNO_3, copper will corrode severely. Copper alloys inhibit growth of organisms and, therefore, are superior where biocides cannot be used.

Copper and its alloys are resistant to most neutral and basic solutions except those containing ammonia, which can cause either SCC or general corrosion. Copper and its alloys do not form a true passive film, although in aqueous solutions near room temperature copper forms an adherent Cu_2O tarnish that acts as a diffusion barrier to slow corrosion reactions. Resistance can be greatly improved by alloying with reactive elements such as Al, Zn, and Sn that form a passive film under the Cu_2O layer.

Table 9-5 lists a few of the more common copper alloys from the hundreds that are commercially available. Tensile strengths vary tremendously, from 220 MPa (32,000 psi) for pure copper to over 1500 MPa (220,000 psi) for beryllium copper.

The Cu-Zn brasses, particularly the high-zinc brasses, may dezincify in stagnant and slow-flowing solutions. Tin brasses, such as admiralty metal and naval brass, are much more resistant to dezincification than ordinary brasses. Inhibited brasses, with 0.02–0.10% P, As, or Sb, are highly resistant to dezincification. At high velocities and impingement situations most copper alloys suffer erosion corrosion because of their low strength. Ammonia plus air and moisture will readily crack them by SCC if stresses are applied or simply if residual stresses are present.

Bronzes, which are copper alloys with Sn, P, Al, Si, or Mn, are much stronger than the brasses and also much more resistant to impingement and to SCC.

Cupronickel, C71500, has the best general corrosion resistance of any commercially important copper alloy. Nickel silvers contain no silver but their high nickel contents give them a silver color and excellent corrosion resistance, especially in seawater.

Titanium and Zirconium Alloys

Titanium Alloys.   Titanium alloys were developed in the 1950s for aircraft and aerospace applications because of their high strength/weight ratio.

TABLE 9-5   Corrosion of Common Copper Alloys in Steam and Salt Water

UNS No.	Alloy	Composition(%)	Corrosion[a] Steam	NaCl
C11000	ETP Copper	99.5 Cu	E	G
C17200	Beryllium copper	98Cu-2Be	E	G
C23000	Red brass	85Cu-15Zn	E	G
C26800	Yellow brass	65Cu-35Zn	F	P
C44300	Inhibited admiralty	71Cu-28Zn-1Sn-0.1As	E	E
C51000	Phosphor bronze	95Cu-5Sn-0.3P	E	G
C62400	Aluminum bronze	85Cu-11Al-4Fe	E	G
C65500	High-silicon bronze	97Cu-3Si	F	G
C71500	Cupronickel	70Cu-30Ni	E	E
C75200	Nickel silver	65Cu-17Zn-18Ni	E	E

Source: Adapted from Davis, 1987.
[a] E is excellent; G is good; F is fair; and P is poor.

Since then, their corrosion resistance to chloride environments and to strong, hot oxidizers has given them an important place in many chemical process operations.

The TiO_2 passive film that forms on the metal in aqueous solutions is the most stable and most protective of all such films. In addition, TiO_2 is an electronic semiconductor so that the passivated metal can be used as an electrode in electrochemical processes, such as the production of Cl_2 gas from molten NaCl and for cathodic protection with impressed current.

Titanium alloys seldom pit but can undergo crevice corrosion that breaks down the passive film. For example, steel gouged into a titanium surface corrodes a very round pitlike hole that is actually crevice corrosion caused by acid formation under the steel. Titanium serves as a cathode when in contact with a corroding metal, such as steel, and becomes charged with hydrogen in aqueous solutions at temperatures above 80°C (175°F). Hydrogen diffusing into the metal embrittles the metal by forming needleshaped, brittle titanium hydrides.

In the laboratory, some titanium alloys have failed by SCC in aqueous chloride solutions when stresses were high and concentrated, but SCC is seldom observed under similar conditions in the field. Several liquid metals do crack titanium, however.

Zirconium.   Zirconium has a very low thermal neutron cross section plus excellent corrosion resistance, allowing it to be used as cladding for uranium fuel and other internals in nuclear reactors. It is also used in chemical processing because of its excellent resistance to strong acids, salts, seawater, and other agents and its outstanding resistance to alkalies.

The ZrO_2 passive film that forms in water and steam is extremely protective but unalloyed zirconium sometimes corrodes irregularly due to impurities in the metal. Zircaloy 2 (1.5Sn, 0.12Fe, 0.10Cr, 0.05Ni) and Zircaloy 4 (1.5Sn, 0.21Fe, 0.10Cr) are superior to unalloyed zirconium in high-temperature water and steam and are formulated to reduce hydrogen pickup in nuclear reactors, since zirconium, like titanium, can be embrittled by hydride formation.

Lead and Zinc Alloys

Lead.   Lead forms a thick, protective coating of lead sulfate in H_2SO_4, making lead and the dilute acid compatible combinations, but the sulfate dissolves in concentrated H_2SO_4. Similarly, lead is very resistant to chromic and phosphoric acids where it forms insoluble chromate and phosphate. Traditionally, lead has been widely used for its high resistance to atmospheres and water but its cumulative, toxic effects on animal life now greatly restrict its use in the environment.

- One unsolved problem at present is how to remove old, scaled, lead-pigment paints from bridges without contaminating the rivers. The solution in many cases, unfortunately, has been to ignore it and just let it quietly rust off.

Lead is still widely used to seal underground power and communication cables, in lead–tin solders, for the electrode plates of lead–acid storage batteries, and in chemical process industries. The metal is soft, and thus subject to erosion–corrosion, but it is low melting and easy to apply as a coating. "Chemical lead," with the best corrosion resistance and strength of any grade of lead, contains small amounts of copper and silver.

Zinc.   Zinc, a highly reactive metal, plays an important role in corrosion protection. It is used extensively for sacrificial anodes in cathodic protection and as a coating on steel to provide both barrier and sacrificial protection. Zinc's excellent resistance to the atmosphere is a result of formation of ZnO on the surface that converts to $Zn(OH)_2$ in the presence of moisture. The hydroxide reacts further with CO_2 to form a basic zinc carbonate that is extremely protective. In hard water an insoluble $CaCO_3$ layer forms on zinc, protecting it even further. Usually, high-purity zinc is used for sacrificial anodes in soils while a 0.2Al-0.05Cd zinc alloy is better in seawater.

9.4 PLASTICS

The three general classes of plastics are

1.   *Thermoplastics.* The thermoplastics consist of long hydrocarbon chains with molecular weights in the thousands. These plastics are relatively weak and soften further as temperature increases. Cross-linking, adding bulky side groups to the chains, or partial crystallization strengthens the polymers.

2.   *Elastomers.* Elastomers are thermoplastics with coiled or kinked chains. High-temperature strength is greatly improved by occasional cross-links (e.g., by vulcanization with sulfur). These thermoplastics can be deformed elastically several hundred percent. In corrosion service, elastomers are mainly used as coatings.

3.   *Thermosets.* Thermosets have been polymerized into three-dimensional, noncrystalline structures that make them strong and rigid. These plastics do not soften when heated and are relatively expensive. Thermosets are all reinforced for industrial applications.

Table 9-6 rates 30 types of commercial plastics for corrosion resistance and also gives tensile strength and percent elongation. The ratio of tensile strength to elongation is a measure of stiffness. Polar solvents include reactive organic liquids, such as formic acid, methanol, water–ethanol mixtures, and acetic acid. Important nonpolar solvents are the ethers, alkanes (e.g., hexane), carbon tetrachloride, benzene, and acetone. Also listed is the temperature of heat distortion as measured by the ASTM D-668 standard test, but keep in mind that the safe temperature for continuous heating is usually a few degrees less.

Some of the properties and uses of the commercial plastics listed in Table 9-6 along with common abbreviations are given here.

ABS: Acrylonitrile–butadiene–styrene copolymers are available in thermoformed sheets or extruded pipes and shapes. They are fairly resistant to weathering, inexpensive, and easily joined.

Acetals: The acetals have a high modulus and good impact strength from −40 to +110°C (−40 to 230°F). Acetals are used for gears, pump impellers, and valve parts.

Acrylics: The acrylics are transparent and soft. These plastics are cast, extruded, and thermoformed and are often used as coatings.

Alkyds: Thermosets of the alkyds are available in molded shapes and they are also used in protective coatings.

Allylic Plastics: Urea–formaldehyde (UF) and melamine–formaldehyde (MF) are used as baked coatings and as cellulose-filled molding resins.

Aramids: The aramids are aromatic polyamides with high strength and excellent resistance to heat.

Cellulosics: The cellulosics, such as cellulose acetate and ethylcellulose, are used as coatings, sheet, extruded pipe, and injection-molded shapes. These plastics are transparent and weather well but their overall chemical resistance is low.

Epoxy Resins (EP): Epoxy resins are thermosets, and are used in moldings, often reinforced with fiberglass. These resins are also used as protective coatings.

Fluoroplastics: The fluoroplastics are heat-resistant thermoplastics, highly crystalline, with excellent corrosion resistance. These plastics are the most versatile and important of the thermoplastics. The major varieties of fluoroplastics are

PTFE (Teflon), polytetrafluoroethylene, is the most difficult fluoroplastic to work. Teflon has the highest temperature stability, is porous, and can fail by fatigue in thermal cycling. High compressive stresses can cause cold flow.

PCTFE (Kel F), polychlorotrifluoroethylene, is used for liners and coatings.

PVDF (Kyner) is polyvinylidine fluoride.

PVF (Tedlar), polyvinyl fluoride has lower heat resistance than other fluoroplastics but is more workable, and is used for coatings or liners.

TABLE 9-6 Chemical Resistance and Physical Properties of Commercial Plastics[a]

	Nonoxidizing Acids		Oxidizing Acids		Aqueous Salt Solutions		Aqueous Alkalies		Polar Solvents		Nonpolar Solvents		Water		Tensile Strength (MPa)	Elongation (%)	Heat Distortion Temperature (°C at 1.82 MPa)
	25°C	90°C	25°C	90°C	25°C	90°C	25°C	90°C	25°C	90°C	25°C	90°C	25°C	90°C			
ABS terpolymer	S	U	Q	U	S	S	S	U	Q	U	U	U	S	U	34.5	60	90
Acetal	U	U	U	U	S	U	S	U	S	U	Q	U	S	U	69.0	30	125
Acrylics (PMMA)	S	U	U	U	S	U	S	U	S	U	U	U	S	U	65.5	4	95
Alkyds	S	U	U	U	S	U	Q	U	S	U	U	U	S	Q	41.4	2	200
Allylic	S	U	U	U	S	Q	Q	U	S	U	U	U	S	Q	58.0	4	200
Amino	Q	U	U	U	S	U	Q	U	S	U	Q	U	S	U	55.2	0.7	130
Aramids	U	U	U	U	S	S	S	Q	Q	U	U	U	S	Q	120.0	5	260
Cellulose acetate	U	U	U	U	S	S	S	U	Q	U	U	U	S	U	51.7	5	140
Ethylcellulose	Q	U	U	U	S	U	S	U	Q	U	U	U	S	U	34.5	10	65
Fluoroplastics *PTFE*, PETFE[b]	S	S	S	S	S	S	S	S	S	S	S	S	S	S	24.1	200	100
Epoxy	S	S	U	U	S	S	S	Q	S	S	S	U	S	S	51.7	5	140
Fluoroplastics *PVDF*, PVF, PE-CTFE	S	S	S	Q	S	S	S	S	S	Q	S	Q	S	S	55.2	200	80
Furan carbon-filled cement	S	S	U	U	S	S	S	S	S	Q	S	Q	S	S	41.4	1.5	80
Nylon, *nylon-66*, nylon-6	U	U	U	U	S	S	S	Q	Q	U	S	U	S	Q	82.8	60	75

Phenolic, mineral-filled	S	S	S	U	S	S	U	U	S	Q	S	U	S	S	41.4	0.5	200
Poly(amide-imide)	S	U	U	U	S	S	U	U	S	S	S	S	S	S	186	12	275
Polyaryl esters	S	Q	Q	U	S	S	S	Q	S	U	S	Q	S	S	68	50	175
Polycarbonate	Q	U	U	U	S	S	U	U	S	Q	U	U	S	S	72	110	130
Polyesters (unsaturated)	Q	Q	U	U	S	S	U	U	Q	U	U	U	S	Q	69	1.5	200
Polyethylenes *LDPE*, LLDPE, HDPE	S	S	Q	U	S	S	S	S	S	Q	Q	Q	S	S	5.5	100	40
Polyimides	Q	U	Q	U	S	S	U	U	S	S	S	S	S	S	96.5	8	315
Polymethyl pentene, *polypropylene*	S	S	Q	U	S	S	S	S	S	S	Q	U	S	S	34.5	100	55
Polyphenylene oxide	S	Q	Q	U	S	S	S	U	S	Q	U	U	S	S	55.2	50	100
Polyphenylene sulfide	S	U	U	U	S	S	S	Q	S	S	S	Q	S	S	74.0	1.1	135
Polystyrene	S	Q	U	U	S	S	S	U	S	U	U	U	S	S	41.4	1.5	90
Polysulfones	S	U	U	U	S	S	S	Q	S	Q	U	U	S	S	82.8	25	175
Polyurethanes, cast	Q	U	U	U	S	S	Q	U	U	U	Q	U	S	S	6.9	200	75
Polyvinyl chloride (plasticized)	S	S	S	Q	S	S	S	S	S	S	U	U	S	S	10.3	200	65
Silicones	Q	U	U	U	S	S	S	Q	S	S	Q	U	S	S	3.5	200	200
Thermoplastic elastomers (styrene–butadiene block copolymer)	S	S	U	U	S	S	S	S	S	S	U	U	S	S	6.9	600	25

Source: Adapted from Seymour, 1982; Fontana, 1987; Schweitzer, 1990.
[a]S is satisfactory; Q is questionable; U is unsatisfactory
[b]Where similar plastics are grouped, mechanical properties are given for the italicized one.

FEP, fluorinated ethylene propylene, is a copolymer of PTFE and hexafluoropropylene. It is often used as piping and as a fitting liner.

PE–CTFE, copolymer of polyethylene and PCTFE, has excellent chemical stability and workability.

PETFE is a copolymer of PTFE and PE.

Furan Plastics: These plastics are used as mortar, usually with a carbonaceous or siliceous filler.

Nylons: Nylons are molded unfilled, with fiberglass reinforcement, or filled with PTFE and MoS$_2$, which act as lubricants for nylon gears, cams, and bearings. Nylons are used for mechanical parts and coatings and have excellent abrasion resistance.

Phenolics: Phenolics are thermosetting and are generally molded, but are also used as mortars and coatings.

Poly(amide-imide): This type is an excellent injection-molded plastic used in pumps, valves, and turbines.

Polyaryl Esters: These esters include polyarylates, polyethylene terephthalate (PET), polybutylene terephthalate (PBT), and PET–PBT copolymers. These thermoplastics are used for containers, pump housings, impellers, valves, gears, and nozzles.

Polycarbonates: Polycarbonates are clear, tough thermoplastics used for glazing, blow-molded containers, and extruded sheet. These plastics have high impact strength but are easily stress cracked.

Polyesters (Unsaturated): Polyesters are often fiberglass reinforced and are widely used where corrosion resistance requirements are not high.

Polyethylenes (PE): These plastics are the lowest–cost polymers, with excellent chemical resistance. They are easily fusion welded and are available in high density (HDPE), low density (LDPE), and linear low density (LLDPE). High density PE is stronger and more chemically and heat resistant than LDPE. Ultrahigh-molecular weight polyethylene (UHMWPE) has outstanding strength and toughness.

Polyimides: The polyimides can be produced either as thermoplastic or thermoset. These plastics are usable from −190°C to +370°C (−375 to 700°F) with some going up to +510°C (950°F) for short periods. The more heat-resistant thermosets are compression-molded into jet engine parts and have excellent creep and abrasion resistance.

Polymethylpentene (TPX): Polymethylpentene is transparent and extremely light. It can be injection-molded into labware and lenses.

Polyphenylene oxide (PPO): This plastic is blended with polystyrene for tanks, pump parts, and machine housing, has excellent heat and dimensional stability but is relatively expensive.

Polyphenylene sulfide (PPS): Polyphenylene sulfide is heat resistant, with outstanding chemical resistance. This plastic can be injection or compression molded and either filled or unfilled. Polyphenylene sulfide is used for pumps and electrical parts.

Polypropylene (PP): Polypropylene is translucent, hard, inexpensive, and light. It is used in appliances, electrical apparatus, and ducts. Polypropylene has a higher modulus and lower thermal expansion than polyethylene.

Polystyrene (PS): Polystyrene is colorless, brittle, with excellent electrical properties but poor heat resistance. Polystyrene is resistant to HF acid but it generally has low chemical resistance.

Polysulfones: Polysulfones have excellent electrical properties and temperature stability. These plastics are used for circuit boards, valves, and motor components.

Polyurethanes (PU): Polyurethanes are used most as foams but are also available as elastomers, moldings, adhesives, and coatings. These plastics have outstanding abrasion resistance.

Polyvinyl chloride (PVC): Polyvinyl chloride is available rigid or plasticized, and chlorinated (PVDC). Rigid PVC is used for piping, ducts, and fans; plasticized PVC is used for flexible tubing and sheeting. These plastics are easily fabricated but have high thermal expansion and poor modulus. Polyvinylidene chloride (PVDC) has better mechanical and chemical properties.

Silicones: The silicones consist of -Si-O- chains instead of the carbon chains that the other polymers have. Silicones have superior heat and moisture resistance.

Thermoplastic Elastomers (TPE): These elastomers are used for hoses, cable insulation, adhesives, linings, and coatings.

9.5 OTHER NONMETALLICS

Ceramics

Ceramics are compounds of metallic and nonmetallic elements. The range of ceramics is quite wide—from glass to earthenware, porcelain, and brick, to concrete. These compounds have outstanding high-temperature stability and chemical resistance, except for HF and caustics. Ceramics are hard, electrical insulators, extremely strong in compression, and very resistant to erosion–corrosion. However, ceramics have much lower thermal conductivities than metals. Because ceramics are brittle, they can withstand only low tensile stress, and are very susceptible to thermal shock.

Glass.   Glass is a noncrystalline ceramic, mainly SiO_2 with other inorganic oxides dissolved in it. The compounds Al_2O_3, B_2O_3, CaO, PbO, MgO, K_2O, and Na_2O are added to enhance mechanical or thermal properties and corrosion resistance. Silica glass (fused silica) is made extremely pure (impurities as low as 1 ppm) for optical uses. It also has excellent corrosion resistance, low thermal expansion, and can be used continuously up to 900°C (1650°F). Glasses of 96% silica, more easily formed and cheaper than fused silica, are made into crucibles, heat shields, laboratory glassware, and tubing where their high thermal shock resistance and good chemical resistance are important.

Aluminosilicate glasses, unlike fused silica, can be processed by all the standard glass-making techniques. These glasses are resistant to water and weathering but have only fair resistance to acids. Typical uses in industry include thermometers and water-level tubes for boilers. Soda lime glasses are the cheapest of the glasses, being used for windows, bottles, and industrial piping. Soda lead glasses have improved refractive index but reduced corrosion resistance. The brilliance of these glasses makes them ideal for crystal ware, and they are also used for shielding from x-rays.

- One fellow in our city installed a glass sewage pipe that ran through his basement game room. Guests could watch the flow.

Borosilicate glasses, such as Pyrex, are the glasses most used by chemical process industries and are the most versatile. A typical glass might be $80SiO_2$-$13B_2O_3$-xNa_2O-yAl_2O_3. Low-expansion borosilicate glasses have excellent chemical resistance and so are widely used in chemical piping. Aluminoborosilicates are used for laboratory apparatus, columns, centrifugal pumps, and industrial piping.

Acid Brick.   Acid brick is made from fireclay or red shale to resist hot acids and erosion–corrosion caused by the acids. The fireclay type has better thermal shock resistance, the best dimensional stability, and does not leach iron into the acid. A high-silicon variety with about 10% more SiO_2 is better in phosphoric acid and is a better thermal insulator. The red shale type of acid brick is used less frequently but is cheaper and more impermeable.

Acid brick is used primarily as a lining for storage tanks, digesters, kettles, and chimneys as well as for floors and trenches where acid spillage may be a problem. For linings, the steel or concrete shell has an intermediate lining of rubber, plastic, lead, or asphalt overlayed with acid brick solidly mortared in place.

Earthenware, Stoneware, and Porcelain.   Earthenware, stoneware, and porcelain are made from mixtures of clay, silica, and feldspar. Earthenware commonly is only clay, fired at a low temperature where little fusion or sintering can take place, so that it is relatively porous (up to 20%) and weak. The clay "soil pipe" used for drains has these properties. Stoneware is fired at a higher temperature to give only about 4% porosity. It is used for chemical tanks and coils, being virtually unattacked by acids. Porcelain is fired at an even higher temperature to make it semivitreous (glassy), nonporous, hard, and stronger. A typical composition for porcelain insulators might be 60% kaolin clay, 20% silica, and 20% feldspar. Porcelain is used for china, semi-vitreous whiteware, acid nozzles, and the like, but more extensively as an enamel coating on steel.

   Concrete.   Being alkaline, concrete is readily attacked by acids. Calcium ions in the concrete leach out in corrosives containing sulfates or they exchange with magnesium ion solutions, turning the concrete into a soft, porous mass. Some organic compounds also degrade concrete. Fluoride surface treatment or sodium silicate washes densify and harden the concrete surface to make it more corrosion resistant. Concrete tanks and pipes must be lined if strong corrosives are involved.

Carbon and Graphite

Carbon is a hard, brittle, porous (up to 30%) material with reasonable strength at room temperature (tensile strength about 7.5 MPa (1100 psi), and compressive strength 43.5 MPa (6300 psi), which increases with temperature until at 1650°C (3000°F) it is one of the strongest materials known. While carbon is a thermal and electrical insulator, graphite is a good conductor. Both carbon and graphite are extremely resistant to thermal shock.

   Impervious Graphite.   Impervious graphite eliminates the porosity of ordinary graphite by saturating it with a chemically resistant polymer, such as a phenolic or epoxy, almost doubling its strength. The impregnated graphite has a maximum operating temperature of only 170°C (340°F), but is among the most chemically inert of materials. It is not corroded by nonoxidizing acids up to their boiling points, by alkalies and salts, nor most organics. Strong oxidizing conditions and halogen gases are damaging.

   Graphite and baked, partially crystalline carbon are used as electrodes in electrochemical and electrometallurgical processes at high temperatures, and for combustion chambers, molds, and crucibles. Impervious graphite is used for chemical process equipment, such as shell-and-tube heat exchangers, pumps, and valves.

Wood

For corrosive environments, only the *inner heartwood* of the following four woods is used, listed in decreasing order of corrosion resistance.

Tidewater red cypress—resists decay and is considered one of the best tank woods.

California redwood—good mechanical properties and weathers well. Widely used for cooling towers.

Long leaf yellow pine—often used for sulfuric acids and sulfates. Excellent
service life if, but only if, continuously saturated with moisture.
Douglas fir—straight–grained, used extensively for brine tanks.

The wood should be kiln dried to make the extractable resins less soluble.
These woods are not affected by water, abrasive slurries, mild acids, salt
solutions, or most organics at their maximum service temperature of about
50°C (120°F), with intermittent service to 65°C (150°F). They cannot take
steam, strong mineral acids, strong oxidizers, or alkali solutions of any
concentration. Wood is sometimes impregnated with coal tar, phenolic resins,
or furan resin, which gives it a corrosion resistance similar to that of the
impregnating agent. Rotting of wood cooling towers, caused by a fungus, is
prevented by impregnation with a biocide.

Wood is used in industry for applications such as tanks, launders, columns,
filter plates and frames, fume scrubbers, and piping. It is often overlooked in
situations where it could outlast metals by many years because it is considered
too old-fashioned.

STUDY PROBLEMS

9.1 Which material would be the *least* suitable under the following conditions? (a) Concentrated caustic solution: Hastelloy B, nickel, or tantalum. (b) Concentrated hydrochloric acid: Hastelloy B, rubber, or 316
stainless. (c) Hot, concentrated HNO_3: nickel, 316 stainless, or titanium.

9.2 A company asks you to recommend a material suitable for their process
but unfortunately they cannot reveal the environment or operating
conditions. They can tell you, however, that steel cracked and that
linings of Teflon, glass, aluminum, and even tantalum were badly
attacked. Suggestions?

9.3 What metals naturally come to mind as suitable for the following
environments? (One metal each.) (a) The internals of a freshwater
distillation unit. (b) Battery acid (37% H_2SO_4). (c) Hot, strong hydrochloric acid. (d) Molten sodium dichromate oxidizer. (e) Storage tank
for 90% vitriol (impure H_2SO_4).

9.4 A manufacturing process involves a hot, agitated solution containing
2.6 g/L of cupric chloride and 10% chromic acid (a powerful oxidizer).
Two metals have been suggested to handle this: 304 stainless steel and
Hastelloy C. Please comment. Any better ideas?

9.5 Corrosion of a cast iron pipe in soil containing H_2SO_4 was mitigated by:
(a) cleaning and painting the affected area with coal tar enamel; (b)
installing a drain tile near pipe to remove much of the acid; (c) installing
a Duriron anode with impressed-current cathodic protection; and (d)

removing wood contacting the pipe to eliminate crevice corrosion. Which of the 8 ways to prevent corrosion do these remedies correspond to?

9.6 Stress-corrosion cracking of welded steel storage tanks containing anhydrous ammonia was prevented by: (a) stress relieving large tanks; (b) purging tanks of all air before filling; (c) adding 0.2% H_2O, since lab tests indicated that as little as 0.1% H_2O prevented cracking. Which of the 8 ways to prevent corrosion do these remedies correspond to?

9.7 Rank the major types of stainless steels according to (a) strength and (b) corrosion resistance.

9.8 In which classifications of stainless steels do the cast alloys *not* fall? Can you explain why?

9.9 Which of the components in the basic alloy of duplex stainless steels would tend to stabilize the austenitic structure or the ferritic structure?

9.10 An alloy nominally 67Ni-33Cu has been used for fasteners in apparatus handling 50% H_2SO_4 at moderate temperatures. Corrosion has been 0.7 mm/y. Can Nickel 201 do a better job? Explain.

9.11 An alloy nominally 67Ni-33Cu has been used for a pump for industrial-grade phosphoric acid. Its corrosion rate is 0.4 mm/y. Could a high-silicon Duriron pump do a better job?

9.12 An Incoloy 825 lining in a furnace has held up very well in 800°C (1475°F) air for 6 years. A second furnace is to be built with Incoloy 800 suggested for the lining because it is considerably cheaper. Would you OK this? Explain.

9.13 Naphthalene with some H_2SO_4 at 90°C (195°F) has proved to be highly corrosive to stainless steels, even Alloy 20. Can any plastic handle this combination?

9.14 A 58% solution of calcium and lithium bromides and chlorides in water at approximately 50°C (120°F) is causing crevice corrosion and pitting of stainless steel. Suggest a plastic coating that should be able to withstand this environment.

Memos

9.15 Memo To: Corrosion Engineer
From: Chet Bailey, Production Supervisor
Since you've been pretty lucky on some of your guesses in the area of corrosion, here's another one for you. In our extraction unit we have H_2S dissolved in water operating between 80 and 150°C with no aeration or agitation. We are temporarily using an old unit designed for sulfuric acid that is made out of Carpenter 20, whatever that is. Don't tell me we shouldn't have done it, just give me some idea of how long

that extractor is going to last. We can lose about $\frac{1}{16}$ in. before it becomes unsafe.

9.16 Memo To: Corrosion Engineer
From: Chet Bailey, Production Supervisor
We've hit a peculiar corrosion problem—not serious, but I was wondering if you'd explain it. A 304L stainless reactor containing 40% HNO_3 and 20% $Zr(NO_3)_4$ is only slightly corroded after 3 years of service. Temperature [120°C] is monitored with a thermocouple that recently went bad. The thermocouple sheath is 405 stainless and was corroded slightly but not bad. The thermocouple was put in a new sheath from the same piece of tubing as the first sheath. We even made sure we insulated the sheath from the reactor, now that you've got us corrosion conscious. *Result*: The sheath corroded away completely in 3 weeks. We can go to a 304 tube, but tell us why 405 worked once when it won't work again.

9.17 Memo To: Corrosion Engineer
From: Milton T. Klein, General Manager
Chet Bailey had a large pump made of Monel 400 that failed very quickly from corrosion. He now plans to order a high-silicon cast iron pump, but I think you should take a look at the situation because these pumps are fairly expensive. The solution being pumped is 30% hydrochloric acid, well aerated, at room temperature. If cast iron seems reasonable, let me know soon because we've got to place the order.

9.18 Memo To: Corrosion Engineer
From: Chet Bailey, Production Supervisor
It has been suggested that we could switch to 50% HNO_3 in the processing section instead of the 80% we've always used. This will save us money on the acid and reduce the corrosion of the aluminum piping. Now it's over 0.1 mm/y. The idea looks good to us but the manager thought we should check with you since you're our corrosion expert.

9.19 Memo To: Corrosion Engineer
From: Milton T. Klein, General Manager
Chet Bailey had a large pump made of Alloy 20 that failed very quickly from corrosion. He now plans to order a Hastelloy B pump, but I think you should take a look at the situation because these pumps are very expensive. The solution being pumped is 35% H_2SO_4, well aerated, at room temperature. If Hastelloy B seems reasonable, let me know soon because we've got to place the order.

9.20 Memo To: Corrosion Engineer
From: Chet Bailey, Production Supervisor
When the contact acid plant was built we installed plain carbon steel pipe between the plant and a 304 stainless steel storage tank. For the

next 3 years we had to replace the corroded steel pipe every 6 months, although fortunately a little iron in the H_2SO_4 didn't cause any trouble in the mill. Since we had noted that the tank was standing up well to the acid, we finally got around to replacing the pipe with 304 stainless. After one year the stainless pipe is leaking! I know it's off-grade in some way because the storage tank is still in good shape. What are your ideas on the subject?

9.21 Memo To: Corrosion Engineer

From: Chet Bailey, Production Supervisor

A Type 316 stainless steel reaction vessel 5 ft in diameter has failed within months with tomato sauce oozing out of a network of cracks around a flange welded to the tank. We've used similar 316 tanks 10 years or more with the same sauce. The sauce has a pH of 3.9 with 2.5% salt and is held at 85°C by steam injection, which causes considerable vibration. Tests show the flange to be Type 304 stainless. What caused this failure and what can be done about it? Should the next vessel be made out of 304, since it was the 316 that leaked, not 304?

10

Protective Coatings

The major purpose of coating a metal is to protect it from a corrosive environment when the metal is otherwise suitable for the service conditions in terms of mechanical and physical properties. Coating metal with good mechanical properties (usually steel) is often more practical in terms of cost and required life than selecting a more corrosion resistant but expensive material.

Protection can be achieved in four ways, with many coatings functioning in more than one way.

1. A barrier coating that prevents the corrosive environment from contacting the metal.
2. A sacrificial coating that corrodes while giving cathodic protection to the underlying metal.
3. An inhibitor coating that slows electrode reactions.
4. An electrically resistive coating that stifles electrochemical corrosion cells. Paints fall into this last category.

No coating can be serviceable for long if the metal surface underneath is rusted or is contaminated with oil, salts, or the like. The key to good protection is good surface preparation, a step even more important than the properties of the coating itself.

Most coatings consist of more than one layer. For example, a phosphate conversion coating might be covered with a paint undercoat and several more paint layers. An electroplated coating may consist of two or three different metals plated on top of each other. Each layer contributes its own properties in providing protection to the base metal.

10.1 METAL COATINGS

The coating metal is either cathodic or anodic to the substrate metal. If cathodic, the coating must be a perfect barrier layer or galvanic corrosion will attack the underlying metal at any pore or scratch in the coating. An anodic coating can provide both barrier and sacrificial protection so that even corrosion of cut edges is prevented. The thicker a sacrificial coating is, the longer the cathodic protection continues.

Before coating, typical surface preparation involves degreasing in solvent vapor or a solvent emulsion, intermediate cleaning in hot alkali, and an acid pickle to remove oxide films left by the previous steps and to lightly etch the surface. Each step is followed by water rinses.

Electroplated Coatings

Metals that are applied as electroplatings for corrosion protection include nickel, chromium, zinc, cadmium, and tin. Copper is electroplated as an undercoat for multiplate systems. The part to be plated is made the cathode in an electrochemical bath containing ions of the plating metal with an anode usually composed of the plating metal that dissolves during the plating. Direct current is applied to the cell so that the cathode becomes negative enough to reduce the metal ions on the cathode surface. Current interruptions or periodic current reversals may be used to smooth the plated layer.

Nickel.   Nickel is often plated over copper for use in atmospheric corrosion resistance and for bright, decorative finishes. Most frequently, nickel plating is used as an undercoat for chrome plating. The thickness, ductility, and hardness of the coat can be controlled by changing current density, temperature, plating time, and bath chemistry.

Electroless Nickel.   Plating can be applied to metal parts by immersing them in a special bath containing an unstable nickel salt whose decomposition is catalyzed by the metal surfaces. The plating's resistance to corrosion is excellent and its wear resistance is also good because the nickel deposit contains 4–12% phosphorus, which hardens it. Coatings are very uniform; plating inside holes and shielded areas is far better than electroplating. However, the coatings are brittle and the chemicals are expensive.

Chromium.   Chromium plating, unlike other plating processes, uses insoluble lead alloy anodes so chromic acid must be added to the bath during electroplating. (Chromium anodes would passivate and stop current flow.) Chromium is a reactive metal but because it passivates so well the plating is cathodic to steel and to an intermediate layer of nickel. The best atmospheric corrosion resistance is obtained from microcracked chromium, electrodeposited under conditions that leave 400–800 cracks/cm (1000–2000 cracks/in.). The microcracks reduce stresses in the plating without penetrating very far; increasing the number of cracks decreases the depth of penetration.

Zinc   Zinc, being anodic to steel, protects steel by serving as both a barrier and sacrificial coating. A thin (5–10 μm or 0.2–0.4 mil) electroplated zinc coating is often used as a paint base, especially on the lower parts of automobile bodies, because it can be applied more smoothly than hot dipping. If zinc is not coated with paint (e.g., on screws, nuts, and bolts) the metal is given a chromate dip to preserve its bright, shiny appearance. Zinc-plated steel is not used in continual immersion because of the galvanic corrosion, nor in contact with foods because of health hazards.

Cadmium.   Like zinc, cadmium is anodic to steel and is plated thinly (5–15 μm or 0.2–0.6 mil) onto steel springs, fasteners, television chassis, and electronic parts because of its good ductility and electrical conductivity. Its atmospheric corrosion resistance is similar to zinc's, it is poisonous, and it is not resistant to common chemicals. Unlike zinc, cadmium plating is not usually painted.

Cadmium can cause liquid metal (and solid metal) embrittlement of high-strength steels above 230°C (450°F). High-strength steels can also be embrittled by hydrogen charged into the metal during electroplating, especially cadmium plating, and must be baked at about 200°C (400°F) immediately after plating. Baking as little as 3 h may be needed for steels with tensile strengths below 1500 MPa (220 ksi) but up to 24 h is required for higher strengths and sections over 25 mm (1 in.) thick.

Tin.   Tin coatings on steel are almost always electroplated, in coats as thin as 0.4 μm (0.02 mil), followed by a flash melting to reduce porosity. Tin is noble to steel in water but it is sacrificial to steel inside sealed tin cans of food with no air present. Being nonpoisonous and seemingly tasteless, it is frequently used for food handling equipment. Tinplate is also widely used in the electronics industry because it solders well and resists tarnishing.

- A large commercial sausage stuffer made of cast iron was used for several years. It was then set outside for 5 years, where it rusted. Cleaned of rust by power brushing and then oiled, it was put back into service but it discolored the meat! The answer: re-tin, in this case by hot dipping to put on a thick layer.

Hot Dip Coatings

Galvanized Steel.   Galvanizing is used wherever steel must remain serviceable in the atmosphere, soil, or water for 10–25 years. Structural steel for electrical transmission towers, culverts, reinforcing steel in precast concrete, bridge structural members, light standards, and marine pilings are typical applications. Hot dip galvanizing puts on a relatively thick layer of zinc that is desirable because the service life is linearly related to the thickness.

Hot dip galvanizing is applied by immersing steel in a molten zinc bath. Diffusion of zinc into the steel bonds it strongly as the two metals form three thin intermetallic compound layers between the zinc layer and the steel, as shown in Figure 10-1. These compound layers tend to make the corrosion very uniform but they are brittle, so for galvanized steel sheet that must be formed after dipping, the compound layers can be suppressed by adding 0.1–0.2% Al to the bath or by electrogalvanizing at room temperature.

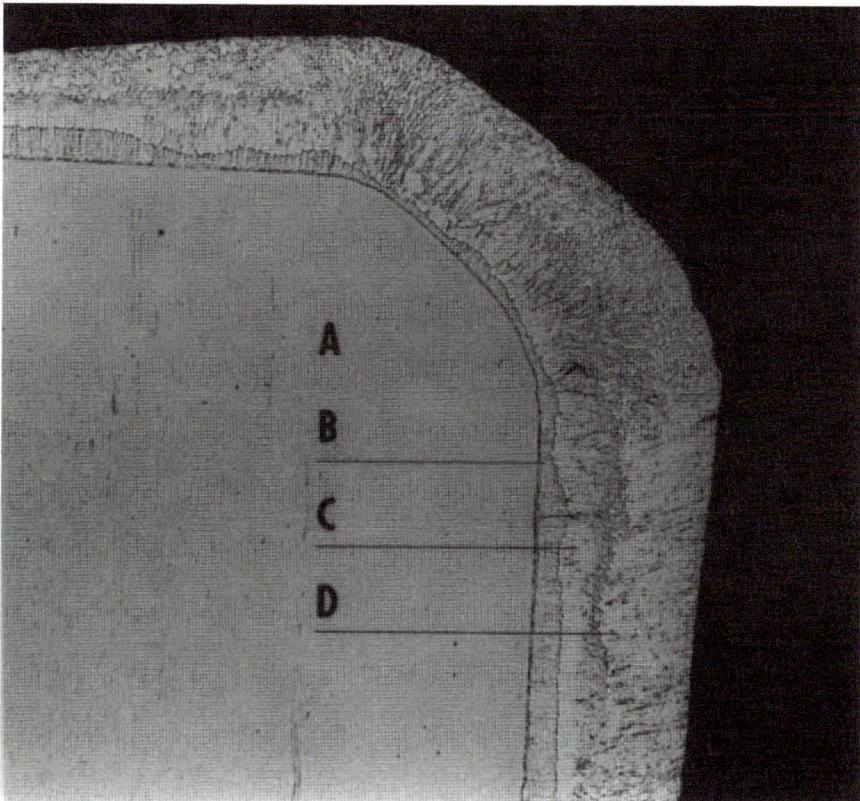

FIGURE 10-1.   Micrograph of a typical galvanized coating, showing the underlying steel (A), the hard Fe-Zn alloys (B and C), and the overlay of zinc (D) Note that galvanized coatings are thicker at corners, giving extra corrosion protection in these vulnerable areas. (Compare with paint coatings, Fig. 10.4.) [Reprinted with permission from *Galvanized Steel on the Move*, Cominco, Ltd., 1985, p. 4.]

Aluminum.   Aluminum-coated (aluminized) steel, made by hot dipping, is used in the atmosphere where it often outlasts galvanizing, but it is not generally used in water.

55Al-Zn.   "Galvalume" (actually 55Al-43.5Zn-1.6Si) is the fastest growing hot dip coating worldwide. The alloy is two to six times more resistant to atmospheric corrosion than galvanized steel, perhaps up to three times better in freshwater, and probably equivalent in seawater.

Lead.   Lead applied to steel by hot dipping, a material called "terne" when the lead is alloyed with a few percent tin, improves corrosion resistance and the steel's ability to be formed, painted, or soldered. It is used principally for automotive gasoline tanks and fuel lines.

Vapor Deposited Coatings

Aluminum.   Aluminum is deposited from its ionized vapor onto high-strength steel aircraft components with a high negative potential applied between the steel and the vaporization source. Complex shapes can be uniformly plated with a dense, thick (up to 25 μm or 1 mil) coating that is very adherent and formable, replacing hot dipping.

MCrAlY.   Alloys of Cr, Al, and Y, where M is Co, Ni, or Fe, are evaporated by electron beam and vapor deposited on gas turbine blades and other turbine parts for resistance to oxidation, sulfidation, and hot corrosion.

Thermal Spray Coatings

The coating material in the form of wire or powder is fed through a gun where it is melted and blown by a stream of compressed gas onto the substrate. As the droplets strike the surface they flatten into platelets that adhere and conform to the surface irregularities, producing a lamellar coat that can be built much thicker than galvanizing. The heat source to melt the feed is acetylene or other fuel gas, an electric arc, or plasma (inert gas heated in an electric arc).

Aluminum and zinc are the metals most commonly applied by thermal spray, in coatings 50–500 μm (2–20 mils) thick. Other metals used for special applications include austenitic stainless steels, aluminum bronze, nickel-base alloys, and MCrAlY. Aluminum spraying has been used on TV towers, offshore oil rigs, ships, fuel storage tanks, railroad hopper cars, and even for kilns operating at temperatures up to 900°C (1650°F). Zinc is sprayed on bridges, on steel in coal mines, the interiors of water storage tanks, and other similar applications.

Cladding

A thin sheet of the cladding metal is bonded to the base metal by forcing them into intimate contact by cold or hot rolling, hot pressing, explosion bonding, or extrusion bonding. Combinations that bond particularly well are Al/Al alloys, Ni/Cu, Au/Cu, Sn/Ni, Cu/steels, Ag/Cu, Sn/Cu, and Au/Ni. An example, illustrating how clad metal can be slit and expanded, is shown in Figure 10-2.

Copper-clad stainless steel, the most commonly clad corrosion barrier material, is used for telephone and fiber optic cable shielding. In acidic or sulfidic soils the stainless steel protects the inner copper layer that is needed for electrical shielding. Aluminum clad with stainless steel is used extensively for automobile trim.

Stainless steel cladding on carbon steel intended for SCC service cannot be properly stress relieved because the difference in thermal contraction during cooling still leaves the cladding in tension.

- Tantalum, which was explosion bonded to steel, showed bad bonding in one 3 × 5-in. spot on every sheet. The work crew had not noticed that each tantalum sheet had a paper label on the back side.

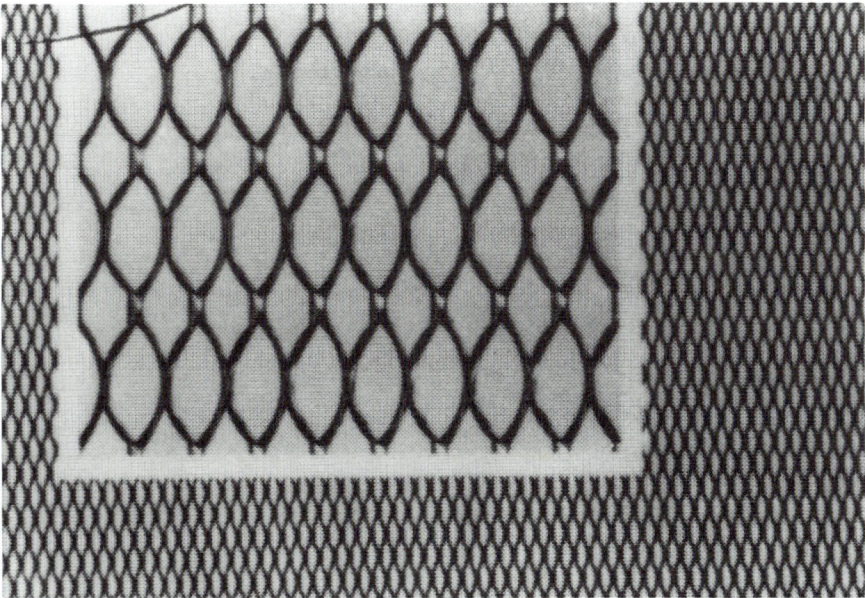

FIGURE 10-2.   Expanded Pt-clad Nb for electroplating anodes. [Reprinted with permission from Robert Baboian and Gardner Haynes, *Metals Handbook*, 9th ed., Vol. 13, Joseph R. Davis, Senior Editor. ASM International, Materials Park, OH, 1987, Fig. 5(*a*), p. 889.]

Alcladding.   Alclad aluminum is the most common example of cladding. A high-strength aluminum alloy sheet is clad on one or both surfaces with a thin layer of pure aluminum or a corrosion-resistant aluminum alloy, such as 3003 (1.2% Mn), which is anodic to the base metal.

Diffusion Coatings

Precleaned parts are packed in a powder mixture containing the coating metal, an inert filler, and an activator. The pack is then held at a high temperature to allow the powder to diffuse into the surface of the parts and form an alloy.

Calorizing.   Calorizing is a method of impregnating steel with aluminum to produce a surface with less than 25% Al. Solid-state diffusion layers up to 1 mm (40 mils) thick can be obtained in this way. Calorized steel is used for molten salt pots, steam superheaters, and furnace parts up to 750°C (1380°F).

Chromized steel.   Steel with a chromium coating, produced similarly to Calorizing, is used for automobile mufflers.

Sherardizing.   Sherardizing reacts steel with diffused zinc, which forms Zn-Fe intermetallic compounds. The process is used to coat small parts, such as nuts, bolts, and washers.

Laser Surface Alloying

The metal to be protected is first coated with a thin layer of the alloying element by electroplating, vapor deposition, thermal spraying, or some other technique, and then surface melted with a laser beam rastored over the entire surface. The high cooling rates either eliminate second-phase precipitates or make them extremely fine, and also prevent any segregation of impurities to grain boundaries (see Fig. 10-3). Stainless steels treated with a laser become much easier to passivate and much more difficult to pit.

10.2 CONVERSION COATINGS

A metal can undergo a surface reaction that converts its surface to an inorganic compound that is extremely adherent, since it formed from the metal itself. This surface is extremely corrosion resistant, since the metal has already reacted.

Anodizing

While magnesium and zinc are sometimes anodized, the process is very common for aluminum. The aluminum is made the anode in an electrochem-

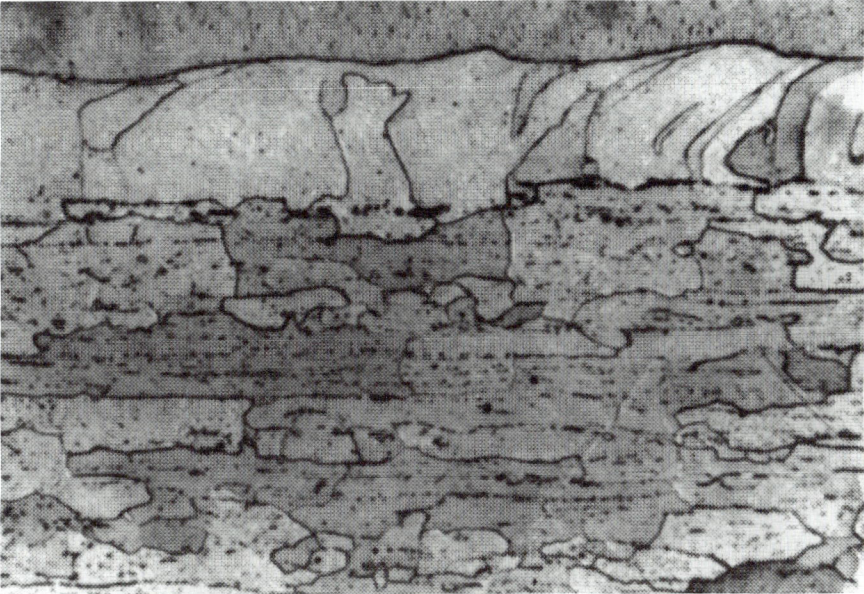

FIGURE 10-3. Coarsening of grain size in a ferritic stainless steel by laser melting. Magnification 425X. (Note the absence of precipitates.) [Reprinted with permission from R.D. Granata and P.G. Moore in *Metals Handbook*, 9th ed., Vol. 13, Joseph R. Davis, Senior Editor. ASM International, Materials Park, OH, 1987, Fig. 12, p. 503.]

ical cell and deliberately corroded to build up a thicker layer of protective oxide than forms in natural environments.

Different types of cell baths are used depending on properties desired for the anodizing. A bath of boric acid and borates produces a very thin and impervious aluminum oxide layer that has the high dielectric constant and oxidation resistance needed for electrolytic condensers and rectifier plates. To make a thicker oxide, the cell electrolyte must slowly dissolve the oxide to leave pores for current to flow. Dilute sulfuric, chromic, or oxalic acid baths produce thick, porous oxide that is then sealed in boiling, deionized water to plug the pores (up to 10^{11} pores/cm^2 or 10^{12} pores/in.^{2}), by converting Al_2O_3 to $Al_2O_3 \cdot H_2O$. If color is desired in the coating, dyeing is done before sealing.

Oils and waxes may be used to seal the anodizing to give better resistance to industrial atmospheres, or dichromate sealants to give added protection in saline environments. Bright automotive trim requires an anodized thickness of only about 6μm (0.25 mil), whereas architectural aluminum will have anodizing up to 30 μm (1 mil) thick.

Phosphate Coatings

Iron, steel, galvanized steel, or aluminum can be sprayed or immersed in a dilute solution of phosphoric acid plus zinc, iron, or manganese phosphates that reacts with the metal to form a gray to black surface layer of insoluble metal phosphate. The phosphate layer provides a bit of corrosion protection, aids cold forming, and increases wear resistance, but primarily it is used as a base for paints on such products as automobile bodies and refrigerator cabinets. The phosphate improves paint adhesion and retards the spread of any corrosion under the paint film. As a dielectric it insulates the local anode and cathode sites on the metal surface. The pores in a phosphate coating are usually sealed with a chromate rinse to improve corrosion resistance at the weakest areas of the coating.

Galvanized sheet steel phosphated for paint adherence provides superior corrosion resistance with its multilayer protection system. Aluminum is easily coated by spraying with zinc phosphate containing fluoride. Medical implants are now being coated with calcium phosphate.

Iron phosphating produces the thinnest, least protective layer but provides a good paint base and is the least expensive phosphating. It is used at the highest acidity to best remove rust, eliminating a pickling step. Zinc phosphates are usually intermediate in thickness, providing either a paint base or a structure to hold oils or wax. Paint applied over zinc phosphate provides good corrosion resistance but the layer is usually too thick for coil coating where painted sheet must be formed. Heavy manganese phosphate coatings are not painted but are porous enough to permit absorption of waxes or oils containing corrosion inhibitors, which is the usual finish applied to steel nuts, bolts, and screws.

Chromate Coatings

Aluminum, zinc, cadmium, steel, and magnesium are often given a chromate treatment, usually by immersion or spray, to produce a protective chromate film on the metal, to seal other coatings of phosphate, oxide, and the like, or to provide an excellent, nonporous bond for paint. The solutions used contain chromic acid or Cr^{6+} ions.

The resulting coatings, ranging in thickness from 10 to 1000 nm (5×10^{-4}–0.05 mil) are primarily a hydrated chromium chromate but contain extensive amounts of other metals and anions. Color varies with thickness, from transparent to iridescent blues and greens, with yellows typical for the thicker coatings. They can also be dyed.

Chromate coatings, being more insoluble in water than other conversion coatings, are exceptionally good for corrosion in immersion conditions. The coatings provide a barrier layer, passivate the metal, and inhibit the cathode process, with corrosion protection being directly proportional to coating thickness.

Cadmium plating must be stress relieved or baked before chromating because the heat would destroy the protective chromate. Galvanizing cannot be successfully painted or phosphated if chromated first. Most chromating of steel is done to produce "tin-free" steel sheet, which is then lacquered for tin cans.

Oxide Coatings

Aluminum is often oxidized with solutions containing chromate or dichromate to produce a thick, strong oxide for better corrosion resistance or as a foundation for lacquer and varnishes. The films produced are a gray color and may be sealed with hot dichromate solution.

In times past, firearms and tools made of steel were blued, producing a surface layer of mainly Fe_3O_4, which has a pleasing appearance and was thought to provide corrosion protection. The steel was heated to about 300°C (575°F) in a bed of hot charcoal or to about 330°C (625°F) in a salt bath until the desired color was obtained. Chemical bluing can be achieved by immersing the metal in hot alkali with an oxidizing agent. In any case, the bluing provides little protection but it is porous enough to hold oil and waxes that can shield the steel from atmospheric attack if regularly applied. Bluing has largely been replaced by phosphate coatings.

Nitriding and Carburizing

Wear resistance of steel is improved by the surface hardening treatments of carburization and nitriding. They diffuse carbon or nitrogen atoms into the surface to harden it while also producing high compressive stresses on the surface. Although carburized or nitrided steels are no more, and in some cases less, corrosion resistant than untreated steel, they are resistant to SCC, which requires tensile stress at the surface to open up a crack.

Rust Converters

A variety of chemical formulations, most containing tannin, are sold for application on rusted steel to penetrate and react with the rust, converting it to nonporous, stable compounds that serve as a barrier layer to protect the steel. They do penetrate and do convert the rust to a partial barrier layer, which in some cases protects the steel. However, they do not remove moisture from the rust and do not counteract the effects of aggressive salts, especially chlorides, in the rust, so corrosion can continue unabated under the converted rust. Rust converters are effective only on clean, uncontaminated rust less than 2.5μm (0.1 mil) thick, in mild service conditions.

10.3 ORGANIC COATINGS AND LININGS

The difference between a polymer coating and a lining is the difference in thickness and, therefore, in service conditions and a difference in mechanism of protection. A coating is thin (by definition <0.5 mm or 20 mils), typically 50–100 μm (2–4 mils) while a lining will usually be around 3 mm (0.1 in.). A lining is thick enough to be a complete barrier between the metal and the environment but a coating is not. Linings are applied to the interior of tanks and vessels, where they are subjected to immersion conditions while coatings are applied to the exterior, where they may be subjected to weathering, condensation, fumes, splashes, and the like but not immersion, which is particularly severe.

- A detached plastic lining plugged an 11-ft Houston sewer 4 years after it was installed. Clean-up cost 1 million, along with major environmental contamination.

Coatings and linings can be applied either in the mill or in the field. They will perform much better if they can be mill applied, mainly because surface preparation is much better in the mill. Also, field application will usually be at least twice as expensive. On the other hand, mill-applied coatings must be much stronger and more resistant to weathering and sunlight to prevent damage during shipping, storage, and handling on the site before construction.

- One engineer was asked to design a telescoping oil drilling derrick that would be constructed in the United States, shipped on the deck of a freighter to the Middle East, and erected in the desert. The design was easy for him but he had great difficulty finding a paint that could withstand rough handling, sea air, intense sunlight, sandstorms, and could be readily repaired.

Surface Preparation

Poor surface preparation causes about three-fourths of all premature coating failures. The very best coatings will not perform well on a poor surface but even poor coatings may be satisfactory on a good surface.

- Engineers are very skeptical of glowing reports on lab and field tests of new coatings because they know that coatings in real service do not behave at all like coatings perfectly applied to perfect surfaces in coupon tests.

Unfortunately for the corrosion engineer the trend these days is toward less surface preparation initially and greater maintenance costs later because of the time value of money and tax advantages in deducting maintenance as an operating cost. The cost of *ideal* surface preparation can run 50–80% of the cost of the entire paint job.

Surface cleaning techniques are listed in Table 10-1. At the present time evidence indicates that wet abrasive blasting is the best surface preparation method, although being new, it has not yet been used as extensively as other techniques. It uses a water jet with chemical inhibitor added to blast an abrasive against the metal surface. This method also avoids the spread of dust that has been a serious problem with the dry-blast methods.

Generally, white metal blast cleaning is necessary for inorganic zinc coatings, a near-white blast cleaning is satisfactory for phenolic, epoxy, vinyl, and organic zinc coatings, while commercial blast cleaning is sufficient for alkyd, epoxy ester, coal tar epoxy, and chlorinated rubber coatings.

- One expert categorically declared, "If wire brushing is all that can be done by way of preparation, then high-performance paints are a waste of money."

Dry blasting with sand or grit roughens the metal surface to give an "anchor pattern" of 40–50 μm (1.5–2 mils) for good coating adherence. Many of the best corrosion-resistant coatings (vinyls, chlorinated rubber, and fluorocarbons) have poor adherence. As a general rule the surface profile (peak height) should be no more than one-half the dry-film thickness to avoid pin-spot rusting at the high spots (see Fig. 10-4).

TABLE 10-1   Surface Preparation Techniques

Technique	Contaminant Removed
Solvent cleaning	Oil, grease, dirt
Vapor decreasing	Oil, grease
Hand tool cleaning	Loose rust, loose mill scale, old paint, dirt
Power tool cleaning	Same as hand tool cleaning
White-metal blast	Oil, grease, dirt, mill scale, rust, old paint
Near-white blast	To 95% white metal
Commercial blast	To two-thirds white metal
Brush-off blast	Oil, grease, dirt, rust scale, loose mill scale, loose paint
Flame cleaning + wire brush	Old paint, dirt, moisture, unbonded mill scale, unbonded rust
Steam cleaning	Acid, oil, grease
Pickling	Mill scale, rust
Water jet	Oil, grease, dirt, salts, loose rust, loose mill scale
Wet abrasive blast	Oil, grease, dirt, rust, mill scale, old paint, salts

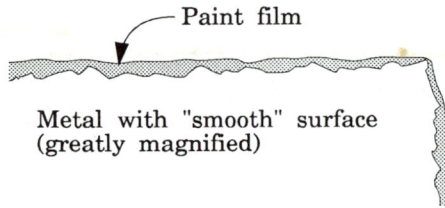

FIGURE 10-4.   Paint coating on metal surface.

Application

Failure to coat the metal sufficiently causes a great many coating failures, second only to poor surface preparation. Crevices simply are not coated, as for example, back-to-back angle irons skip welded. Floor gratings are almost impossible to coat completely. Since they may rust badly enough to become dangerous to walk on, many chemical plants have switched to plastic gratings. A full, even coating of paint must be applied; just coloring the metal gives no protection.

- One inspector found a painter spray painting a 4.5-m (15-ft) high tank from the ground—in a 40-km/h (25-mph) wind! Erecting the proper scaffolding would have taken a lot of time.

For steel in corrosive environments, it is generally agreed that the minimum paint coating should be 3 coats with 125 μm (5 mils) total thickness, which gives 50–75 μm (2–3 mils) at the edges. Coating failures commonly occur at weld spatters and sharp edges where the paint film is the thinnest. That is the reason for the ingenious design of the KTA test panel, used to field test coatings (Fig. 10-5). Paint should not be applied below 10°C (50°F), when the surface temperature is below the dew point, or at relative humidities above 85%.

Principles of Protection

Linings are thick enough to be impermeable barriers but coatings (a term that includes paints, varnishes, and lacquers) are thin and porous. Both moisture and oxygen penetrate coatings, and that is all it takes to set up corrosion. However, corrosion will not occur if the pores through the paint are so fine and circuitous that the resistivity of the electrolyte path becomes

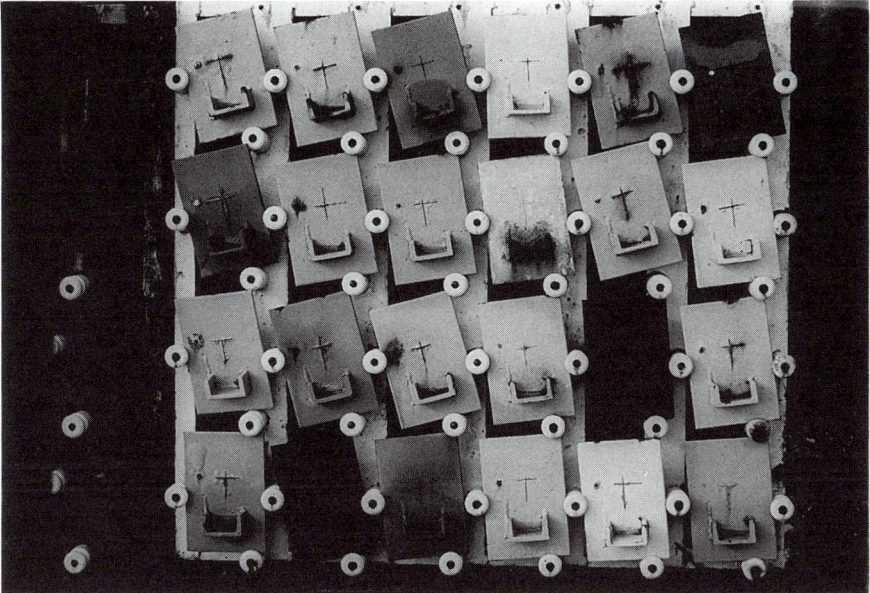

FIGURE 10-5.   The KTA test panels that include round and sharp edges, corners, pits, welds, a crevice behind the channel, a moisture pocket, with coating damage by cut and punch. Supplied with the original mill scale or blasted to white metal. [Courtesy of KTA.]

very large. Also, local charged regions on the polymer molecules attract and hold ions moving through the pores. An electrical resistance of 10^8 Ω/cm^2 (10^9 $\Omega/in.^2$) is sufficient to stop corrosion. Low-resistance porous coatings that hold moisture against the metal, creating crevice corrosion, are worse than no coating at all.

Multiple, thin layers are much more effective than one thick layer because only a few pores in one layer will connect with pores in the next. A steel boat hull in seawater, for example, will typically have four to seven coats of paint. The U.S. Navy uses a five-coat system plus cathodic protection.

Some coating types contain corrosion inhibitors, such as red lead or zinc chromate, that interfere with the anode reaction. The zinc-rich paints contain zinc dust, which becomes a virtually complete layer of zinc metal when the paint dries. Galvanic protection by the zinc slows corrosion and prevents localized attack. The Statue of Liberty and the gantries at the Kennedy Space Center have now been painted with inorganic zinc silicate, which should greatly extend their lives.

Zinc-rich paints must be applied as the primer directly to the steel; they cannot be applied over other paint, even old, oxidized zinc-rich paint, because they will not make good electrical contact with the steel for galvanic protection. Zinc-rich paints can actually accelerate corrosion of steel under wet thermal insulation above 60°C (140°F) where the zinc passivates.

- *Warning*: In case of fire, molten zinc will embrittle steel.

Formulation

Paints consist of a vehicle, pigments, and additives. The vehicle contains a liquid binder that will react with oxygen in the air or contains components that react with each other to form a continuous solid film. Emulsions of water with acrylic or styrene binders are becoming popular to reduce volatile organics (health hazard). Pigments suspended in the vehicle give it color, reinforce the film, and protect it from damage by sunlight. They may also be inhibitive or galvanic (Zn). Additives stabilize the binder, control flow, and may include driers, plasticizers, and wetting agents. Varnishes are clear coatings with the pigment omitted. Lacquers have a vehicle containing a volatile solvent that evaporates to leave a solid film.

Table 10-2 lists the common types of coatings and linings, their corrosion resistance, and typical uses.

"Paint-Over-Rust" Paints.   These paints are sold mainly to the home handyman. Some famous brands of paint advertise or imply that they can be applied over rust. These paints usually have oil-based vehicles (often fish oil) that penetrate rust well. They look good but they do not eliminate the water or salts under the rust. These paints may be satisfactory for clean, uncontaminated, rusted steel but they do not provide long-term service in severe environments.

Corrosion Under Coatings

Corrosion under coatings begins after water, oxygen, and ions have penetrated the polymer film to the metal and set up minute electrochemical cells between local anodes and cathodes on the metal surface. This results in a deterioration of the metal–coating bond.

The following types of corrosion occur under organic coatings.

1. *Blistering* is the accumulation of water or gas formation at the metal–coating interface.
2. *Early rusting* is found under thin latex coatings before they dry thoroughly, especially on cool metal surfaces exposed to high moisture conditions.

3. *Flash rusting* forms on blast-cleaned steel, where crevices have been formed or galvanic cells set up with imbedded steel grit.
4. *Anodic undermining* is delamination by a galvanic cell between the base metal and a metal coating or by applied potential.
5. *Filiform corrosion* shows threads of corrosion products formed after water from high humidity penetrates the coating.
6. *Cathodic delamination* occurs from cathodic protection that generates a high pH on the metal surface, attacking the oxide or polymer at the metal–polymer interface.

10.4 GLASS AND CEMENT COATINGS

This section describes the ceramic glasses and cement coatings used to protect metals from corrosive liquids. Refractory ceramics resistant to high-temperature gases, discussed in Section 15.4, are entirely different in composition and characteristics.

Porcelain Enamels

Glass is used to line hot water heaters and chemical reactors, storage tanks, valves, and pumps in the chemical, food, and pharmaceutical industries, especially where strong acids must be handled.

Steel, cast iron, and aluminum sheet are coated with glass, often containing crystalline ceramic powders, to produce an attractive and corrosion-resistant surface layer with a typical thickness of 50–100 μm (2–4 mils). Porcelain enamels are most often used for major appliances, sanitary ware, cookware, architectural panels, and signs. Because they offer only barrier protection they must be free from defects.

Two coats are generally applied to steel: a dark-colored ground coat of borosilicate glass that bonds well to metal and a clear or pigmented cover coat, formulated to resist the specific corrosive environment that will be involved. For example, range oven liners must resist food acids and cleaners at pH values from 2 to 10 and temperatures up to 315°C (600°F). For aluminum, with its natural oxide layer that adheres well to glass, only one layer of porcelain enamel is applied.

To coat steel it must be low in dissolved gases, particularly hydrogen, or they will diffuse out during the firing stage and leave pinholes in the glass. The steel surface is cleaned by sandblasting or pickling, then coated with a slurry of powdered glass by dipping or spraying, and fired in an oven at 800–870°C (1470–1600°F), which melts the glass and bonds it to the steel.

Glass is not resistant to hydrofluoric acid, concentrated phosphoric acid, or strong caustics. It is brittle and easily damaged by impact or thermal shock

TABLE 10-2   Organic Coatings and Their Properties

Coating Type	Excellent Resistance to	Not Resistant to	Comments	Typical Uses
Acrylics	Weather, solvents, UV light, alkalies	Immersion (thermoplastic types); acids	Adhesion poor-fair. Tend to be brittle	Automotive finishes, appliances, Al siding
Alkyds	Weather	Acids, alkalies, solvents	Inexpensive, good adhesion, flexibility	Primers, exterior enamels
Chlorinated rubber	Acids, water, oil, weather	Solvents, heat, flexing	Easy recoating. Moderate cost	Ship bottoms, swimming pools, industrial coatings
Coal tar pitch	Water, acids, alkalies, oils	Sunlight, heat above 40°C(100°F)	Hard and brittle at low temp	Underground pipe lines, immersion
Epoxy	Acids, alkalies, salts, solvents, water, weather	Oxidizing acids	Versatile. May chalk in sunlight	Maintenance paints, automotive primers, appliances
Coal tar epoxy	Water, salts, soil	Oxidizers, strong solvents, low-temp abrasion, UV light	Applied to bare steel without primer	Tank linings, sewage plants
Fluorocarbons	All chemicals, weather, heat to 200°C (400°F)		Expensive, adhesion difficult	Coil coatings, siding

Neoprene	Saltwater, sunlight, oils, abrasion	Solvents	High dielectric constant	Tank linings, gasoline hoses
Phenolic	Oils, waters, solvents	Alkalies, oxidizers	Expensive, difficult to repair	Can and tank linings, ships
Silicones	Heat to 315°C (600°F)	Solvents	Expensive, brittle	Stove parts, roasters
Urethane	Abrasion, weather	Oxidizers, strong acids, alkalies	Durable	Aircraft finishes
Vinyls	Acids, abrasion, weathering, waters	Solvents, heat above 65°C (150°F)	Tough, flexible	Can and tank linings
Zinc, inorganic	Solvents, weather, abrasion, heat to 370°C (700°F)	Acids, alkalies, oxidizers	Requires good surface prep	Primer on bridges, offshore, ships, cars
Zinc, organic	Heat to 370°C (700°F) Similar to binder	Similar to binder	Easier surface prep. than inorganic Zn	Primer on steel Repairs on galvanizing

Sources: Fontana, 1987; Van Delinder, 1984; Wood, 1982; Moses, 1978; Treseder, 1980; Carboline Protective Coatings Reference Handbook.

and difficult to repair. Special formulations can handle strong acids up to 230°C (450°F) and mild alkalies up to pH 12 at 90°C (200°F). Thirty-year tests have established the ability of porcelain enamels to withstand weather while retaining color and high gloss, essential characteristics for architectural uses and signs. The coatings also are completely inert to organic solvents.

- A corrosive organic chloride had to be transported by tanker truck but no metal could be found to contain it. The engineer finally went to a glass-lined tank trailer with a specially designed suspension system as soft as a baby buggy's. It rode very smoothly until the truck turned, causing the trailer to roll over, gently.

Cement Coatings

Four major types of cementitious coatings are generally applied to steel for corrosion resistance.

1. Potassium silicate.
2. Modified potassium silicate containing some sodium silicate.
3. Calcium aluminate.
4. Portland cement.

The silicates undergo polymerization reactions during solidification, whereas the last two types react with water to form insoluble, solid hydrates. All types form somewhat permeable barrier layers but also form alkaline conditions on the surface of the steel.

Coatings can be applied by casting, spraying (Gunite or Shotcrete processes), or troweling. Pipes are coated internally by rolling them with a thin mortar mix inside. Pipes are also coated externally with cement to protect them from soil and water corrosion and sometimes to prevent buoyancy of submerged gas pipelines. All cementitious coatings can be damaged by mechanical or thermal shock but they are easily repaired. Portland cement is inexpensive, with a thermal expansion closely matching steel's. Chemical resistance of these coatings is shown in Table 10-3.

Cement coatings are not resistant to strong alkalies, hydrofluoric acid, or fluoride salts. Acids slowly permeate the coatings so it is common practice with linings to put an interliner of a corrosion-resistant elastomer between the steel and the cement coating. Modified silicates in long exposure to sulfuric acid form some voluminous sodium sulfate that may crack or spall the coating.

Cementitious linings are widely used for wastewater treatment and throughout the chemical industry in storage tanks for hot or cold water, oil, and chemicals, and in gas stacks and scrubbers.

TABLE 10-3   Chemical Resistance of Cementitious Coatings

Chemical resistance	Silicate– Silica	Modified Silicate	Calcium Aluminate	Portland
H_2O resistance	E	E	E	E
H_2SO_4	E	E	X	X
HCl	E	G	X	X
H_3PO_4	G	G	P	X
HNO_3	E	E	X	X
Organic acids	E	E	F	P
Solvents	E	E	G	G
NH_4OH	F	E	F	G
NaOH	X	F	F	F
$Ca(OH)_2$	X	F	F	G
Amines	X	G	F	G
Resistant at pH	0–7	0–9	4.5–10	7–12

Source: Adapted from Hall, 1986.
E is excellent; G is good; F is fair; P is poor; X is not recommended.

STUDY PROBLEMS

10.1 In applying organic coatings to metal (a) What is the purpose of sandblasting? (b) Why do painted edges and corners fail first? (c) Can coatings protect if moisture penetrates? Explain.

10.2 Why is electrical resistance the single most important measure of the corrosion resistance of a paint coating on metal?

10.3 "A coat of paint on metal is a barrier between the metal and the environment." Criticize this statement and explain.

10.4 A bronze cleat is fastened on a steel ship deck that is splashed with seawater. How could you apply paint to reduce this corrosion?

10.5 Complete the following: (a) The most common cause of premature failure of organic protective coatings is _____.
(b) Paints protect metals by providing a barrier to _____.
(c) Coatings often fail first at sharp edges because _____.

10.6 An automobile is given a phosphate coating followed by two coats of enamel. (a) What is the purpose of the phosphate? (b) Why use *two* coats of enamel? (c) If rust spots begin showing through within a year, what is the most likely cause of failure? (d) Why does the enamel contain pigment?

10.7 If a paint film is so porous that it cannot prevent oxygen or water from reaching the metal substrate, how can it serve as a barrier to corrosion?

10.8 A steel tank with an internal copper heating coil was slightly rusted by the solution. To reduce iron pick-up, the tank was lined with a compatible liquid-applied coating. In almost no time the tank wall perforated and much of the coating was undercut and lifted off the tank. Explain why.

10.9 What mechanism(s) of protection is involved for a chrome-plated automobile bumper? Explain.

10.10 Extremely thin layers of chromium plating do not develop high stresses and so may not form microcracks. How would their corrosion resistance compare with thicker, microcracked chromium?

10.11 Zinc-coated steel bolts given a chromate dip illustrate what mechanism(s) of protection? Explain.

10.12 Compare the corrosion protection offered by a 0.4-μm electroplated zinc coating with 5-mil, hot-dipped galvanizing.

10.13 If aluminum is twice the price of zinc, compare the protection offered per dollar for Galvalume and galvanizing in industrial atmospheric corrosion.

10.14 What is the point of using an anodizing electrolyte for aluminum that will dissolve the oxide? Explain.

10.15 Why do you think anodized aluminum for automotive trim has a much thinner coating than for architectural purposes?

10.16 Automotive sheet steel passed through iron phosphate and chromate baths has been given corrosion protection by what mechanism(s)?

10.17 Weld spatters are little beads and tits of metal that adhere to the metal surface apart from the weld. Since they are quite small, why should they offer any problem in painting?

10.18 The KTA test panel has a short channel welded to a plate coupon with the weld deliberately incomplete. Sketch a welding arrangement that would cause the *least* corrosion.

10.19 Compare glass linings with portland cement linings for (a) resistance to acids, (b) resistance to alkalies, (c) resistance to impact, and (d) ease of repair.

10.20 A solution of 80% phenol with some amylene, sulfuric, and sulfurous acids at 100–110°C (212–230°F) has cracked a 316 stainless steel digester. What could we line it with?

11

Corrosion Inhibitors

An inhibitor is a chemical added to the corrosive environment in small amounts to reduce the corrosion rate. Some inhibitors interfere with the anode reaction, some with the cathode reaction, and some with both. They are usually used to prevent general corrosion but most are not effective in preventing localized attack, such as crevice corrosion, pitting, or SCC.

Inhibitors generally interact with the metal surface in some way: to form a passive film, to form a barrier film of adsorbed inhibitor that may be only a monolayer or less, or to form a thick barrier layer of reaction products or inhibitor.

In considering the use of inhibitors the corrosion engineer must understand what problems can be solved with inhibitors, which inhibitors can be used in the particular environment without interfering with other metals or with the process, and whether or not inhibitor additions will be economically practical.

- How would you go about selecting an inhibitor for cast iron municipal water mains? Answer: Forget it. (1) Cast iron corrodes very slowly in water; (2) citizens only tolerate additives in their drinking water for health reasons, and reluctantly even then; (3) the expense of inhibiting all that water would be prohibitive.

Inhibitors have a critical concentration that must be reached or exceeded for them to be effective, and in some cases to prevent them from making corrosion worse. Consequently, the inhibitor additions must be frequently monitored and the corrosivity of the environment checked with coupon tests or electronic probes, such as the linear polarization type.

While examples of corrosion inhibitors are given in this chapter, almost

235

all good inhibitors are proprietary (secret) mixtures of several chemicals that make them much more effective. They are marketed under a bewildering variety of trade names with only the chemical companies that make them knowing where they can and cannot be used. Some companies even use the same name for different formulations marketed in different parts of the country. The engineer *must* consult a representative from one of the reputable inhibitor companies who can prescribe a suitable product for the particular corrosion conditions he/she faces.

11.1 PASSIVATORS

Passivators are anodic inhibitors that shift the corrosion potential of the metal sharply in the positive (noble) direction. These inhibitors deactivate anodic sites on the metal surface by causing the local current density to exceed the amount needed for passivation. Passivators are used in the neutral pH range, primarily to treat cooling waters.

Passivators slow corrosion by several mechanisms.

1. They stabilize the passive film, reducing the corrosion rate.
2. They repassivate the metal if the film is damaged.
3. They assist in film repair, some passivators forming insoluble compounds that plug pores in the film.
4. They prevent adsorption of aggressive anions, such as Cl^-, by competitive adsorption of inhibitive anions.

But they are dangerous if not used in sufficient amounts, because a metal *almost* completely passivated has a very large cathode/anode area ratio that will concentrate all attack on unpassivated regions to create pits. The critical concentration required usually varies from 10^{-3} to 10^{-5} mol/L. For example, chromate requires about 3×10^{-4} mol/L (500 ppm Na_2CrO_4) to inhibit steel in dilute NaCl solutions. The critical concentration of inhibitor increases if chloride increases, temperature increases, pH decreases, or the amount of dissolved oxygen decreases in the corrosive solution.

• How about inhibiting hydrochloric acid with a passivator? Answer: Get real.

The concentration of inhibitor must exceed the critical concentration everywhere in the system and must be maintained at that level. This means that systems with crevices, with dirt or grease, with rust or lime scale in them can actually be damaged by additions of passivators.

• You cannot play "catch-up" by tossing in a drum of inhibitor after you find corrosion in your system.

Direct Passivators

Also called "oxidizers," direct passivators have anions that react with the metal surface and become incorporated in the passive film to strengthen it, complete it, and repair it. Chromates (CrO_4^{2-}) and nitrites (NO_2^-) are the best oxidizers, passivating steel even in deaerated solutions. Molybdates (MoO_4^{2-}) and tungstates (WO_4^{2-}) are less effective and require some oxygen in the solution to assist in exceeding the critical current density.

Unfortunately, the good oxidizers cannot be used in situations where they could conceivably contaminate the environment or contact human beings. Chromate is carcinogenic and nitrite decomposes to carcinogenic nitrosoamines, although bacteria gradually decompose them to harmless compounds.

Examples of where oxidizers have been particularly effective or ineffective include:

a. Reinforced concrete: Calcium nitrite is the only inhibitor shown to be effective as an additive to the cement mix for protection of steel reinforcing.
b. Automotive antifreeze: Sodium molybdate added in combination with other inhibitors protects all the cooling system metals: iron, aluminum, copper, brass, and solder.
c. Sour gas wells: Oxidizers are exhausted converting hydrogen sulfide to elemental sulfur; completely ineffective.

Indirect Passivators

Indirect passivators are not oxidizers but they improve the adsorption of dissolved oxygen onto the metal surface to enable the oxygen to actually passivate the metal. These passivators only work if the environment contains dissolved oxygen or an oxidizing inhibitor, such as nitrite or molybdate. Indirect passivators are alkaline compounds that create a high OH^- concentration that reacts with H^+ ions adsorbed on cathode sites, cleaning them off so oxygen can adsorb. Both inorganic and organic indirect passivators are effective. Inorganics most often used are alkaline compounds, such as $NaOH$, Na_3PO_4, Na_2HPO_4, Na_2SiO_3, and $Na_2B_4O_7$ (borax). For example, Calgon, a mixture of glassy phosphates, is frequently used to protect steel in water. Organic indirect passivators include sodium benzoate and sodium cinnamate, which have the advantage that they do not cause pitting if the chloride ion concentration should get too high. Corrosion rates increase, of course, but the corrosion is uniform attack.

• Could you use a passivating inhibitor in a system containing both steel and copper? (*Hint*: Copper does not passivate but it can be corroded by solutions containing oxygen.)

Only in recent years has the synergistic effect of a combination of inhibitors been generally recognized, although the inhibitor companies have long taken advantage of it. Adding a direct passivator along with Na_3PO_4, an indirect passivator, inhibits corrosion of steel in water slurries by over 99%, for example. Different inhibitors may even affect different surface sites.

- If 1 mg/L of a passivating inhibitor is sufficient to passivate a system, would doubling the concentration be better? Answer: Yes! While you cannot improve much on 100% passivation, you need a safety factor. On the other hand, you do not want to add so much oxidizer that the system goes transpassive.

11.2 BARRIER INHIBITORS

Inhibitors can act as a barrier layer between the metal and environment by adsorbing as a film over the metal surface or by precipitating a compound scale on the surface. These barrier inhibitors usually cover the entire surface, both anode and cathode areas, although some affect only cathode areas or only anode areas. It is possible to identify the types of sites affected because anodic inhibitors shift the metal's corrosion potential in the positive direction, cathodic inhibitors shift it negatively, and general inhibitors have little effect on it.

Adsorption inhibitors that block the entire surface must have at least a monolayer of coverage, but if they block only a certain type of site the coverage will be much less—as little as 4% of a complete monolayer in one case.

Organic Adsorption Inhibitors

Organics that coat metal with an oily surface layer will protect it. These inhibitors are commonly used in acids, although a few function in neutral or alkaline solutions. In adsorbing on the metal, they replace adsorbed water molecules and prevent water from solvating metal ions or prevent H^+ ions from adsorbing at cathode sites where reduction to $H_2(g)$ could occur. Organic adsorption inhibitors will adsorb on any surface—metal, porous rust, dirt—whatever is offered.

- A dirty system can soak up a fortune in inhibitor without slowing corrosion appreciably.

Physical Adsorption.   The electrostatic attractive forces between organic ions or dipoles and the electrically charged surface of the metal cause physical adsorption, although a bit of covalent bonding also takes place.

Chemisorption.   Charge sharing or charge transfer during the adsorption is called chemisorption. It is slower than physical adsorption but improves significantly with temperature. Chemisorption of organic molecules is specific for certain metals.

The adsorbing species must have a dipole in the molecule: a sulfur atom, as with sulfoxides, sulfides, and mercaptans; nitrogen, as with aliphatic, aromatic, or heterocyclic amines; a double or triple bond, or the like, in a hydrocarbon chain that is typically about 18 carbon atoms long.

Several different mechanisms may be responsible for the action of organic adsorption inhibitors.

1. The inhibitor forms a physical barrier, decreasing diffusion of reactant.
2. The inhibitor may reduce metal reactivity by adsorbing on active sites.
3. The inhibitor changes the surface potential, which affects electrode reactions.
4. The inhibitor may even enter into an electrode reaction, forming a stable surface complex.
5. In alkaline solutions that form natural oxide or hydroxide films, the inhibitor may increase the film stability by becoming part of the film, repairing pores, or removing corrosion products from the surface.

Virtually all the better organic adsorption inhibitors are proprietary blends: molecules that adsorb strongly on the metal surface, other molecules that adsorb well to the first type in order to give lateral strength to the film, molecules that improve solubility in the aqueous phase, molecules that help match vapor pressures of the organics with the aqueous phase, plus other additives, too. Years of experimenting have gone into developing these formulations.

Halides actually assist organic amine molecules in adsorbing, with I^- being the best, followed by Br^- and Cl^-, with F^- having little effect. The halide ions adsorb, lower the surface potential, and increase adsorption of the slightly positive nitrogen groups.

- You would not combine a cathodic inhibitor with an oxidizer. Why not? The oxidizer passivates the metal, but slowing the cathode reaction might prevent passivation. Do you see any problem with switching from one inhibitor type to another?

Organic filming inhibitors do have problems, however. The sulfur-containing types decompose slightly and if not added in sufficient concentration can stimulate iron corrosion. Similarly, the nitrogen types form small amounts of ammonium compounds that attack copper and its alloys. The organics, in interfering with the cathode reaction in acids, may encourage hydrogen penetration into the metal.

- You would not use an anodic inhibitor if crevice corrosion is the problem. Why not? (*Hint:* Consider how well molecular chains of 18 carbon atoms would penetrate into crevices.)

Inorganic Precipitation Inhibitors

A number of compounds including many of the indirect passivators react with the environment to precipitate insoluble products that give general coverage to the metal or in some cases form preferentially at anode or cathode sites. Bicarbonate ions (HCO_3^-) for example, form insoluble carbonate in alkaline solution. Phosphates are the most widely used, precipitating ferrous and ferric phosphates ($FeHPO_4$ and $FePO_4$) on steel. Silicates can even inhibit steel already coated with rust—a rare phenomenon.

Vapor-Phase Inhibitors

Vapor-phase inhibitors are used to protect critical machined parts from corrosion during shipment and storage. For example, electronic equipment is sealed in a plastic sack with a slip of paper informing the purchaser that a vapor-phase inhibitor protects the metal. The inhibitor is in the printer's ink on the paper. An entire warehouse storing coils of metal for fabrication may contain a vapor-phase inhibitor in the atmosphere.

The amount of inhibitor present is extremely small but it is volatile enough to vaporize and then condense on the metal surface to form a protective film. The vapor pressures of many inhibitors are quite low; dicyclohexylamine nitrite, widely used in industry, has a vapor pressure of only 3×10^{-7} atm at 25°C (77°F), for example. This organic nitrite reacts with moisture on the metal to passivate it, but most other vapor-phrase inhibitors protect by the mechanisms used by liquid-phase organic adsorption inhibitors. Diethylamine, injected into sour gas pipelines, reacts with the H_2S in the gas and forms a protective polysulfide film on the steel.

- A very few individuals develop serious allergies to some inhibitors if they work daily in an atmosphere containing them, even though the quantities involved are almost undetectable.

Table 11-1 lists a variety of common inhibitor types. Keep in mind that only the principal reactive ingredients are stated.

TABLE 11-1   Corrosive Systems and Their Inhibitors

Environment/Inhibitors	Conditions	Metals Protected
Water, Potable		
$Ca(HCO_3)_2$	10 mg/L	Steel, cast iron, etc.
Polyphosphate	5–10 mg/L	Fe, Zn, Cu, Al
$Ca(OH)_2$	To pH 8.0	Fe, Zn, Cu
Na_2SiO_3	10–20 mg/L	Fe, Zn, Cu
Water, Cooling		
$Ca(HCO_3)_2$	10 mg/L	Steel, cast iron, etc.
Na_2CrO_4	0.1%	Fe, Zn, Cu, Al
Na polyphosphates	pH 6.5–7.5	Fe, Cu alloys
MBT[a]	If no Cl_2	Cu, brass
Tolyltriazole		Cu
Zn phosphonates	pH 6.5–9	Fe
$NaNO_2$	0.05%	Fe
NaH_2PO_4	1%	Fe
Morpholine	0.2%	Fe
Water, Boiler		
NaH_2PO_4	10 mg/L	Fe, Zn, Cu
Polyphosphate	10 mg/L	Fe, Zn, Cu
Morpholine	variable conc	Fe
Hydrazine	O_2 scavenger	Fe
Ammonia	Neutralizer	Fe
Octadecylamine	Variable conc	Fe
Brines		
$Ca(HCO_3)_2$	10 mg/L	Fe, Zn, Cu
Na_2CrO_4	0.1%	Fe, Zn, Cu
Na benzoate	0.5%	Fe
$NaNO_2$	(NaCl 5%)	Fe
Brines, Oil Field		
Na_2SiO_3	0.01%	Fe
Na_2SO_3 or SO_2	O_2 scavenger	Fe
Quaternary amines	10–25 mg/L	Fe
Imidazoline	10–25 mg/L	Fe
Rosin amine acetate	5–25 mg/L	Fe
Coco amine acetate	5–15 mg/L	Fe
Formaldehyde	50–100 mg/L	Fe
Seawater		
Na_2SiO_3	10 mg/L	Zn
$NaNO_2$	0.5%	Fe
$Ca(HCO_3)_2$	pH dependent	All
$NaH_2PO_4 + NaNO_2$	10 mg/L + 0.5%	Fe
Engine Coolants		
Na_2CrO_4	0.1–1%	Fe, Pb, Cu, Zn
$NaNO_2$	0.1–1%	Fe
Borax	1%	Fe
Glycol–Water Coolant		
Borax + MBT[a]	1% + 0.1%	All
Na benzoate + $NaNO_2$	5% + 0.3%	All
MEA[b]		All
DEA[c]		All

TABLE 11-1   Corrosive Systems and Their Inhibitors *(continued)*

Environment/Inhibitors	Conditions	Metals Protected
HCl Acid		
Ethylaniline	0.5%	Fe
MBT[a]	1%	Fe
Pyridine +		
phenylhydrazine	0.5% + 0.5%	Fe
Rosin amine +		
ethylene oxide	0.2%	Fe
H_2SO_4		
Phenylacridine	0.5%	Fe
Diethylaniline	Pickling	Fe
HNO_3		
$K_2Cr_2O_7$		Fe
$KMnO_4$		Fe
KI		Fe
H_3PO_4, conc		
NaI	200 mg/L	Fe
Most Acids		
Thiourea	1%	Fe
Sulfonated castor oil	0.5–1%	Fe
As_2O_3	0.5%	Fe
Benzotriazole		Cu
Vapor Condensate		
Morpholine	Variable conc	Fe
Ammonia	Variable conc	Fe
Ethylenediamine	Variable conc	Fe
Cyclohexylamine	Variable conc	Fe
Cement		
$Ca(NO_3)_2$	Variable conc	Fe
Drilling Muds		
Na_2CrO_4	pH > 8	Fe
Vapor-Phase Inhibitors		
Cyclohexylamine		
carbonate	30 g/m^3	Fe, Al, Sn, Zn, solder
Dicyclohexylamine		
nitrite	10 g/m^2	Fe, Al, Sn
Amylamine benzoate	Variable conc	Fe
Diisopropylamine		
nitrite	Variable conc	Fe
Methylcyclohexylamine		
carbonate	Variable conc	Fe
Coating Inhibitors		
$ZnCrO_4$ (yellow)	Variable conc	Fe, Zn, Cu
$CaCrO_4$ (white)	Variable conc	Fe, Zn, Cu
Pb_3O_4 (red lead)	Variable conc	Fe

Sources: Davis, 1987; Kapusta, 1988; Van Delinder, 1984.

[a]Mercaptobenzotriazole is MBT.

[b]Monoethanolamine is MEA.

[c]Diethanolamine is DEA.

11.3 POISONS

Poisons inhibit the cathode reaction—hydrogen or oxygen reduction. In acids, the combining of hydrogen atoms on cathode sites to form molecular hydrogen (H_2) is poisoned by elements of Group VA: P, As, Sb, and Bi. Addition of arsenic trioxide to acid, for example, gives the following reaction:

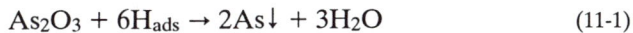

$$As_2O_3 + 6H_{ads} \rightarrow 2As\downarrow + 3H_2O \qquad (11\text{-}1)$$

plating out arsenic on cathode sites. Sulfide and selenide ions also poison the cathodes by adsorbing on them, but their solubilities are too low to make them generally useful as inhibitors, and H_2S and H_2Se gases are dangerous.

All of these cathode poisons are toxic to humans but they also can damage metals by causing hydrogen blistering or hydrogen embrittlement in steels and other susceptible metals. Embrittlement can be removed after pickling in acid by baking out the hydrogen as is done after electroplating (described in Section 10.1).

Oxygen reduction can also be poisoned. Brass in seawater or freshwater is prone to dezincification, where the cathode reaction is formation of OH^- ions from dissolved O_2. Inhibited admiralty brass contains 0.1% of As, Sb, or P to prevent dezincification. As the brass corrodes, the O_2 reduction raises the pH at the surface and forms $As(OH)_4^-$ ions that adsorb on the cathodes (surface sites with Cu atoms) and prevent O_2 adsorption.

- Would adding a cathodic poison to an acid cleaning solution be effective? Answer: Only if you do not want it to clean.

*11.4 POLARIZATION WITH INHIBITORS

Passivating inhibitors are anodic inhibitors because they depend on the shape of the anode polarization curve (Fig. 11-1) and because they shift the corrosion potential in the noble (+) direction. As the figure shows, insufficient passivator increases corrosion instead of reducing it, and it is even possible for excessive amounts of passivator to increase corrosion by raising the corrosion potential into the transpassive region.

Anodic barrier inhibitors adsorb at anodic sites on the surface, thus reducing the exchange current density of the anode reaction. They may also increase the activation polarization of the anode in making it more difficult for metal ions to collect their solvation sheath of water molecules that they must have to escape into the solution. The effect on the polarization diagram

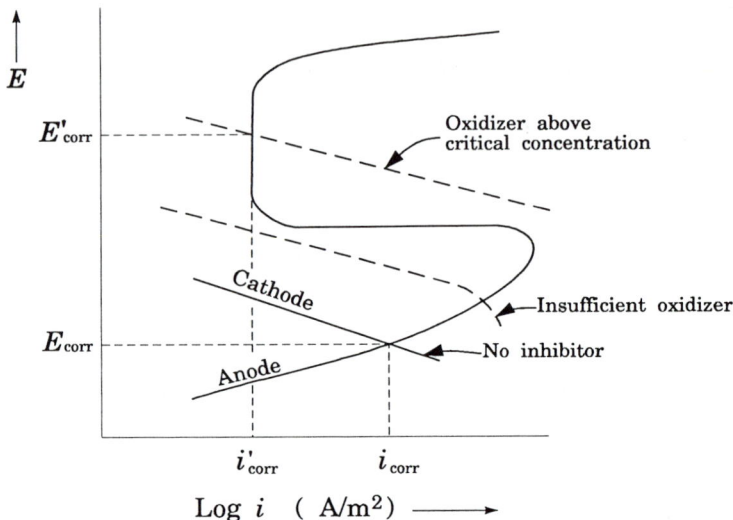

FIGURE 11-1.   Effect of a passivator on corrosion kinetics.

is shown in Figure 11-2a. The corrosion potential shifts in the positive direction as with an oxidizer.

A cathodic inhibitor reduces the cathode exchange current density and often increases activation polarization (Fig. 11-2b) as, for example, by interfering with the step involving combination of two H atoms to form $H_2(g)$. The corrosion potential shifts decidedly in the negative direction.

A general inhibitor affects both anode and cathode, Figure 11-2c, with only a minor effect, either + or −, on the corrosion potential.

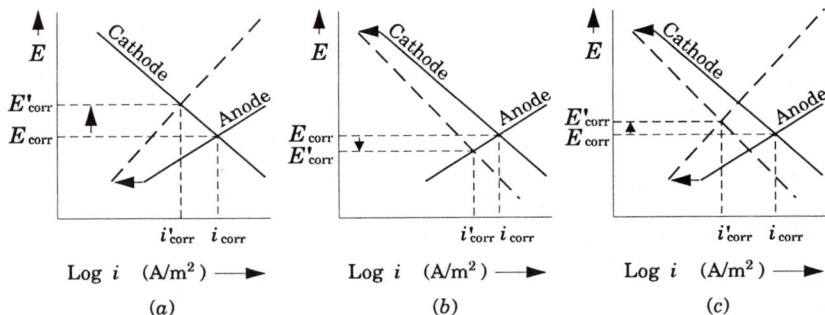

FIGURE 11-2.   Effect of adding (a) an anodic inhibitor, (b) a cathodic inhibitor, or (c) a mixed inhibitor. Dash lines show inhibitor addition.

11.5 SCAVENGERS

Scavengers eliminate dissolved oxygen from a closed system with neutral or alkaline pH. With no oxygen for a cathode reaction, corrosion stops. Sulfite is a common, inexpensive scavenger, widely used in water flooding for oil production:

$$Na_2SO_3 + \tfrac{1}{2}O_{2soln} \rightarrow Na_2SO_4 \qquad (11\text{-}2)$$

Boiler waters are often vacuum degassed or stripped with steam to remove most of the oxygen and then scavenged to reduce the oxygen to less than 10 $\mu g/L$ (10 ppb). If the scavenger is slow acting, it helps form a dense, protective magnetite (Fe_3O_4) scale on steel rather than Fe_2O_3 or a porous Fe_3O_4. However, a scavenger is not always effective in preventing pitting.

- Phosphites were patented in 1936 for O_2 scavenging of boiler waters but they were so powerful that they reduced the magnetite protective scale, exposing bare steel.

Hydrazine (N_2H_4) is another scavenger often used for boilers, especially for high-pressure, high-temperature systems. It is slow acting at low temperatures, although some proprietary preparations sold as concentrated aqueous solutions include catalysts. It has an advantage in leaving no solid residue.

$$N_2H_4 + O_2 \rightarrow N_2 + 2H_2O \qquad (11\text{-}3)$$

Scavengers remove oxygen, not H^+ ions, and so are ineffective in acids.

- Scavengers are often added with other corrosion inhibitors, but not passivators. Why not? Answer: The two types would work at cross-purposes—one trying to prevent oxidation and the other trying to oxidize the metal.

11.6 NEUTRALIZERS

Neutralizers reduce corrosion by reducing the concentration of H^+ ions in solution. Neutralizers are even added to neutral solutions because the H^+ ion concentration is much greater at high temperatures due to the presence of CO_2, which forms carbonic acid (H_2CO_3). Neutralizers control small amounts of HCl, CO_2, SO_2, H_2S, organic acids, and the like that are likely to concentrate at some point in a system and react with water to form H^+ ions.

If the acid condenses from a vapor, the neutralizer must be present when the acid forms, which means that it must have the same condensation properties as the acid. A list of some common types of neutralizers is given in Table 11-2.

Conclusions

The corrosion engineer needs to know how his/her inhibitors work, their limitations, and their pitfalls. He/she must keep a check on the equipment used to make inhibitor additions and monitor inhibitor effectiveness continually. It is also very important that he/she should develop a good working relationship with the technical people in the inhibitor companies.

STUDY PROBLEMS

11.1 A sour gas line averaged 2.5 leaks per month, mainly at pits along the bottom of the pipe where liquid water flowed. We injected 0.3 gal/day of Tol-Aeromer A (an organic sulfophosphate that is an anodic indirect passivator) into the line, reducing leaks to 0.5 per month. Could we improve this record further by (a) increasing the amount of passivator, (b) decreasing it, or (c) eliminating the 1% O_2 from the gas? Why?

11.2 A metal passivates but its corrosion rate is still too high. Which would be best: an anodic adsorption inhibitor, a cathodic adsorption inhibitor, or a general adsorption inhibitor? Explain.

11.3 The corrosion rate of your stainless steel equipment is extremely high

TABLE 11-2   Typical Neutralizing Inhibitors

Inhibitor	Systems
NH_3	Boilers, cooling water, freshwater, vapor phase
MEA[a], DEA[b]	Glycol cooling systems
Borate	Glycol cooling systems (automotive)
CaO	Fresh waters
Cyclohexylamine	Vapor-phase inhibition
Amylamine benzoate	Vapor-phase inhibition
NaOH	Petroleum refining
Na_2CO_3	Petroleum refining
Alkylamines, proprietary	Petroleum refining, crude oil production
Polyamines, proprietary	Petroleum refining, sour (H_2S) water

Sources: Davis, 1987; Kapusta, 1988.
[a]Monoethanolamine is MEA.
[b]Diethanolamine is DEA.

because it is in the transpassive region. Which would be better: an anodic adsorption inhibitor, a cathodic adsorption inhibitor, or a general adsorption inhibitor? Explain.

11.4 You are asked to suggest a suitable inhibitor type to protect high-strength steel engine components from corrosion by acid condensate contaminated with seawater. Briefly explain why the following types would or would not be suitable. (a) A passivating inhibitor, (b) a poison, and (c) a scavenger.

11.5 We have a 430 stainless steel evaporator containing 1–10% H_2SO_4. Crevice corrosion is creating serious problems. What type(s) of inhibitor could we use? Explain. (Do not waste time discussing the types that will *not* work.)

11.6 You are considering a corrosion inhibitor for stainless steel tubing carrying 4% H_2SO_4 with large amounts of NaCl. The tubing has been cleaned but may have some active pits. For each of the following inhibitor types that would be *un*satisfactory, explain why you would reject it. (a) Direct passivator, (b) indirect passivator, (c) anodic adsorption inhibitor, (d) cathodic adsorption inhibitor, and (e) scavenger.

11.7 We have a 316 stainless steel gasoline distillation column with no corrosion except in crevices. Temperature 140–220°C (350–425°F), no aeration, 130 mg/L of Cl2. We also have three drums of inhibitor: ECH-1 is a cathodic adsorption inhibitor, ECH-2 an anodic adsorption inhibitor, and ECH-3 an indirect passivating inhibitor. Which do we dump in, and why?

11.8 The inhibitor ANAF 107 is described in the company literature as a "cathodic adsorption inhibitor." (a) How can you tell from its action in your system that it is a *cathodic* inhibitor? (b) Will ANAF 107 work in an already badly rusted system? Explain.

*11.9 On a single polarization diagram, sketch: (a) the situation in which stainless steel is corroding severely in an industrial acid; (b) the same situation with low corrosion when an indirect passivating inhibitor is added. Label the diagram completely, showing electrode reactions.

*11.10 Sketch a polarization diagram (labeled fully) that shows how an indirect passivating inhibitor works.

*11.11 You should not use a cathodic inhibitor on a passivated metal. Sketch a polarization diagram (labeled fully) that shows why.

11.12 Unfiltered river water, carrying some silt, is used as coolant for your plant. The steel pipe from the river has pitted through on the bottom in several places. Someone suggests adding an inhibitor at the intake to protect the pipe and the plant cooling system. Briefly explain why the following inhibitor types would or would not be suitable: (a) oxidizers, (b) organic filming inhibitors, or (c) scavengers.

11.13 A plant is using a borax inhibitor to protect aluminum equipment. This compound, an indirect passivator, is working perfectly at present, but in the future it is possible that changes may be made in plant operation or design. What changes should be avoided?

11.14 Sodium sulfite, a good scavenging inhibitor for steels, has not been found to be effective in protecting aluminum in strong acids. In fact, it sometimes makes things worse. Any idea why?

Memos

11.15 Memo To: Corrosion Engineer
From: Chet Bailey, Production Supervisor
Our plant cooling system consists mainly of brass internals, which have corroded rather severely in the water-antifreeze mixture, judging from the amount of crud we saw when we looked into it. We have tried to catch up with the problem by pouring in a couple of drums of inhibitor, but it doesn't seem to be working because we have had two leaks within the last week and the crud is still there. The inhibitor was something a salesman recommended, based on mercaptobenzothiazole. Give us some help—we've got a problem here that's getting worse by the day.

11.16 Memo To: Corrosion Engineer
From: Chet Bailey, Production Supervisor
Side reactions at the ethylene plant have finally been traced to the organic adsorption inhibitor they've been using. Some snake-oil salesman tells me we should use sodium vanadate, which he says is a passivating inhibitor. But will it passivate the carbon steel sufficiently at 110°C? I'm desperate enough to give it a try. Do we run service tests or lab tests?

12

Cathodic and Anodic Protection

Cathodic protection converts all anodic areas on a metal surface to cathodes so that corrosion ceases. The protected metal has positive current flowing onto it from the electrolyte everywhere on the surface so that no current flows off. This result can be achieved in two distinctly different ways.

1. By connecting a sacrificial anode to the metal that is to be protected.
2. By applying an electric current from a separate power source, a technique called impressed-current cathodic protection.

Anodic protection, on the contrary, makes the entire metal surface anodic—so anodic that the metal completely passivates. Obviously, then, this technique is limited to metals that can form protective passive films. Since passivated metals still corrode at a low rate, anodic protection almost, but not completely, stops corrosion.

12.1 CATHODIC PROTECTION

The metal to be protected is made the cathode in an electrochemical cell, where the anode is either a more corrodible, sacrificial metal, or an inert electrode with impressed dc current. Corrosion in soils and most aqueous environments can be completely prevented at a cost far less than for inhibitors or coatings.

Cathodic protection protects against both general corrosion and localized attack, such as intergranular corrosion, pitting, and dealloying. It can stop SCC but not hydrogen-induced cracking. In fact, cathodic overprotection can cause hydrogen embrittlement. Cathodic protection can eliminate corrosion fatigue, but not straight mechanical fatigue failure.

Cathodic protection *cannot* be used in the following situations.

1. Above the water line, in vapor. Electric current will not flow through a gas, at least not at the moderate voltages used in cathodic protection.
2. In nonconducting liquids, such as oil. An electrolyte is one of the essential ingredients of an electrochemical cell.
3. In electrically screened areas. This is an important limitation that will be described in the next paragraph.
4. In extremely corrosive environments. While theoretically possible, cathodic protection would be impractical because of horrendous power costs.

Shielding.   Shielding, or screening, prevents the flow of protective current through a low-conductivity electrolyte, where the current can find a much better conducting metallic path. For example, in Figure 12-1 the current flowing from the sacrificial anode protects all the tank wall area below the waterline, as well as the screen at the bottom. Very little current flows down the pipe beyond the screen because it divides proportionally with the conductance of the paths, most of it choosing the high-conductance path through the metal back to the anode. (To give some idea of the differences in conductivity, at 25°C (77°F) a saturated NaCl brine has a conductivity of 0.25 S/cm while steel's is 60,000 S/cm.) [The siemans (S) is now the unit of conductance, replacing Ω^{-1} and mhos.]

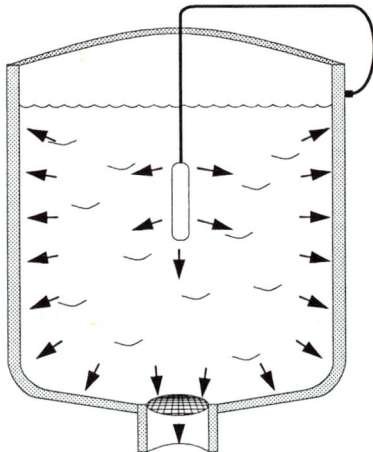

FIGURE 12-1.   Cathodic protection of a vessel by a sacrificial anode with shielding of the pipe by a metal screen. Arrows show positive current flow.

Other examples of shielding would be

a. A bundle of pipes in an electrolyte. Although the outside surfaces of the bundle would receive protective current, no current would penetrate the narrow path of electrolyte to the center of the bundle.
b. A pipe buried in a landfill of cinders. Cinders, being fairly good electrical conductors but also very inhomogeneous, will catch all the current, which then flows to the pipe through a few high-conductance paths. Much of the pipe surface receives no protective current. Cathodic protection in this case just increases the cathode/anode area ratio.
c. Protective current flowing down a pipe. Cathodic protection current will only protect down a pipe for a length equal to three or four diameters. Within that distance almost all current has found some metal to flow through.

Actually, the "distance effect" in galvanic corrosion is another example of shielding: The current wants to return to a metallic path as quickly as possible.

Nonconducting materials, such as concrete or urethane barrier coatings, do not shield the metal from cathodic protection current.

12.2 SACRIFICIAL PROTECTION

Figure 12-1 shows the electrical arrangement of a sacrificial anode used to protect the inside of a tank containing an electrolyte. The anode in the center of the tank is electrically connected to the tank by an insulated wire. The anode is made of a metal more corrodible than the tank so that a galvanic cell is set up with the tank as the cathode. If the insulated wire and its connections remain intact, cathodic protection continues until the anode metal corrodes away. In fact, even if the anode corrodes off from the wire and falls to the bottom of the tank, it will still cathodically protect the tank, although protection will be less at greater distances from the anode.

Sacrificial protection is inexpensive because it has the obvious advantage that it does not need any external power source. It also gives a fairly uniform distribution of current, but that current output is low and the only way to increase it is to use more anodes. Little or no maintenance is required except to replace the anode, which should be necessary only after several years of service. But because the system is cheap and simple, there is usually no way to monitor the amount of protection being given.

• The head of a major cathodic protection company ran into a client who exclaimed, "Those sacrificial anodes you put in 2 years ago to protect our underground piping–they're great! We accidentally dug one up the other day and it was just like new. They'll last forever!" (But they aren't supposed to.)

The sacrificial anodes used are magnesium, zinc, or aluminum. Although pure magnesium is sometimes used, the AZ63 alloy (6Al, 3Zn, 0.2Mn) is more commonly specified to reduce pitting of the anode. A high-potential alloy containing 1% Mn is also popular. Because of local cell action and corrosion products that form on its surface, a magnesium anode has an efficiency of only about 50%; that is, only one-half its corrosion current flows to the protected structure. The potential of the standard magnesium anode is −1.6 V versus a $Cu/CuSO_4$ reference electrode, with consumption at about 7.9 kg/A · y (17.5 lb/A·y). Anodes are usually placed within a few meters of the protected structure because of the low voltage.

Zinc anodes are usually 99.99% Zn or, for seawater, an alloy with a few tenths of a percent Al and Cd. The potential of zinc is less than the potential of Mg, at about −1.1 V versus $Cu/CuSO_4$, but the efficiency is excellent, around 95%. Consumption is about 10.7 kg/A · y (23.5 lb/A · y).

- Submarines are designed to be dry-docked once every 5 years. Zinc anodes must last that long.

Aluminum passivates in most soils but can be used as a sacrificial anode in seawater if passivation is prevented by alloying with zinc plus Hg, In, or Sn. The potential is about the same as for zinc (−1.1 V) with an efficiency of 95% and a consumption rate of 3.0–9.4 kg/A · y (6.5–20.7 lb/A · y) depending on the alloy and the conditions.

In soil, special backfills are commonly used with sacrificial anodes. The anodes often come packaged in a sack with bentonite clay and gypsum to give better conductance and a more uniform environment for improved anode efficiency.

Sacrificial anodes are used where

1. Current requirements are relatively low.
2. Soil resistivity is relatively low, usually under 10,000 Ω-cm.
3. Electrical power is not available.
4. Short-term protection is needed (a sacrificial anode system is seldom designed for a life of more than 20 years).
5. "Interference" problems could occur. Stray currents (Section 5.10 and Fig. 5-16) put out by impressed-current cathodic protection anodes can corrode other metal in the vicinity if they are not part of the same system.

12.3 IMPRESSED-CURRENT CATHODIC PROTECTION

The sketch of an impressed-current cathodic protection system shown in Figure 12-2 illustrates one main difference from sacrificial protection: An

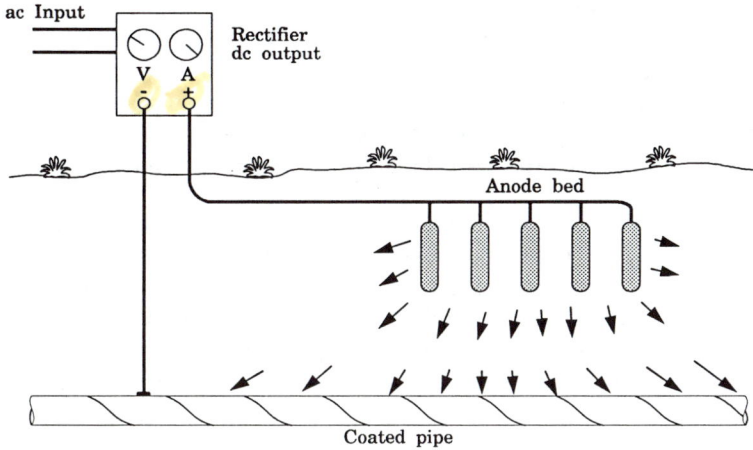

FIGURE 12-2.   Diagram of impressed-current cathodic protection system for a buried pipe-line.

external dc power source is required, with the negative wire connected to the protected structure. The other main difference is that sacrificial anodes are not used; the anodes are chosen to be as noncorroding as possible.

• A cathodic protection rectifier had to be moved to allow widening of a road. An ordinary electrician, unfamiliar with dc current, reconnected the unit backwards—positive wire to the pipe. *Result*: three leaks appeared almost immediately.

Impressed current has several important advantages over sacrificial anodes.

1. Impressed current is versatile. A wide range of voltage settings allows the protection current to be adjusted precisely.
2. It is effective in high-resistance soils, even solid rock with resistivity over 100,000 Ω-cm.
3. Knowledge of the status of the system is continually available. The protection current can be read on an ammeter at the power source.
4. Maintenance can be performed at rectifier stations. A bad anode is easily found and replaced. Contrast that with walking along a pipeline with a magnetic detector trying to find sacrificial anodes.

On the other hand, impressed current can have some disadvantages compared with sacrificial anodes.

1. It requires a source of electric power.
2. A high potential develops close to the anodes that may cause interference on nearby structures.
3. Overprotection is possible, reducing protective rust scale, disbonding coatings, or charging protected steel with hydrogen.
4. Finding a suitable location for the anode groundbed can be difficult.

- An impressed-current "cathodic protection" scheme has been marketed that allegedly protects the underside of an automobile from corrosion. Two permanent anodes connected to the car's battery are supposed to distribute current through a thin film of moisture on the steel. If shielding prevents current from flowing down a pipe more than 3 or 4 diameters, how far from the anodes would you estimate the protection current would travel?

Anodes for impressed-current protection are more noble than the metal they are designed to protect because sacrificial anodes would corrode away at a disastrous rate. However, scrap iron (old pipe, railroad rails) is sometimes used for temporary anode beds even though it dissolves at the rate of 9.1 kg/A · y (20 lb/A · y). Much more common are anodes of graphite or Duriron (14.5% Si cast iron), which while not as good electrical conductors as most metals, corrode much less: graphite under 0.5 kg/A · y (1 lb/A · y) in most soils and Duriron 0.1–0.2 kg/A · y (0.2–0.4 lb/A · y). In recent years sintered magnetite (Fe_3O_4) and sintered Pt-group oxides on titanium have become very popular because their dissolution rates are virtually nil.

Buried anodes are usually placed some distance from the protected structure—50 m to 50 km (50 yards–30 miles), wherever a good low-resistance location for the anode bed can be found. If stray currents might cause interference with other metal in the vicinity a "distributed anode" system is used, with anodes 1 or 2 m from the structure. A backfill of coke breeze (carbon powder) is commonly used around the anodes to increase their effective size and ensure good contact with the soil.

The anode reaction is not corrosion of the anode material but oxidation of the environment. For example,

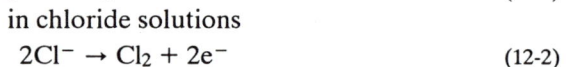

in water

$$2H_2O \rightarrow O_2 + 4e^- + 4H^+ \tag{12-1}$$

in chloride solutions

$$2Cl^- \rightarrow Cl_2 + 2e^- \tag{12-2}$$

The dc power supply necessary to force current from the anodes is a rectifier if a power line is available. The cathodic protection rectifier (Fig. 12-3) consists of two parts: an ac step-down transformer with taps that take

FIGURE 12-3.   Rectifier for impressed-current cathodic protection. [Courtesy of JA Electronics Manufacturing Co., Inc.]

off voltages of different amounts to adjust power output, and a rectifier to change ac to dc by means of selenium or silicon plates.

In remote areas where power lines are not available, batteries, motor generators, thermoelectric generators, fuel cells, solar cells, or windmills may be used. The last two methods usually charge batteries to provide continuous protection current.

Ships have long been protected by sacrificial anodes around their vital areas: the rudder and screw propeller. Modern ships are now going to impressed-current cathodic protection of the entire hull. Figure 12-4 shows two permanent anodes, several of which are arranged on each side of the ship.

The amount of protection current required depends on the corrosion conditions which can vary considerably. A well-coated pipe in high-resistivity soil may need only 5 mA/m^2 (0.5 mA/ft^2) while a bare pipe in low-resistivity clay may need 2 A/m^2 (200 mA/ft^2). As a general rule of thumb the current required will be about one and one-half times the corrosion rate. Since 1 A/m^2 is approximately 1-mm/y corrosion for many metals including steel, a rough estimate can be made of the current requirement.

The voltage of the protected structure determines whether protection is

FIGURE 12-4.   Impressed-current anodes on the hull of an ice breaker. [Courtesy of David Lingnau, Gulf Canada Resources Ltd.]

sufficient or not. For steel in soil, lowering the steel-to-soil voltage everywhere to at least -0.850 V, with respect to a $Cu/CuSO_4$ reference electrode, provides complete protection. Since this is not always possible, a shift of 300 mV in the negative direction when cathodic protection is applied is usually sufficient. Other more-disputed voltage criteria are sometimes used if neither of these two potential changes can be made.

It has become standard practice to combine cathodic protection with protective coatings to achieve a long-life system at minimum power costs. Current distribution is more uniform and, since current only flows to flaws in the coating, the current requirements are greatly reduced. Typically, the cost of cathodic protection will be only about 10% of the cost of protecting bare metal. Nevertheless, in some ship ballast tanks and on some offshore oil-drilling rigs, impressed current protects bare steel because the cost of maintenance painting is too high.

- Extensive experiments have recently shown the feasibility of pulsed dc cathodic protection for oil well casings. Pulsing at 1–5 kHz gives better protection at the bottom of the well and causes less interference.

12.4 ANODIC PROTECTION

Anodic protection, a relatively new technique, is used to passivate storage tanks and process vessels containing highly corrosive electrolytes. The environment must be uniform so that all the metal passivates, which rules out environments, such as soils, that vary greatly in composition.

The ease with which metals passivate decreases in the sequence:

Ti, Al, Cr, Be, Mo, Mg, Ni, Co, Fe, Mn, Zn, Cd, Sn, Pb, Cu, Ag

Anodic protection can be dismissed for the very reactive metals Mg, Zn, and Cd because they should never be used in highly corrosive environments, and for Cu, Cu alloys, and Ag because they seldom form protective passive films. All the other metals can theoretically be given anodic protection, although in practice the method has been used mainly for steel and stainless steels.

Three pieces of equipment are needed for anodic protection: a potentiostat, a cathode, and a reference electrode, shown in Figure 12-5. The potentiostat is an electronic device to hold a constant potential. Powered from an ac line, it includes a rectifier to provide dc output. The desired potential for passivation is set on the potentiostat, which then compares this potential with the potential of the tank, as measured versus the reference electrode. If the two potentials differ, the current between tank and cathode is automatically adjusted by the potentiostat until the difference is eliminated.

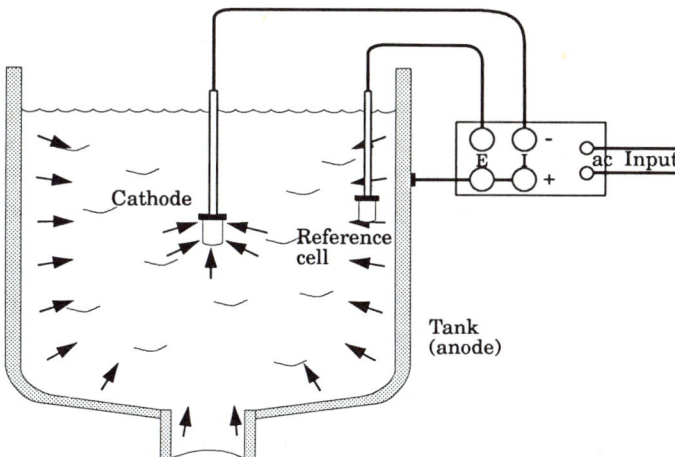

FIGURE 12-5.   Anodic protection of a tank.

When anodic protection is first applied, the tank potential is usually far below the desired potential so that increasing amounts of current must flow from the tank walls until the critical current density for passivation is exceeded. The potentiostat and wiring must be strong enough to handle these large currents. After passivation the currents flowing in the cell are several orders of magnitude smaller but the potentiostat must be sensitive enough to control them.

The reference electrode is often a calomel half-cell but it may be any stable electrode that will not be affected by the environment. This is a weak spot in the system because if the reference goes bad the tank will be controlled to the wrong potential and will not passivate. Sometimes two reference electrodes are used, with the potentiostat continually comparing them, so that if the potentials begin to differ an alarm sounds.

The choice of cathode metal is not critical; even old steel cable has been used, although carefully engineered designs prefer a corrosion-resistant metal that will not contaminate the solution, with Hastelloy C being the most widely used.

Galvanic Cathodes.   Anodic protection with galvanic cathodes has been successful in a few cases, with no potentiostat or external power source involved. A noble cathode of platinum, graphite, or passivated titanium is fastened directly to the tank so that the tank polarizes enough to passivate.

Anodic protection has several advantages over cathodic protection.

1. It can be used in extremely corrosive environments where power costs of cathodic protection would be prohibitive.
2. The corrosion rate of the protected system is directly proportional to the current density applied so that the corrosion rate can be continually monitored.
3. The current requirements are low once the metal has passivated. However, a great deal of current is required to passivate. For example, 316 stainless steel requires 6500 A/m^2 (600 A/ft^2) to passivate in 67% H_2SO_4 at room temperature but only 1 mA/m^2 (0.1 mA/ft^2) is needed to maintain passivation.
4. Operating conditions can be determined precisely by lab experiments, whereas cathodic protection requires "fine-tuning" adjustments over periods of months to get the proper protection.
5. Current distribution or "throwing power" is excellent. A single cathode can handle large anode areas with no shielding problems.

   • A single Hastelloy C cathode rod in the center of a shell-and-tube heat exchanger for H_2SO_4 can protect *all* the steel tubes by anodic protection. Try that with cathodic protection!

6. Anodic protection can be used to *reduce* the passive potential and hold it below the critical pitting potential. However, it is still anodic protection, not cathodic protection, because anodic polarization is being controlled.

Some of the disadvantages of anodic protection have already been mentioned but they must be kept in mind.

1. Anodic protection will not work if the metal does not passivate well. Steel in hot acid or in strong chloride solutions corrodes too rapidly even in the passive potential range.
2. Corrosion is never completely stopped, unlike cathodic protection.
3. Metal in the vapor zone does not passivate, although a passive film *may* remain on an emptied tank for up to several hours.
4. High currents (and high corrosion rates) are required to passivate or repassivate.
5. Anodic protection costs almost as much to protect a small system as a large one.
6. All protection equipment must work properly or disaster can strike. A potential a little too high (transpassive) or a little too low (active) will destroy the metal that is supposed to be protected.
7. Anodic protection cannot be used in an inhomogeneous environment, such as soil or some unstirred solutions.
8. Anodic protection will not work in a nonelectrolyte, such as oil or air.
9. It cannot be used for a mixture of metals. Corrosion of some might be increased at the protection potential.

- Anodic protection can be a key feature of a process. It is essential to one sulfuric acid plant that processes waste SO_2 from a large smelter. If the anodic protection stops, the entire smelter must immediately shut down.

In addition to widespread use of anodic protection for steel and stainless steels in H_2SO_4 plants, the method is almost universally used to protect steel kraft digesters in the pulp and paper industries where the environment is $NaOH + Na_2S$ at 175°C (350°F). The life of the kraft digesters is extended sevenfold so the system pays for itself in about 2 years. Phosphoric acid plants have anodic protection of stainless steel tanks largely to prevent the danger of H_2 accumulating from corrosion. Carbon steel anodically protected in aqueous NH_3 fertilizer solutions has half the cost of aluminum or stainless steel used previously. Nitric acid plants are using anodic protection to protect nickel and nickel alloys.

• A steel tank for sulfuric acid was rated unrepairable but anodic protection was applied anyway, extending its life by over 5 years.

*12.5 ELECTROCHEMICAL THEORY

Figure 12-6 illustrates the situation of iron corroding in an acid (the upper two solid diagonal lines) with the corrosion potential and corrosion current indicated by their intersection. For cathodic protection a new anode is introduced: a sacrificial anode more reactive than the iron or an impressed-current anode with applied current forced from it. The power required for protection is the product of the applied current and the potential at which the protected structure is being held.

For simplicity the protection potential is shown at the equilibrium potential of the iron anode so that all net current attempting to leave the anode is exactly opposed by an equal current attempting to flow onto the anode, but in real life it is not necessary to balance the two exactly. With the current density on a logarithmic scale, the protection current could be several orders of magnitude greater than the original corrosion current. For this reason cathodic protection in very corrosive environments requires so much electrical power or consumes sacrificial anodes so rapidly that it becomes impractical.

Because cathodic protection converts the protected metal wholly to a cathode it increases the exchange current of the cathode reaction. However, the exchange current usually has little effect on the protection current because of concentration polarization.

Figure 12-7 shows the polarization diagram of a passivatable metal that

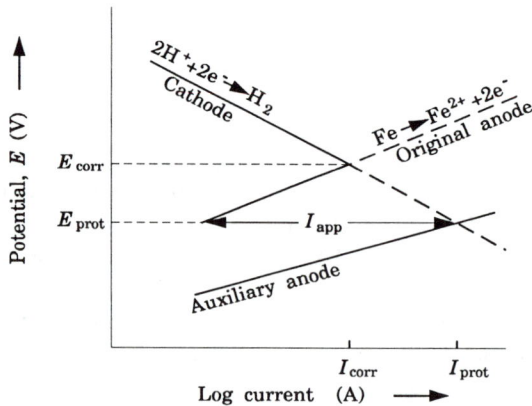

FIGURE 12-6. Polarization diagram for cathodic protection of iron in acid.

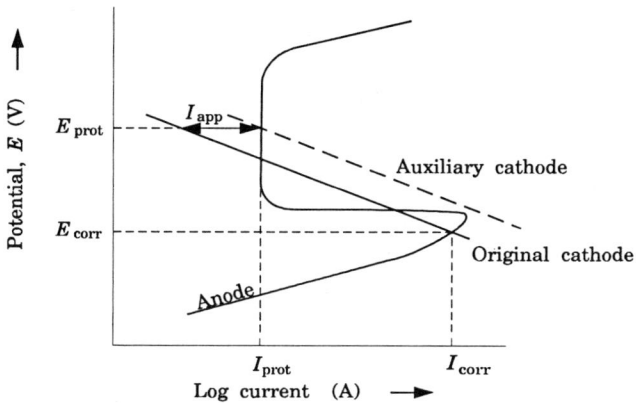

FIGURE 12-7.   Polarization diagram for anodic protection.

corrodes in the active potential region (E_{corr}) with no protection applied while with anodic protection the potential is raised to E_{prot}, which passivates the metal. The corrosion rate is directly proportional to the protection current density i_{prot}. Power required for protection is the product of the protection potential, E_{prot}, and the applied current but since the metal has passivated, the applied current density is so low that the power required to maintain passivity is also extremely low.

STUDY PROBLEMS

12.1   A kiln for cement blocks is corroding badly at the bottom. Condensed steam fills the bottom up to about 3 cm during the firing. Protective coatings cannot take the temperature. Can you describe a cathodic protection system that would prevent the corrosion? Give details of equipment, electrical connections, and precautions.

12.2   Cathodic protection is to be applied to a steel beer fermenter tank lined with a baked phenolic. It is known that the dissolved iron level in the beer must be kept below 3 mg/L (3 ppm) and the applied current density must be controlled closely or the lining may blister and strip. Adjustment must be made frequently because resistance changes as the wort becomes beer. (a) Describe the equipment you would use. (b) Describe exactly how it would be connected. (c) The tank sits on a moist concrete floor that corrodes its bottom. Will cathodic protection inside the tank increase, decrease, or not affect this external corrosion?

12.3 Anodic protection is used only for certain metals in certain environments. Explain these limitations.

12.4 A steel pipeline is cathodically protected with a zinc electrode. Explain how the life of the *zinc* will be affected if (a) the moisture content of the soil decreases; (b) the zinc electrode is moved further from the pipe, from 2 m originally to 20 m; (c) the plastic coating on the pipe deteriorates, exposing more steel.

12.5 Sketch simple graphs for anodic protection showing (a) how the current required for protection changes with time and (b) how the corrosion potential changes with time.

12.6 Steel equipment used for sulfuric acid manufacture is to be electrically protected from internal corrosion. (a) What type of protection would you select? Explain. (b) Identify electrodes and power control equipment.

12.7 A village standpipe (tall tank for drinking water) is to be electrically protected from internal corrosion. (a) What type of protection would you select? Explain. (b) Identify electrodes and power control equipment.

12.8 The cooling water for a heat exchanger is taken from the St. Lawrence River, and varies from freshwater to seawater depending on the tide and the season. Cathodic protection is decided upon to stop the severe corrosion. (a) What type do you recommend? Why? (b) List all equipment required (except wires) and specify materials to be used.

12.9 You have installed an anodic protection system in a 304 stainless steel storage tank containing 15% H_2SO_4. Although the system as originally installed was drawing 4.4 A/m^2, after a short time it drops abruptly to 0.73 A/m^2. What could have caused this change? What adjustment should you make in the system?

12.10 An underground, bare copper steam line was laid in a trench backfilled with coke breeze. This backfill is expensive and is normally used only around impressed current anodes but it was used in this case to provide thermal insulation and a neutral environment. The line was protected with magnesium anodes that appeared to be working, but severe leaks developed in 1 month. How would you improve protection?

12.11 Cathodic protection personnel of gas companies continually watch for accidental electrical connection between water lines and gas lines inside dwellings. For example, a coat hanger hung on a water pipe may contact a gas line. Explain the reasons for their concern.

12.12 A cast iron pipeline is to be cathodically protected with zinc. Sketch the electrical setup, showing anode, cathode, direction of current flow, and any other equipment needed. How could current be adjusted? How could voltage be adjusted?

*12.13 Chromium metal will have a potential of -0.74 V in neutral water. With the help of the Pourbaix diagram (Fig. 2-11), explain the effect on chromium of (a) anodic protection and (b) cathodic protection.

Memos

12.14 Memo To: Corrosion Engineer
From: Milton T. Klein, General Manager
I had Chet Bailey work out the specifications for the new processing plant but we're already in trouble. His plans called for 4-in. copper waterlines (insulated from the cast iron city main, of course) and they've sprung a leak after being in the ground only 2 months. The men who dug up the line say it is corroded on the outside. Much of that area is a drained bog with the soil mixed with underlying sandy clay; the utility company tells me the soil has a pH of 2.0 and a resistivity of 247 Ω-cm.
Will corrosion reduce with time or will things get worse? Should we coat the lines with something? Put on some kind of cathodic protection? What can we do to ensure a continual water supply? Give it some thought and jot down any ideas you can give me.

12.15 Memo To: Corrosion Engineer
From: Chet Bailey, Production Supervisor
At the meeting with Klein the other day when we were discussing protecting the sulfuric acid storage tanks, you mentioned anodic protection. This was a new one on me. I tried looking it up in a book but all I could find was a bunch of S-shaped curves. Could you explain the theory of it briefly and simply so that even I could understand what you guys are talking about? I would really appreciate it. But remember, I'm no corrosion engineer and don't pretend to be one, so don't try to snow me.

12.16 Memo To: Corrosion Engineer
From: Milton T. Klein, General Manager
Chet Bailey has a large carbon steel tank that he plans to use for leaching a siliceous ore pulp with 50% solids. They will add 5 g/L of ferric sulfate with sulfuric acid to get an initial pH of 0.5, heat to 50°C and stir rapidly. Since this combination is probably both corrosive and abrasive, it occurs to me that we may need to protect the tank in some

way. A zinc coating? Cathodic protection? Anodic protection? The least expensive? No protection?

Give it some thought and jot down any ideas you have that could help us.

12.17 Memo To: Corrosion Engineer

From: Chet Bailey, Production Supervisor

We have a titanium-lined reactor corroding at nearly 1 mm/y that something has got to be done about. We could line the tank with a good plastic or rubber coating or install a cathodic protection system. Some hot-shot salesman tried to sell me a new super cathodic protection, which was supposed to have lower power costs than ordinary cp and could even protect the tank walls when the process liquor level dropped. But I'm naturally suspicious, as you know by now, and would prefer to stick to a proven protection method. You've been pretty lucky on your guesses so far, so I leave it to you. What do we do?

12.18 Memo To: Corrosion Engineer

From: Chet Bailey, Production Supervisor

In light of the underground corrosion problems we've had with galvanized steel around here, Clyde Young, my assistant, suggests that cathodic protection should be put on the new water line that will be installed to the steam plant. Plans call for a 6-in. galvanized pipe 500 ft long. Should a protection system be installed when the line is new or later after we see how bad the corrosion is? What kind of system? What about this anodic protection I've heard about? We'd much appreciate any advice you can give us on this.

12.19 Memo To: Corrosion Engineer

From: Chet Bailey, Production Supervisor

We have a large underground storage tank for diesel fuel that seems to have corroded. It gave no trouble for several years but now the fuel is seeping into the nearby creek. We're going to have to dig up the steel tank so the question is Do we patch it? Replace it with something better? Do we have to put on some sort of protective coating? Do we cathodically protect it? We have a new cathodically protected pipeline running near the tank, so could we extend that system to protect the tank, too?

Since you've been getting mighty lucky on your guesses lately I'll leave the decisions to you.

13

Designing for Corrosion

Mistakes in plant design are the most frequently cited (58%) cause of corrosion failure in chemical-process industries. Design mistakes are even blamed slightly more than incorrect materials selection. Although "get a better material" might seem to be the answer to any corrosion problem, it becomes obvious, as a failure is investigated, that in most cases the design could be improved.

Unfortunately, for corrosion engineers most plants are built at minimum cost with the idea that troubles will be met as they develop by upgrading the equipment as needed. Even so, careful design can overcome many of the shortcomings of materials.

Consumer goods, such as automobiles, outboard motors, or whatever, are offered in a highly competitive market where price is a major factor. In times past the manufacturers dodged corrosion problems by giving short-term guarantees. The manufacturers regarded corrosion as a process that operated in their favor, guaranteeing frequent replacement. Consumer groups changed all that with their unfavorable publicity, making consumers much more concerned about corrosion and other design problems. When manufacturers began trying to improve the corrosion resistance of their products, better designs solved many problems at no additional cost.

- Design engineers must understand corrosion or they become part of the problem. "As long as design engineers are around, corrosion engineers will always have jobs," one corrosion engineer remarked sarcastically.

All design engineers know that they are supposed to make allowances for corrosion by increasing wall thickness, and so on, but a generation ago that was all they needed to know, or so they thought. The following points identify the major corrosion principles necessary for a successful design. They are arranged according to types of corrosion: chemical corrosion, electrochemical corrosion, and corrosive–mechanical interaction.

13.1 ALLOW FOR UNIFORM ATTACK

The design engineer selects a material with a satisfactory corrosion rate in the environment involved. Handbook data, results of lab tests, pilot plant and field tests, or previous service are considered. Any metals showing localized corrosion are eliminated at the start. Then, knowing the required life of the equipment and assuming uniform corrosion, the engineer calculates the corrosion allowance to be added to the metal thickness, plus a safety factor, because real life is not as precise as mathematics.

Remember that modern engineering equipment does not have to last as long as the pyramids. Corrosion at a reasonable rate can be tolerated unless the corrosion products themselves create a problem: contaminate the desired product, plug up flow lines, or the like.

13.2 MINIMIZE ATTACK TIME

Corrosion takes time. The design engineer tries to reduce the time that the environment will contact metal. If dirt and sediment can collect they hold moisture next to the metal long after the other metal surfaces are dry. Figure 13-1 shows how braces of channels and angle iron should be positioned to prevent liquids from sitting and corroding.

Corrosion is often particularly severe on sheltered surfaces where dirt and salts are not washed off and evaporation of moisture is slow. Avoid structural

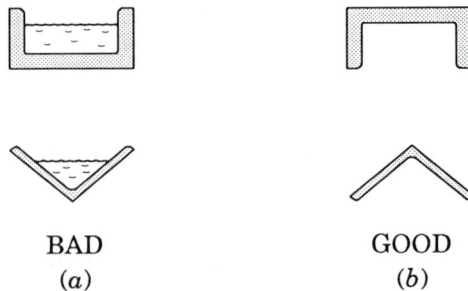

BAD   GOOD
(a)   (b)

FIGURE 13-1.   Arrangement of steel shapes (a) that hold moisture and dirt and (b) those that do not.

shapes (Fig. 13-2) that collect dirt and moisture with no way to orient these shapes to prevent this accumulation.

Drainage Problems

Find a way to allow water to drain off metal surfaces. Figure 13-3 illustrates how structural members and armored electrical cable should be oriented to allow moisture to drain from the metal.

Drainage holes (Fig. 13-4) can eliminate liquid entrapment, but the holes must be located so that the draining liquid does not drip or splash onto other

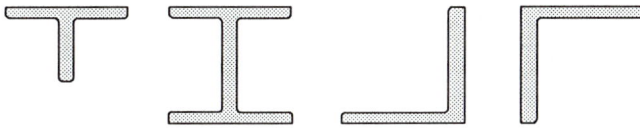

FIGURE 13-2.   Structural shapes that collect dirt and moisture.

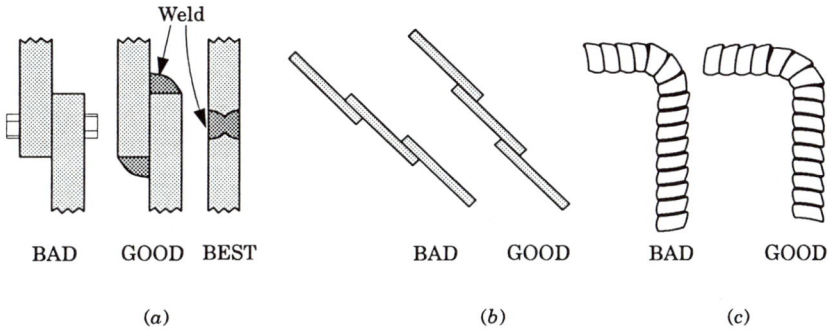

FIGURE 13-3.   Orientation of structural members and armored electrical cable to prevent crevice corrosion.

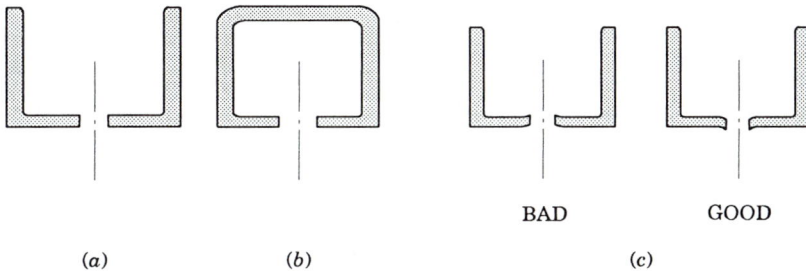

FIGURE 13-4.   Drainage holes in open and partially closed structures.

metal. Horizontal surfaces should be sloped or bevelled if possible, as indicated in Figure 13-5.

Tanks and reaction vessels must be designed with drains allowing them to be emptied completely (Fig. 13-6). Be sure that the slope of the tank bottom is sufficiently steep that suspended solids and sediment will also be removed, but not so steep on large tanks that the swirling motion of the liquid during drainage will cause erosion–corrosion.

Pipes should be supported so that they slope uniformly without low spots that will not drain and that will allow sediment to collect (Fig. 13-7).

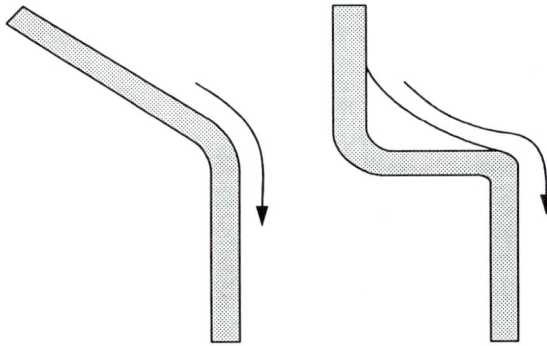

FIGURE 13-5.   Slopes or modified profiles permit liquid flow.

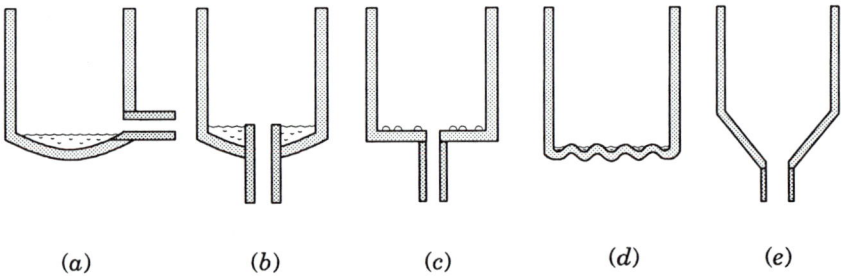

(a)   (b)   (c)   (d)   (e)

FIGURE 13-6.   Tanks should be designed so that they can be emptied completely. Examples (a–d) show poor designs. (e) Good design.

FIGURE 13-7.   Low spots in a pipe can never be drained.

Susceptible Metals

In particular, corrodible metals and microstructures must be given special care. For example, empty drums and cylinders may contain a small amount of liquid that remains to corrode the bottom, as Figure 13-8 shows. The galvanic cell setup at the weld can be prevented by locating the weld above the bottom.

Dissimilar metals must be joined so that liquids run off the junction to prevent the initiation of galvanic corrosion (Fig. 13-9).

Concentrating Liquids

Spills, splashing, spray, and intermittent flow leave a thin film of liquid that may become more corrosive with time. Condensed moisture and spills will run down the walls of a tank to the bottom, where droplets will sit, absorbing O_2, and slowly evaporating while concentrating dissolved corrosives. A drip skirt should be welded on (Fig. 13-10). The bottom of the drip skirt corrodes badly but that is only a cosmetic problem.

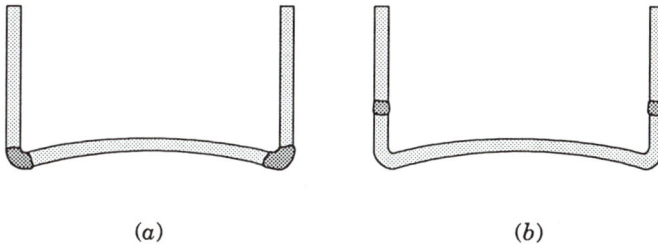

(a)   (b)

FIGURE 13-8.   (a) Weld at tank bottom will corrode if liquid residue or condensation collects. (b) Improved design.

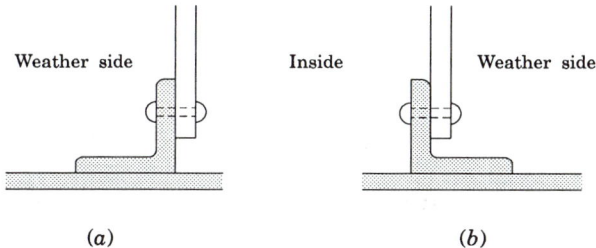

(a)   (b)

FIGURE 13-9.   Aluminum boat superstructure fastened to steel deck. (a) Crevice corrosion and galvanic corrosion possible and (b) better design.

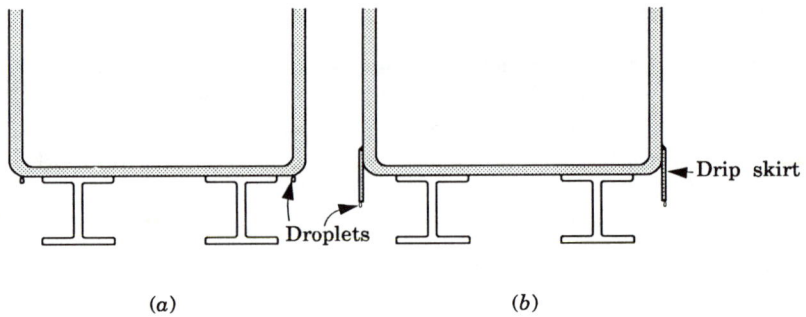

FIGURE 13-10.   Tank (*a*) without and (*b*) with drip skirt.

The design for liquid flow into a container should prevent liquid from running down the wall (Fig. 13-11*a*), where it can wet the wall and absorb oxygen, or where the concentrated solution can contact the metal before mixing. Splashing of liquid up on the wall (Fig. 13-11*b* and *c*) can be prevented by feeding it into the center of the tank (Fig. 13-11*d*).

Do not let solutions concentrate on a metal surface. In Figure 13-12*a* the splashing solution can dry and concentrate on the hot coil, in contrast to the coil in Figure 13-12*b*, which is completely submerged.

Liquid allowed to sit in a partially emptied tank can dissolve oxygen or moisture from the air and increase its corrosiveness. Blanketing with inert gas may be necessary as the tank is emptied.

A pipe often corrodes along the bottom because sediment collects, or it corrodes at the top because of splashing from a partially filled pipe. In either case its life can be almost doubled by rotating it 180°.

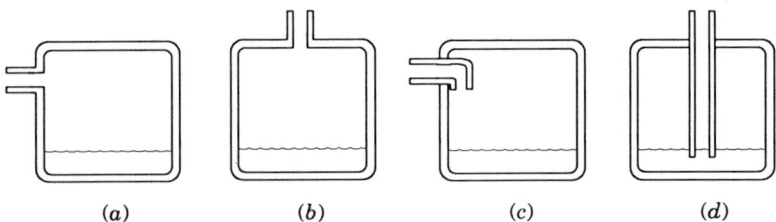

FIGURE 13-11.   Designs for adding concentrated liquid to a vessel. (*a–c*): liquid runs down or splashes on tank walls and (*d*) good design.

Heating coil

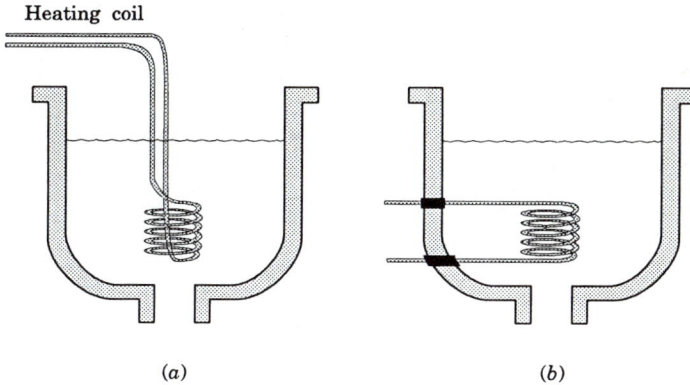

(a)   (b)

FIGURE 13-12.   (a) Splashing solution will concentrate on a hot heating coil and (b) good design.

Vapor Problems

Frequently, the vapor of a corrosive solution attacks the metal much worse than the liquid does. Sidewalls and roofs of storage tanks and reaction vessels or large-diameter horizontal pipes often corrode severely in the vapor. Their design should include venting, a vacuum removal of the vapor, or a condenser return to the vessel.

Excluding air and preventing air entrainment by a fully air-tight system is quite effective with neutral or alkaline corrosive solutions. Designs may also call for sealing out liquids; box girders, for example, should be continuously welded to seal out moisture and humidity, although some engineers despair of getting perfect welds and just put in drain holes. Proper design must prevent gaps in aluminum sheathing or breaks in plastic wrapping around thermal insulation on piping that would permit moisture-laden air to enter the insulation and initiate corrosion or SCC.

Corrosive vapors in the outside atmosphere can sometimes be combatted by locating buildings upwind from nearby pollution-producing industries, storing equipment and supplies where they will be sheltered from prevailing winds that carry corrosive fumes, and locating chimney stacks downwind from plant structures. Corrosion by sea air is much reduced by siting buildings inland from the prevailing winds, even as little as 150 m (500 ft).

13.3 RESTRICT GALVANIC CELLS

Do not use dissimilar metals together, is not likely to be practical advice. However, a design involving metals close to each other in the galvanic series can function successfully with little effect from galvanic corrosion. Keep in mind that situations beyond the design can create galvanic corrosion: The connection of new steel to old steel will attack the new metal because the old metal has become more noble by forming a thick, protective rust coating. Use of graphite-containing lubricants and packing can cause galvanic corrosion of bearings and valves.

The Electrodes

With galvanic corrosion, overcome the urge to protect the corroding metal with paint unless the cathode metal is painted also. Increasing the cathode/anode ratio at the inevitable flaws and damaged spots in the paint film only concentrates the attack. Instead, try to minimize the cathode/anode area ratio. With welds, for example, specify filler metal the same or slightly more noble than the base metal and require that welds be oriented so that the least surface area of the weld is exposed to the corrosive environment (Fig. 13-13).

Make sure that the design does not call for welding that can sensitize stainless steel. Figure 13-14a shows a design that will carburize stainless steel by welding it to carbon steel.

As with welds, fasteners (rivets, bolts, screws, etc.) should be at least as noble as the much larger area of metal they fasten. Figure 4-6 illustrates this principle.

All galvanic corrosion can be stopped if the two metals can be electrically insulated from each other. Figure 13-15 shows ways of insulating one metal from another. However, insulating washers, and the like, cannot be used at

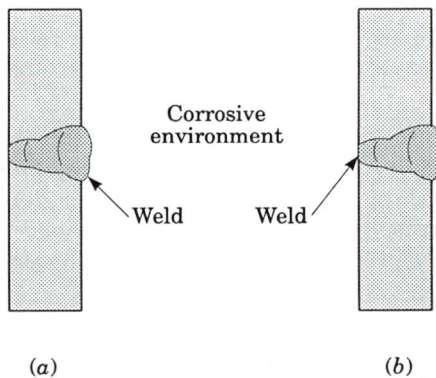

(a)   (b)

FIGURE 13-13.   Weld to expose minimum amount of weld metal to a corrosive environment: (a) poor and (b) good.

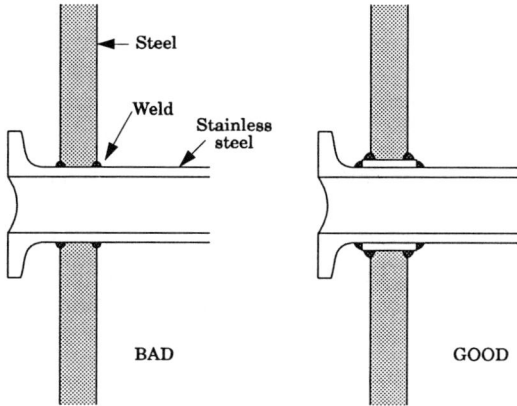

FIGURE 13-14. Prevent (*a*) contamination of corrosion-resistant metal; (*b*) better design.

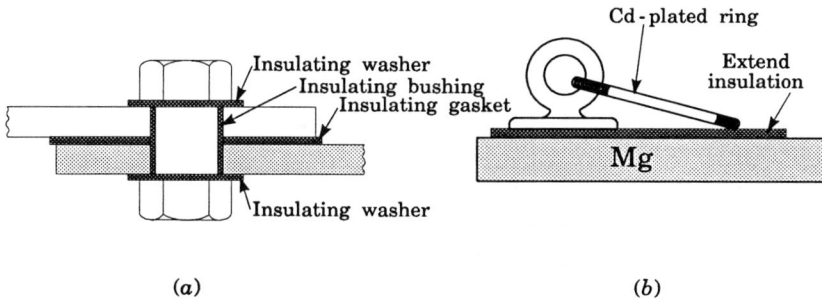

(*a*)   (*b*)

FIGURE 13-15. Methods of insulating metals. (*a*) Two plates insulated from each other and from the fastener. (*b*) Cadmium-plated ring prevented from touching magnesium plate.

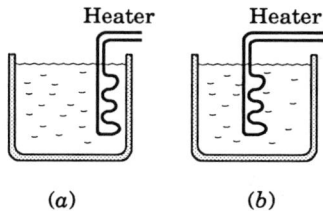

(*a*)   (*b*)

FIGURE 13-16. (*a*) Galvanic corrosion by copper heater in steel tank and (*b*) corrosion reduced by the distance effect.

extremely high stresses or temperatures, or in corrosives that will attack the insulating material.

The Electrolyte

Take advantage of the distance effect in galvanic corrosion by placing anode and cathode as far apart as possible, especially in low-conductivity solutions. The heater in Figure 13-16 should be put in the center of the tank to get it as far from the tank wall as possible, if the two metals cannot be insulated from each other. Figure 13-17 shows how a steel insert, necessary for wear resistance, can be used in a magnesium plate, even though condensation from the atmosphere will contact both metals.

In Figure 13-18 the distance effect can reduce corrosion if the middle section is made of a metal that is intermediate in corrodibility between the other two sections. This technique separates the highly noble metal that serves as the principal cathode from the very reactive metal that is doing most of the corroding. Or, the intermediate section can be the most corrodible if the wall thickness is made extra heavy (see Fig. 13-18). This same approach can also be used to mitigate concentration cells and differential temperature cells.

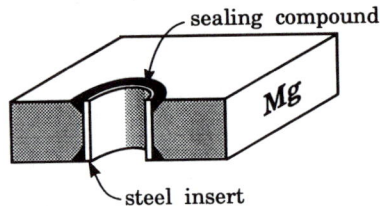

FIGURE 13-17.   Steel insert in magnesium plate. The distance effect is used to protect the magnesium from galvanic corrosion.

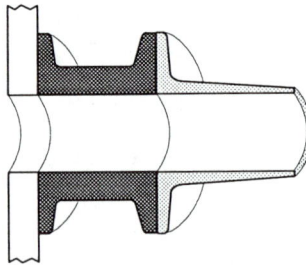

FIGURE 13-18.   Exchangeable connector for joining dissimilar metals.

Where different metals must be used but cannot be insulated, preventing the environment from contacting *one* of the metals prevents galvanic corrosion, as shown in Figure 13-19.

- *Note*: Coating the more noble metal is safer, since absolute perfection is rare in this world.

Liquid should never be allowed to flow over a noble metal and then on to a very reactive metal; even water dripping from copper onto aluminum severely corrodes the aluminum. Copper ions carried by the water will plate out on the aluminum and set up a galvanic cell.

If possible, turn galvanic corrosion into an advantage to protect vital components, as in the use of heavy steel tube sheets to protect thin-walled copper tubing in heat exchangers.

13.4 PROTECT AGAINST ENVIRONMENTAL CELLS

Crevices

Geometric crevices often occur when two pieces of metal are put together, especially if rough machining marks on the surface prevent intimate mating.

Avoid deposits on the metal by making sure that designs do not include ledges, pockets, flanges, or obstructions that will provide a location for entrained solids to settle out of the liquid. The design should ensure that flow rates are rapid enough to keep metal surfaces washed but not so rapid that erosion–corrosion could be initiated.

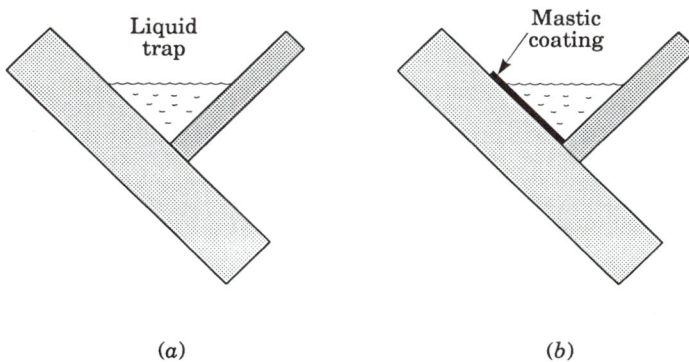

(a)   (b)

FIGURE 13-19.   (a) Galvanic corrosion and (b) prevented by isolating one metal from the environment.

If crevices must be in the system, seal them. Weld, caulk, or solder rather than use fasteners, such as rivets or bolts. Welding must be continuous, not skip or spot welds (Fig. 13-20), and butt welds are better than welded lap joints (Fig. 13-3a).

Where crevices cannot be sealed by any practical method they must be protected. Galvanizing both bolt and nut protects their threads from crevice attack. Painting metals with a zinc-rich or zinc chromate primer before assembling them prevents crevice corrosion for years. Although painting the metal outside a crevice may not seal the crevice, it does reduce the cathode area of the corrosion cell. Inhibited greases provide temporary protection while also lubricating threaded fasteners.

Porous Materials

A design that places absorbent nonmetallic materials against metal sets up potential crevice corrosion conditions. In many cases the nonmetallic is not designed to be wetted, but the designer must assume that sometime, somehow, wetting will occur. Absorbent gaskets, fiberglass blankets, sponge rubber, plastic foam, cardboard, wood, cloth-backed tape, and concrete all potentially can create crevices, especially if they can act as a wick. Do not assume that an adhesive used to fasten nonmetal to metal will seal the metal surface.

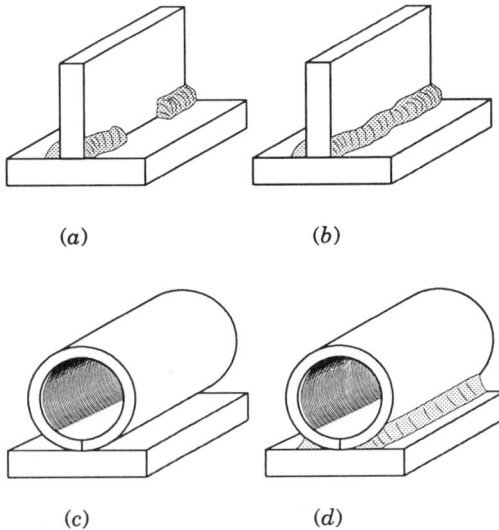

(a)   (b)

(c)   (d)

FIGURE 13-20.   Crevices and ways to seal them. (a) Skip weld, (b) replaced by continuous weld, (c) space between tube and plate, and (d) sealed with mastic.

- "Blessed are they who expect trouble for they surely will not be disappointed."

Metals sitting on concrete pose a special problem because the concrete is porous and holds moisture. The metal should sit off the floor so that spills and cleaning water contact the metal for only a brief time. Concrete should slope away from metal for rapid drainage. Figure 13-21a illustrates these principles. Thin-walled tanks should be supported so that they do not contact concrete at all (Fig. 13-21b).

If horizontal tanks must sit on a concrete saddle, as in Figure 13-22, a support plate should be welded to the vessel to give an extra thickness allowance for the expected corrosion. A metal saddle would be even better.

Figure 13-23 shows a very practical design for thermal insulation of a tank

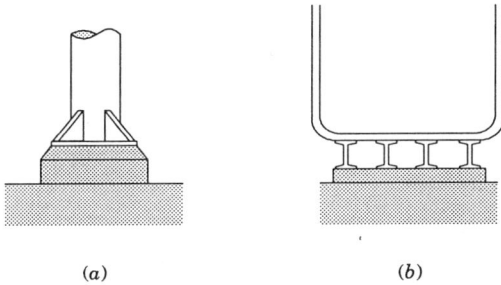

(a)   (b)

FIGURE 13-21.   (a) Supporting post with bottom plate fastened above floor level. (b) Support tanks above floor level.

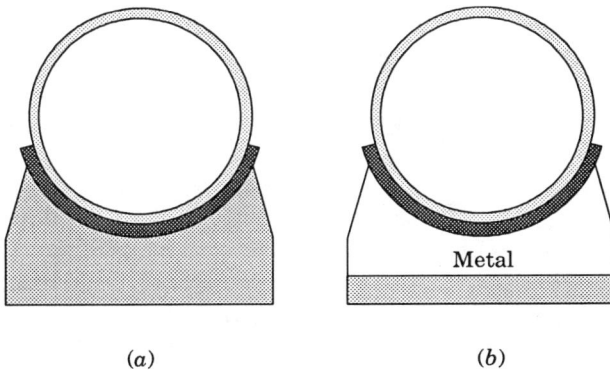

(a)   (b)

FIGURE 13-22.   (a) Support pad protects horizontal tank from concrete saddle, (b) better design: metal saddle with support pad.

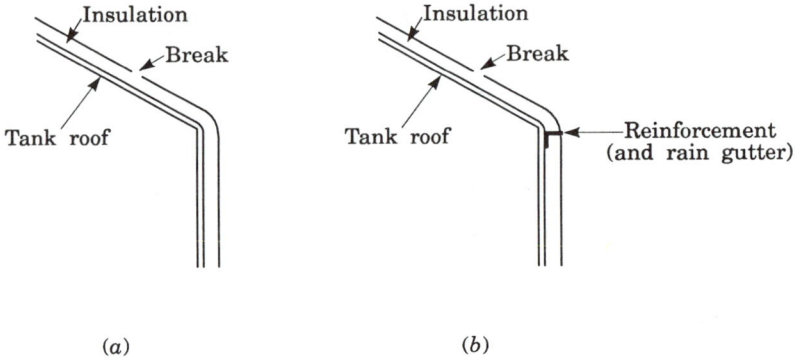

FIGURE 13-23.   Details of tank insulation. (*a*) Leak in roof will wet side wall insulation and (*b*) improved design.

for outside storage. The designer assumed that somewhere the roof would leak. To prevent water from running down the tank walls, drainage was provided for the roof insulation with a steel reinforcement separating roof from walls.

Temperature Differences

Hot spots on a metal surface are often created by impingement of flame or hot gases but can also occur when chemicals are mixed. Local heating means high corrosion rates and may set up differential temperature cells or create stresses by differences in thermal expansion. If boiling or evaporation occurs, the corrodent is concentrated on the metal surface at the hot spot, leading to increased corrosion or even SCC.

Try to heat the metal uniformly. In Figure 13-16 centering the heater would reduce galvanic corrosion but also would heat the tank more evenly. Provide good circulation of fluid to keep the temperature differences at a minimum by forced convection. If a tank must be heated externally, the heater should cover as much of the exterior surface as possible.

Where temperature differences are necessary, as in heat exchangers, ensure that the temperature gradient is uniform. Insulating ferrules in the front ends of the tubes not only guard against erosion–corrosion but also even out heat transfer somewhat. Dirty or scaled surfaces reduce heat transfer and are likely to nucleate boiling. In some cases boiling can be suppressed by imposing back pressure on the liquid.

Cold surfaces cause almost as many corrosion problems as hot surfaces. Moisture from hot gases or humid air condenses on cold metal and corrodes it even if the gases themselves are not harmful. Figure 13-24 shows how rapid heat transfer through noninsulated supports can cause corrosion inside a reaction vessel.

Air ducts and engine mufflers often have to be constructed with insulation between double walls to prevent condensation on the metal. An inlet tube carrying cold solution into a tank with hot products may need an extra-thick wall to withstand corrosion from condensed vapor.

13.5 AVOID CORROSIVE–MECHANICAL INTERACTION

Erosion–Corrosion and Cavitation

Good design can eliminate erosion–corrosion by reducing velocity, turbulence, and impingement. The obvious way of reducing fluid velocity in a pipe is to use a larger pipe, but other erosion–corrosion situations require more ingenuity. For example, in Figure 13-25 erosion–corrosion has been observed

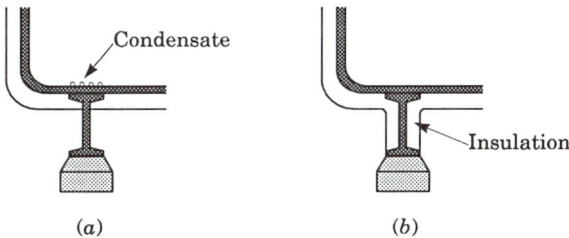

FIGURE 13-24.   Insulated reaction vessel containing moist gases corrodes if the legs act as cooling fins.

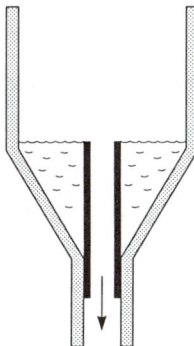

FIGURE 13-25.   Removable drainpipe in reaction vessel.

to occur in large reaction vessels as the liquid swirls around while draining. A removable drainpipe allows drainage in two stages to reduce the velocity in the second stage.

Turbulence is often caused by projections into a pipe or abrupt changes in cross section. The design should eliminate changes or make them more gradual. Streamline inlets and outlets where rapid changes in velocity must occur. For heat-exchanger tubing, specify ferrules for the inlet ends.

Impingement can be minimized by careful design. Bends in piping should be gradual with large radii; branch piping should connect at low angles (Fig. 13-26b). Flow from inlet pipes should be directed away from tank walls (Fig. 13-26d and e). Suspended solids should be filtered from liquids, if possible, and water droplets should be removed from high-velocity steam.

Cavitation can be prevented by reducing hydrodynamic pressure differences or by injecting bubbles of an insoluble gas into the liquid. Bubbles absorb the shock and cushion the implosions.

If erosion–corrosion cannot be prevented, try to move it. An example is prevention of inlet tube corrosion by extending the tubes about 10 cm (4 in.)

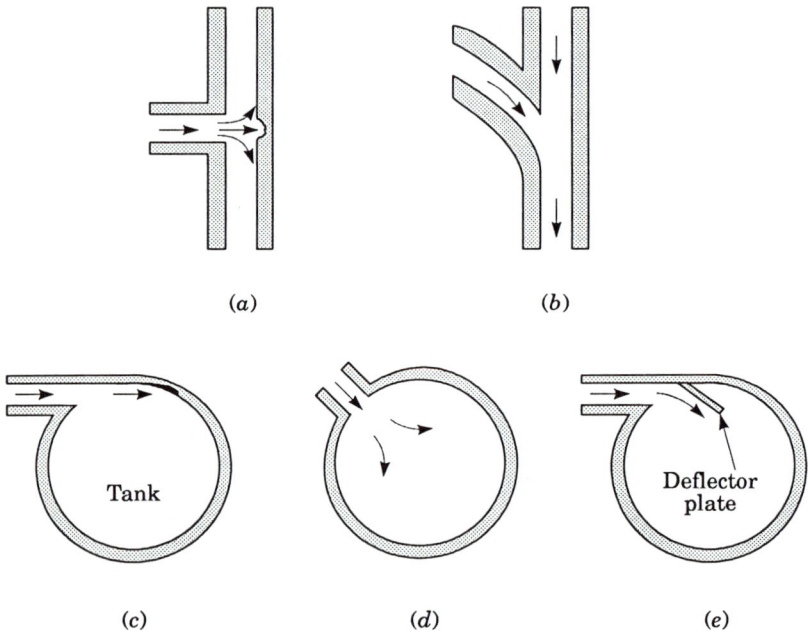

FIGURE 13-26.   (a) Impingement of flow, (b) preventing impingement by streamlined connections, (c) impingement on tank wall, (d) directing flow away from wall, and (e) deflector plate.

beyond the tube sheet. If a heat exchanger is designed so that its flow can be reversed, the life of the tubes can be nearly doubled under erosion–corrosion conditions. Also remember that replaceable impingement plates or baffles will transfer erosion–corrosion to metal that can be sacrificed (Fig. 13-26e).

Stresses with Corrosion

The dangers of SCC and corrosion fatigue are always present but careful design can reduce their likelihood by keeping stresses low. Residual stresses from cold work are well known to cause trouble—the design engineer specifies the lowest possible hardness that will still be suitable for the service.

Local stresses can set up an electrochemical corrosion cell as shown in Figure 13-27a for a welded tank with an extra-thick bottom. The extra thickness at the weld was ground off on the inside of the vessel, which became the location of severe attack, aided by the slightly noble weld metal. The replacement bottom was fabricated larger than the tank, with excess metal ground off on the outside of the tank. It served quite satisfactorily.

- The design engineer has little control over assembly stresses caused by workmen with the philosophy of, "Don't force it. Use a bigger hammer."

Welds create a heterogeneous HAZ (heat-affected zone) susceptible to corrosion. When that is combined with high stress, a dangerous situation is set up. In Figure 13-28 the "bad" design, a hold-over from old riveting design, puts high stresses on the welds, while the "best" design moves the welds entirely away from the high-stress area. The number of welds should also be kept to a minimum.

Another example of separating the corrosion location from the high-stress

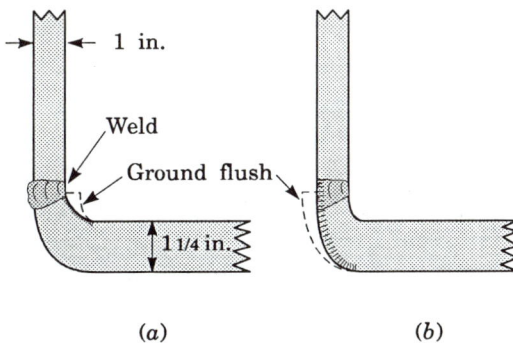

(a)   (b)

FIGURE 13-27.   Grinding can cause local stress cells inside a tank but not outside.

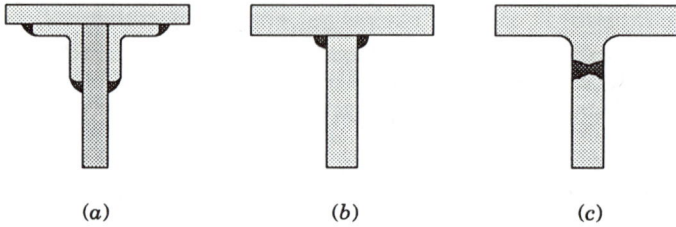

FIGURE 13-28.   High stress should not be put on welds in a corrosive environment. (*a*) Bad, (*b*) better, and (*c*) best.

location is in the redesign of axles for railroad cars (Fig. 13-29). Fretting corrosion between the shrink-fitted wheel and the axle initiated corrosion fatigue. To prevent this problem, the ingenious designer moved the flexing stress out of the wheel and onto the axle.

Operating stresses concentrate at stress raisers, such as notches and grooves, until the stress intensity may be high enough to initiate SCC. Use of generous fillets (Fig. 13-30*b*), smooth surface finish, keyways with sledge endings and fillet radii (Fig. 13-30*d*), and similar designs distribute stresses more evenly. Also, prevent excessive vibration, a problem often found with heat-exchanger tubes, by putting in more tube supports.

The designer must consider stresses created by corrosion products. A covering of 5 cm (2 in.) of concrete is usually considered sufficient to protect reinforcing steel in moderate environments but in high chlorides a thicker covering is better. Stress-corrosion cracking in heat exchangers and evaporators often initiates at crevices where chloride can concentrate and where corrosion products in pits generate high tensile stresses.

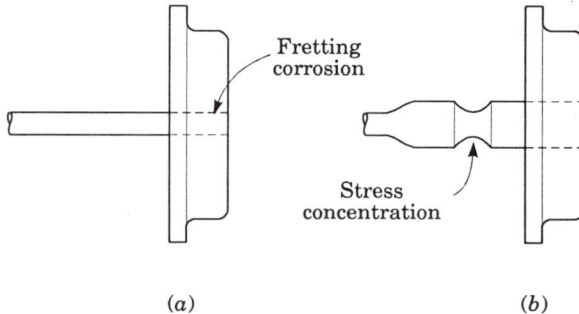

FIGURE 13-29.   Redesign of railroad axle to prevent fretting corrosion and corrosion fatigue.

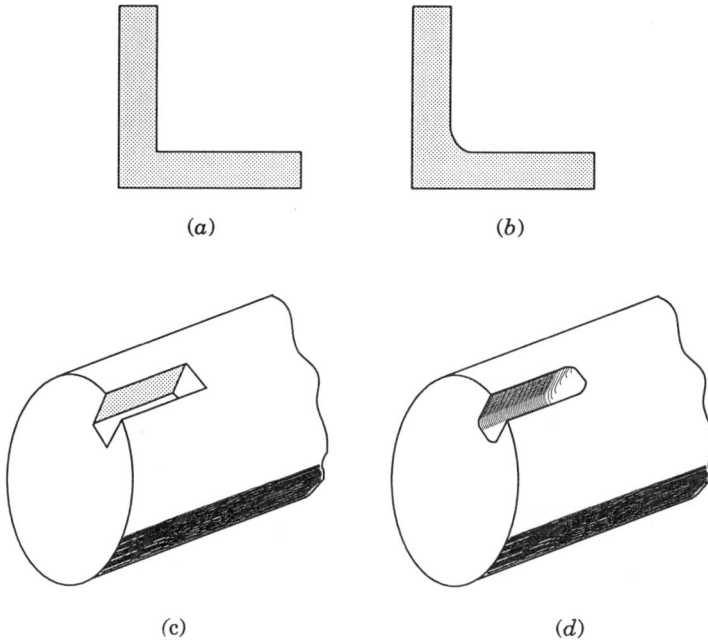

FIGURE 13-30.   Prevent stress concentration by (*a–b*) fillet and (*c–d*) rounded keyway.

While compressive stresses created on metal surfaces by shot peening, swaging, and so on, do prevent crack initiation, bear in mind that the compressive stresses are balanced by equivalent tensile stresses just below the surface. If pits reach the tensile region, cracks will initiate and extend very rapidly. Rolled threads on bolts and nuts have surface compressive stresses and rounded grooves, both of which help prevent cracking. In contrast, the sharply angled grooves on cut threads are common initiation sites for cracking.

13.6 DESIGN FOR INSPECTION AND MAINTENANCE

A design must provide ready accessibility to critical areas for repairs, replacement of parts, and cleaning. Figure 13-18 depicts a connector between two incompatible metals. The connector is going to corrode but it is easily replaceable. Pipes and small vessels often must be cleaned with flexible brushes but the brushes cannot double back on themselves to get to blind spots, around ledges, and into crevices. In large equipment designs, manways are commonly incorporated to allow inspectors and workmen to enter.

- Remember that a corollary of Murphy's law is, "If any place in your equipment is absolutely inaccessible, that's going to be the trouble spot."

The design should include provision for monitoring corrosion, especially in critical locations. Electronic probes or fluid-tight valves where coupons are inserted will provide information about the corrosion.

- Try not to let a failure be your first indication of trouble. Avoid surprises (if it's not your birthday).

Maintenance usually involves painting. A common design mistake is failure to leave sufficient space between surfaces for a painter to get in a brush or spray from a paint gun (Fig. 13-31). The spacing between surfaces must be at least one-third of the height and not less than 45 mm. With large equipment remember that enough room must be left for the painter.

Tubing is easier to paint than I or H shapes, while lattice construction is almost impossible to paint well. Sharp corners, weld spatter, and rough surfaces receive very little protection by paint because the film is particularly thin at these points (Fig. 10-4). The design should call for rounded corners and edges since painted metal invariably rusts at sharp edges first.

Try to make the design foolproof. No human being is infallible and some seem *particularly* fallible. Design with these people in mind.

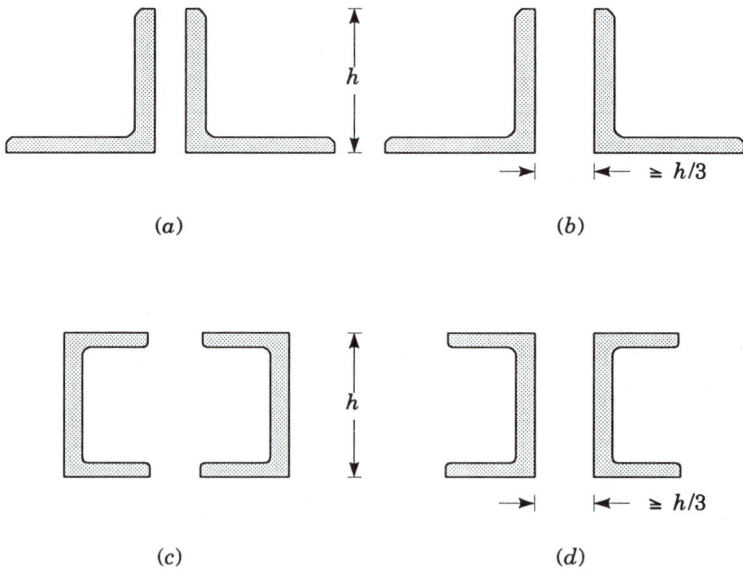

FIGURE 13-31.   Profiles that cannot or can be painted.

- A plant designed for versatility had a $CaCl_2$ brine line connected through a valve to a steam line. One day someone opened the connecting valve, of course, just because it was there. An entire steam turbine was corroded beyond repair.

STUDY PROBLEMS

13.1 Compare the designs for fluid flow in Figure 13-32. Discuss their shortcomings. (a) Branch piping design and (b) heater tube bundle design.

13.2 The cross section of a reaction vessel is shown in Figure 13-33. Point out good features of the design and suggest improvements.

13.3 A diesel locomotive exhaust system has corroded severely from acid condensate. How can it be redesigned to extend its life?

13.4 Explain any design flaws in the storage tank, Figure 13-34.

13.5 What corrosion problems are the following design recommendations intended to overcome? (a) Install a drip skirt. (b) Increase pipe diameter. (c) Install legs on the tank to raise it off the floor. (d) Install controls to hold solution temperature constant. (e) Insulate the legs of a reactor as well as the reactor.

13.6 A boiler has failed by SCC and is to be redesigned. List at least three design modifications that should be considered.

13.7 Choose any commonly used method of preventing corrosion and describe a situation where its use would be inappropriate, for other than economic reasons. Explain.

13.8 A small copper pipe in your plant began spurting a long stream of hot acetic acid at an elbow. Inside the pipe the surface is quite smooth with some rippled appearance. What changes in *design* could you recommend to prevent further trouble here?

13.9 A manufacturer complains about corrosion of brass-coated steel grommets on truck tarpaulins. Corrosion products weaken the canvas and cause it to tear. How could proper design improve the situation?

13.10 A manufacturer plans to use steel screws to fasten aluminum panelling on trailers. How could proper design reduce the possibility of corrosion?

13.11 An instruction booklet held together with a metal staple must be used for 5 years in a very humid climate. How many ways can you improve the design so that the metal staple can survive the corrosion?

13.12 Your boss asks you to come up with a scheme to save his truck radiators from further corrosion. These trucks are in the far north, use any kind of cooling water available, and are infrequently serviced. Here is your big chance, but explain it fully to him.

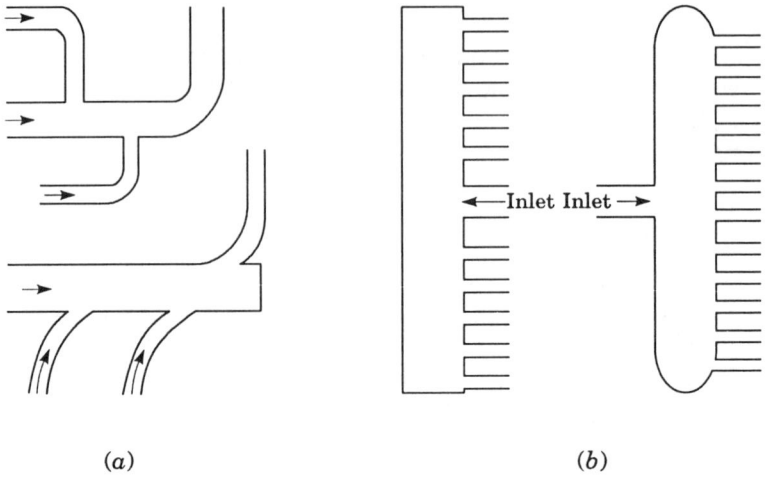

FIGURE 13-32. Suggested designs involving fluid flow. (*a*) Branch piping design. (*b*) Heater tube bundle design.

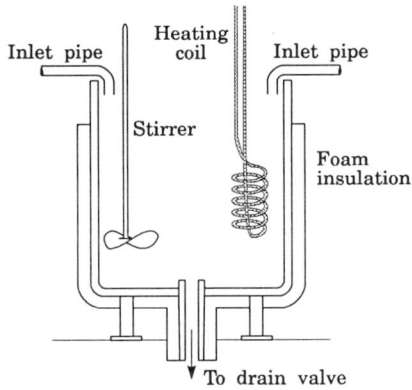

FIGURE 13-33. Proposed design of reactor.

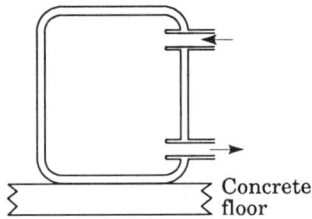

FIGURE 13-34. Design of storage tank.

13.13 If a new Alaska pipeline is built, how would you protect it from corrosion? What materials and equipment would you use?

Memos

13.14 Memo To: Corrosion Engineer
From: Chet Bailey, Production Supervisor
We have to connect a heavy-walled cast iron pipe to a copper pipe. They will be carrying dilute acetic acid at room temperature and very moderate velocity. Clyde Young, a junior engineer under me, says we ought to put a ceramic coupling between the two pipes to prevent corrosion. Is that right?
Also, both pipes are suspended from the roof I-beams by steel straps. Could we have some kind of corrosion problem from the air? Now that we've got a hot-shot corrosion engineer, it's time you get to work.

13.15 Memo To: Corrosion Engineer
From: Chet Bailey, Production Supervisor
Yesterday an inspector found our two vertical deaerator vessels to have extensive cracks transverse to circumferential welds, nozzle welds, and manway welds with several cracks through the wall. Both vessels were built 18 years ago out of SA 515 Grade 70 carbon steel. They use water softened with sodium zeolite. I don't know what operating pressure they were designed for, but there was frequent water hammer. Now we've got to try to weld those cracks up and get back in operation quickly. Since you're supposed to be so smart, tell me why they cracked. I don't think your fancy computer is going to give you the answer, though.

13.16 Memo To: Corrosion Engineer
From: Chet Bailey, Production Supervisor
We have a weak (0.05 N) solution of sodium carbonate at room temperature stored in a large steel tank. There's no agitation or anything like that to cause trouble. And the tank has been sitting there 11 years and looks almost as good as when I put it in—no corrosion. I knew plain carbon steel, cold-rolled and welded, could handle a solution like that.
However, they've been bothered by corrosion of the filter on the outlet. The filter is woven steel wire mesh and lasts about 6 months; also we get complaints about the rusty color of the solution. The steel in the filter has a higher carbon content than the tank, but should it make all that difference? It's about time you earned your pay around here. I can stop the rusting as soon as I find a supplier of stainless steel filters, but tell me what they have been doing *wrong*.

13.17 Memo To: Corrosion Engineer

From: Milton T. Klein, General Manager

When the main shop was built we had a cooling tower installed on the roof for the office air conditioning. Copper tubing carrying water to the tower held up OK but the aluminum tower had to be replaced in less than 1 year. Then somebody got the idea of connecting the water line to a 4-ft coil of aluminum tubing, which is connected to the tower. The aluminum coil corrodes and has to be replaced every 6 months, but the tower has been fine for 8 years. One of my engineers has pointed out that we can save the aluminum tubing by having it coated inside with plastic.

Give it some thought and jot down any ideas you have that could explain this problem and could help us.

Thanks, M.T.K.

13.18 Memo To: Corrosion Engineer

From: Chet Bailey, Production Supervisor

For several years we have been successfully using Type 347 stainless steel for a 20-m cooling coil carrying conc $CaCl_2$ brine, around pH 7, at -2 to 5°C. The coil is flanged to a mild steel vessel and cools river water. We have also had no trouble with 347 heat exchanger tubes for exactly the same brine but at somewhat higher temperatures.

More cooling was needed so we put in a 100-m 347 coil. Now this coil has failed by heavy pitting on the inside. What might have gone wrong and what can we do about it?

13.19 Memo To: Corrosion Engineer

From: Milton T. Klein, General Manager

One of the light poles in the parking lot blew over last night and destroyed two cars but fortunately no one was hurt. The pole had corroded severely at its base. Every pole (carbon steel) is anchored with long bolts in concrete footers. The steel nuts and bolts extending above the flange plate of the pole are covered with a steel box to give the base a pleasing appearance (corporate visitors use this lot, too). Under the boxes everything is a mass of rust: nuts, bolts, and poles. Could our high humidity or industrial air pollution be responsible? The boxes and the poles above the boxes are OK, with paint still intact. What is the best way to prevent more poles from falling and still have them look good?

Regards, M.T.K.

14

Oxidation: Metal–Gas Reactions

Engineering metals react with air. If they react slowly they are usable, but at high temperatures many metals react disastrously because chemical reaction rates increase exponentially with temperature. The principal reactant in air is oxygen, so all gas–metal reactions have come to be called "oxidation," using the term in its broadest sense. The reacting gas may be water vapor, hydrogen sulfide, chlorine, and so on, but the reaction mechanisms are essentially the same as for reaction with oxygen.

At temperatures below 100°C (212°F) metals usually have a thin film of moisture adsorbed on the surface, so that they react with oxygen by the common electrochemical processes of uniform or localized corrosion. At temperatures above 100°C (212°F) the gas either reacts directly with the metal or the reaction becomes controlled by solid-state diffusion through an oxide layer on the metal.

*14.1 THERMODYNAMICS OF OXIDATION

A typical oxidation reaction might be

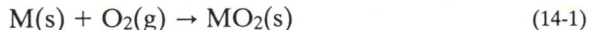

$$M(s) + O_2(g) \rightarrow MO_2(s) \tag{14-1}$$

with the metal M forming a solid oxide MO_2, MO, M_2O_3, or whatever, and the notations (s) and (g) indicating solid and gas phases. As with all chemical reactions the driving force is ΔG, the Gibbs energy that must be *negative* if oxidation proceeds and will become zero at equilibrium. For this reaction ΔG can be expressed in terms of the standard Gibbs energy change ($\Delta G°$) as

$$\Delta G = \Delta G^\circ + RT \ln (a_{prod}/a_{react}) \qquad (14\text{-}2)$$

in which a_{prod} and a_{react} are the activities of products and reactants. Activities of solids are usually invarient; that is, their activities equal 1, and for the high temperatures and moderate pressures of most gas–metal reactions the activity of oxygen can be approximated by its pressure, p_{O_2}. The driving force for the reaction is then

$$\Delta G = \Delta G^\circ + RT \ln 1/p_{O_2} \qquad (14\text{-}3)$$

At equilibrium, where $\Delta G = 0$

$$\Delta G^\circ = RT \ln p'_{O_2} \qquad (14\text{-}4)$$

The driving force for oxidation then is

$$\Delta G = RT \ln (p'_{O_2}/p_{O_2}) \qquad (14\text{-}5)$$

where p'_{O_2} is the equilibrium oxygen pressure and p_{O_2} is the actual O_2 pressure of the gas. Thus gold, for example, would form Au_2O_3 with an equilibrium O_2 pressure of 10^{19} atm at 25°C (77°F), so that the environment would have to exceed an oxygen pressure of 10^{19} atm before ΔG becomes negative and reaction would occur. Consequently, gold does not oxidize. On the other hand, aluminum oxide has an equilibrium O_2 pressure of 10^{-184} atm at 25°C (77°F), so that aluminum will oxidize in just about any real gas—or vacuum.

14.2 OXIDE STRUCTURE

The Pilling–Bedworth Ratio

The first basic understanding of oxidation mechanisms came in 1923 when N. B. Pilling and R. E. Bedworth divided oxidizable metals into two groups: those that form protective oxide scales and those that do not. They proposed that a scale will be unprotective if the oxide layer on the metal occupies a smaller volume than the volume of metal reacted. The Pilling–Bedworth ratio is the volume of oxide divided by the volume of metal that formed the oxide.

If the ratio is less than 1, as it is for alkali and alkaline earth metals, the scales are usually unprotective, being porous or cracked, because the volume of oxide on the metal surface is insufficient to cover the entire surface as it replaces the reacted metal. If the ratio is greater than 1, the scale shields the metal from the gas so that further reaction can occur only after solid-state diffusion, which is slow even at high temperatures.

If the ratio is much greater than 2, and the scale is growing at the metal–oxide interface, large compressive stresses build up in the scale so that *if* the scale grows thick it will likely buckle and spall off, leaving the metal unprotected.

For example, in the oxidation of iron

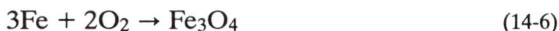

$$3Fe + 2O_2 \rightarrow Fe_3O_4 \qquad (14\text{-}6)$$

the Pilling–Bedworth (PB) ratio is

$$\text{(Volume of 1 mol Fe}_3\text{O}_4)/\text{(Volume of 3 mol Fe)} =$$
$$(44.7 \text{ cm}^3)/(3 \times 7.10 \text{ cm}^3) = 2.10$$

Example Problem:

Aluminum, atomic mass 26.98 g/mol and density 2.70 g/cm³, oxidizes to Al₂O₃, molar mass 101.96 g/mol and density 3.97 g/cm³. Calculate the Pilling–Bedworth ratio.

Solution:

Oxidation is

$$2Al + \frac{3}{2}O_2 \rightarrow Al_2O_3$$

$$\text{PB ratio} = \frac{\text{volume of 1 mol Al}_2\text{O}_3}{\text{volume of 2 mol Al}}$$

$$= \frac{(101.96 \text{ g/mol})/(3.97 \text{ g/cm}^3)}{(2 \text{ mol}) (26.98 \text{ g/mol})/(2.70 \text{ g/cm}^3)} = 1.28$$

Although the Pilling–Bedworth ratio is a gross oversimplification, which is not always correct, it provides information about the protective nature of metal oxides as a rough rule of thumb. Volume ratios for common oxide–metal systems are listed in Table 14-1.

The Defect Structure of Ionic Oxides

Ions can move through crystalline oxides via either Schottky defects or Frenkel defects or both. Schottky defects (Fig. 14-1*a*) are combinations of cation vacancies and anion (oxygen) vacancies in the correct ratio to maintain

TABLE 14-1   Melting Points and Pilling–Bedworth Ratios of Selected Oxides

Oxide	Melting Point (°C)[a]	Oxide/Metal Volume Ratio	Oxide	Melting Point (°C)[a]	Oxide/Metal Volume Ratio
αAl_2O_3	2015 (3660°F)	1.28	αMn_3O_4	1705 (3100°F)	2.14
γAl_2O_3	$\gamma \rightarrow \alpha$	1.31	MoO_3	795 (1465°F)	3.27
BaO	1923 (3493°F)	0.69	Na_2O	subl 1275 (2330°F)	0.57
BaO_2	450 (840°F)	0.87	Nb_2O_5	1460 (2660°F)	2.74
BeO	2530 (4585°F)	1.70	Nd_2O_3	~1900(~3450°F)	1.13
CaO	2580 (4675°F)	0.64	NiO	1990 (3610°F)	1.70
CaO_2	d 275 (527°F)	0.95	OsO_2	d 350 (662°F)	3.42
CdO	~1400 (~2550°F)	1.42	PbO	888 (1630°F)	1.28
Ce_2O_3	1692 (3078°F)	1.15	Pb_3O_4	d 500 (930°F)	1.37
CeO_2	~2600 (~4700°F)	1.17	PdO	870 (1600°F)	1.59
CoO	1935 (3515°F)	1.74	PtO	d 550 (1020°F)	1.56
Co_2O_3	d 895 (1643°F)	2.40	Rb_2O_3	489 (912°F)	0.56
Co_3O_4	$\rightarrow CoO$	1.9°	$P\epsilon O_2$	d 1000 (1830°F)	2.16
Cr_2O_3	2435 (4415°F)	2.02	Rh_2O_3	d 1100 (2010°F)	1.87
Cs_2O	d 400 (750°F)	0.47	SiO	1700 (3090°F)	1.72
Cs_2O_3	400 (750°F)	0.50	SiO_2	1713 (3115°F)	2.15
CuO	1326 (2419°F)	1.72	SnO	d 1080 (1975°F)	1.26
Cu_2O	1235 (2255°F)	1.67	SnO_2	1127 (2061°F)	1.31
Dy_2O_3	2340 (4245°F)	1.26	SrO	2430 (4405°F)	0.65
Er_2O_3	infus	1.20	Ta_2O_5	1800 (3270°F)	2.47
FeO	1420 (2590°F)	1.78 (on αFe)	TeO_2	733 (1351°F)	1.38
αFe_2O_3	1565 (2820°F)	2.15 (on αFe)	ThO_2	3050 (5520°F)	1.35
αFe_2O_3		1.02 (on Fe_3O_4)	TiO	1750 (3180°F)	1.22
γFe_2O_3	1457 (2655°F)	2.22 (on αFe)	TiO_2	1830 (3325°F)	1.76
Fe_3O_4	d 1538 (2800°F)	2.10 (on αFe)	Ti_2O_3	d 2130 (3865°F)	1.47
		1.2 (on FeO)	Tl_2O_3	717 (1323°F)	1.30
Ga_2O_3	1900 (3450°F)	1.35	UO_2	2500 (4530°F)	1.97
HfO_2	2812 (5064°F)	1.62	U_3O_8	d 1300 (2370°F)	2.71
HgO	d 500 (930°F)	1.32	VO_2	1967 (3573°F)	2.29
In_2O_3	d 850 (1560°F)	1.23	V_2O_3	1970 (3580°F)	1.85
IrO_2	d 1100 (2010°F)	2.23	V_2O_5	690 (1275°F)	3.25
K_2O	d 350 (660°F)	0.45	WO_2	~1550 (~2825°F)	1.87
La_2O_3	2315 (4200°F)	1.10	WO_3	1473 (2683°F)	3.39
Li_2O	~1700 (~3100°F)	0.57	W_2O_5	subl ~850 (~1560°F)	3.12
MgO	2800 (5070°F)	0.80	Y_2O_3	2410 (4370°F)	1.13
MnO	1785 (3245°F)	1.77	ZnO	1975 (3585°F)	1.58
MnO_2	d 535 (995°F)	2.37	ZrO_2	2715 (4920°F)	1.57
Mn_2O_3	d 1080 (1975°F)	2.40			

Sources: Adapted from Kubaschewski and Hopkins, 1962; Weast, 1988.
[a]Decomposition is d, sublimation is subl, and infusible is infus.

electrical charge balance in the oxide. A Frenkel defect is a combination of a cation vacancy and an interstitial cation, illustrated in Figure 14-1b. Ions can move around in an oxide when the appropriate type of ion diffuses into a neighboring vacancy or, with Frenkel defects, when cations diffuse interstitially.

Generally, metals oxidize much faster than can be accounted for simply by diffusion through Schottky and Frenkel defects because metal oxides are seldom if ever stoichiometric.

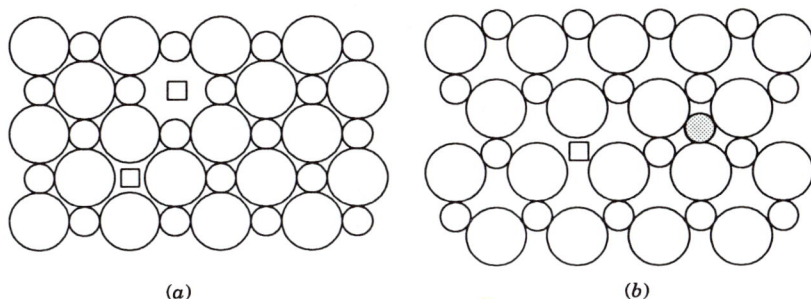

(a)   (b)

FIGURE 14-1. Defects in ionic crystals. (a) Schottky defect and (b) Frenkel defect. Ion vacancies are indicated by squares and an interstitial cation by a shaded circle.

Defect Structure of Semiconductor Oxides

Protective oxide scales are electronic semiconductors as well as having nonstoichiometric defect structures that allow mass transport through the scale. Semiconductors are classified by the way current moves through them as p-type, n-type, or amphoteric.

The p-Type Oxides. The p-type oxides are metal deficient, with cation vacancies in their crystal structure. They also have some Schottky and Frenkel defects that contribute to the total ion diffusion through their structures.

A good example of a p-type oxide is NiO, depicted schematically in Figure 14-2, which has a small fraction of its cations with +3 valence along with enough cation vacancies in its structure to maintain charge balance. The cation vacancies allow nickel ions to diffuse from the metal surface through the NiO scale to the outer oxide surface where they can react with oxygen ions to thicken the scale. For every two nickelic ions (Ni^{3+}), one cation vacancy must be present, so the actual formula of the oxide is $Ni_{<1}O$.

The Ni^{3+} ion has an extra valence electron taken from a lower energy level in its $3d$ subshell and, therefore, is said to have an "electron hole," a low-energy empty site for an electron. Current is carried through the oxide when electrons move through these holes. An electron hole, which is nothing more than a missing negatively charged electron, is considered to have a positive charge, so the oxide is called a p-type (positive-type) semiconductor. Electrons are transferred from the metal, via electron holes, to the oxygen adsorbed on the outer surface of the NiO scale.

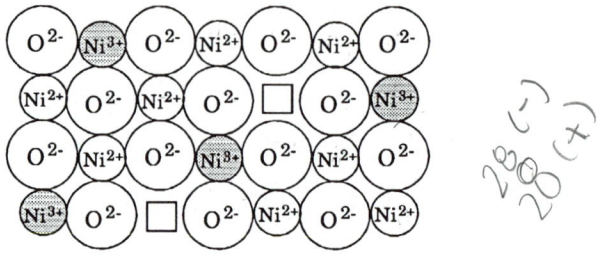

FIGURE 14-2.   Ionic arrangement in p-type NiO scale. Cation vacancies are indicated as open squares. The Ni^{3+} ions are shaded.

One p-type oxide, UO_2, is an anion-excess type with an actual formula of $UO_{2.2}$ to $UO_{2.4}$. Growth occurs by interstitial diffusion of O^{2-} ions through the oxide scale to the metal surface.

The n-Type Oxides.   The n-type oxides can be either metal excess or anion deficient. Beryllium oxide (BeO) typifies the metal-excess type (Fig. 14-3) because the Be^{2+} ion is small enough to exist interstitially in the BeO structure. Some Be^{2+} ions diffuse outward interstitially through the BeO scale to react with oxygen on the outer surface. As the beryllium metal releases a Be^{2+} ion to the scale layer it also gives off two electrons to maintain charge balance. These electrons, having no low-energy holes to go to, move rapidly through empty high-energy levels of the Be^{2+} ions in the oxide until they reach adsorbed oxygen on the outer oxide surface where they are captured to make O^{2-} ions. Since conduction is principally

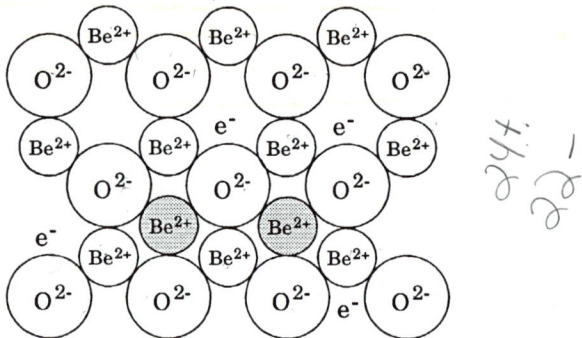

FIGURE 14-3.   Ionic arrangement in n-type cation-excess BeO. Interstitial cations are shaded; free electrons are indicated as e^-.

by the rapid movement of free electrons, BeO is an n-type semiconductor (negative charge carriers). As with cation-deficient p-type oxides, the cations diffuse outward through the scale so that new oxide forms at the scale–gas interface.

Anion-deficient n-type semiconductors, typified by ZrO_2 (actually $ZrO_{<2}$) shown in Figure 14-4, have anion vacancies as a result of a fraction of the cations supplying fewer electrons than needed for stoichiometry. While most zirconium ions are Zr^{4+} ions that determine the crystal structure, a few are Zr^{2+}. To maintain charge balance, the number of anion (O^{2-}) vacancies must equal the number of Zr^{2+} ions in the scale. Mass transport is by the diffusion of O^{2-} inward through anion vacancies to the metal surface where new oxide grows. Since current is carried through the scale by free electrons, the oxide is an n-type semiconductor.

Amphoteric Oxides.   Amphoteric oxides have either a cation or anion deficiency in their structure. An example is lead sulfide (PbS), which has minimum conductivity at its stoichiometric composition. If the composition is cation-deficient $Pb_{<1}S$ it is p-type but if it is anion-deficient $PbS_{<1}$ it is n-type.

Similarly, intrinsic semiconductors, such as CuO, have electron holes in their valence band, which have been created by free electrons that have escaped to a higher energy conduction band, so they are both n- and p-type.

Amorphous Oxides

When a metal surface first begins to oxidize, especially at low temperatures, some oxides are noncrystalline. These oxides contain more oxygen than

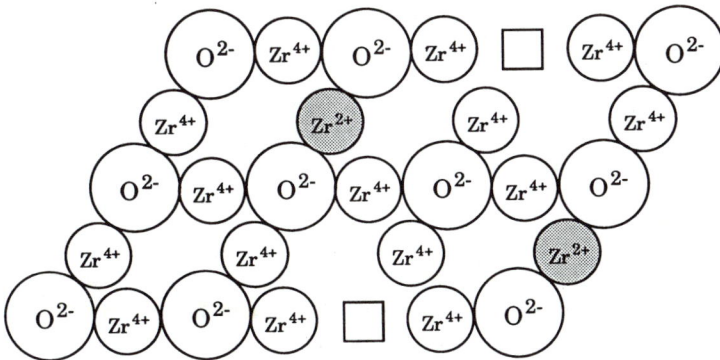

FIGURE 14-4.   Ionic arrangement in n-type anion-deficient ZrO_2. Anion vacancies are indicated as open squares; Zr^{2+} ions are shaded. Free electrons can escape from Zr^{2+} ions.

metal in their formulas, with three or four O^{2-} ions clustered around each metal ion. The triangular or tetrahedral arrangements produce random ring structures that allow large anions or O_2 molecules to move through the oxide more easily than small cations. The amorphous oxides tend to crystallize after a time. Examples are SiO_2, Al_2O_3, Ta_2O_5, and Nb_2O_5.

On the other hand, oxides with M_2O and MO formulas are apparently always crystalline and allow small cations to diffuse readily through their structures. Examples are NiO, Cu_2O, and ZnO.

Epitaxy

As the first layers of oxide grow on a metal crystal they try to align their crystal structures in a way to best fit with the metal structure. This epitaxial relationship aligns certain planes and directions in the two crystals. For example, either (111) or (001) planes of Cu_2O grow parallel to the (001) plane of Cu with the <110> directions in the Cu_2O parallel to the <110> directions in the copper.

As the epitaxial oxide grows, it builds up stresses because it is slightly distorted in trying to fit the metal substrate. At some point, however, the epitaxial oxide becomes so thick (usually 50–100 nm) that increasing stresses force the oxide to become polycrystalline and the epitaxy is gradually lost.

Preferred Orientation

Polycrystalline oxide scale does not grow evenly because some crystals are oriented better for growth and grow rapidly at the expense of more poorly oriented crystals, which gradually disappear. As the oxide develops a pre-ferred-orientation texture its surface becomes roughened by the uneven growth.

14.3 KINETICS OF OXIDATION

Linear Oxidation

If the oxide does not protect the metal surface the oxidation is reaction controlled and does not slow down with time. The Pilling–Bedworth ratio may be less than 1, the oxide may vaporize or melt, or stresses may cause the oxide to spall or crack. The oxidation rate is then constant.

$$\frac{dx}{dt} = k_L \tag{14-7}$$

where x is the mass or thickness of the oxide, t is the time, and k_L is the linear rate constant, which depends on the oxide formed, the gas composition and pressure, and the temperature. When this equation is integrated it becomes

$$x = k_L t \qquad (14\text{-}8)$$

if $x = 0$ at $t = 0$, showing that the mass of oxide formed increases linearly with time. This result is illustrated by the straight line of Figure 14-5.

Logarithmic and Inverse Logarithmic Reaction Rates

Thin protective oxide films forming at low temperatures follow either logarithmic or inverse logarithmic kinetics equations for thicknesses up to the order of 100 nm. The rate-controlling step is the transport of ions or electrons across the film in the presence of a strong electric field. The logarithmic equation is

$$x = k_e \log (at + 1) \qquad (14\text{-}9)$$

where k_e and a are constants chosen to fit the experimental data as closely as possible. The inverse logarithmic equation is

$$\frac{1}{x} = b - k_i \log t \qquad (14\text{-}10)$$

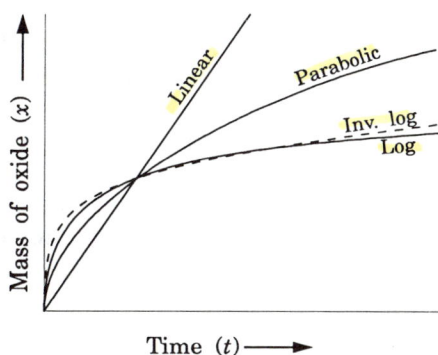

FIGURE 14-5.   Oxidation kinetics for linear, parabolic, logarithmic, and inverse logarithmic rates, drawn so that they all have produced the same amount of oxide after an initial time period.

where b and k_i are experimental constants that can be chosen to match logarithmic curves quite closely, as Figure 14-5 shows. With the thin films that form in the early stages of oxidation the differences between the two equations are almost undetectable experimentally.

Thin-Film Mechanisms

Oxidation begins by oxygen chemisorbing on the metal surface until a complete two-dimensional oxide layer covers the metal. After the monolayer forms, discrete nuclei of three-dimensional oxide begin appearing at structural defects, such as grain boundaries, impurity particles, and dislocations. Aided by surface diffusion, the nuclei grow rapidly outward at an ever-increasing rate until they form a complete film three or four monolayers thick (in the order of 1 nm). The oxidation rate then slows abruptly.

In the second stage of oxidation, visible crystalline nuclei appear uniformly spaced over the surface, with their arrangement depending on crystal orientation. They grow outward uniformly to cover the whole surface.

• "Temper colors," the colors of oxidized metal, show the thickness of the oxide film. For example, copper oxidized until it is a violet color has a film 48 nm thick. A dark blue color appears if the film is 50–52 nm.

Many theories have been advanced to explain the mechanisms of oxidation when oxide films 1–100 nm thick cover the metal in the early stages of oxidation or at low temperatures. The measured rates show that the films are too thin for the slow step to be solid-state diffusion.

The best-established theory of thin-film oxidation is one proposed by N. Cabrera and N. F. Mott, which applies to films up to about 10 nm thick. In this theory electrons from the metal pass easily through the oxide film by quantum mechanical tunneling to reach oxygen adsorbed at the oxide–gas interface where they form oxygen anions. A potential difference of about 1 V then exists between the layer of anions at the outer surface of the oxide and the metal. For a film 1 nm thick this gives a field strength of 10^7 V/cm, powerful enough to pull ions through the film. The rate-controlling step is either the transfer of ions into the film or the movement of those ions through the film. As the film grows thicker the field strength decreases until its effect becomes unimportant and the usual parabolic diffusion mechanism takes over.

Parabolic Kinetics

Protective scales (oxide layers thicker than 100–200 nm) usually obey the parabolic growth law. It describes the common situation where the oxidation is controlled by ion diffusion through a compact barrier layer of oxide. As the oxide grows thicker the diffusion distance increases and slows the oxidation because the diffusion rate is inversely proportional to the oxide thickness

$$\frac{dx}{dt} = k_p/x \tag{14-11}$$

which when integrated gives

$$x^2 = 2k_p t \tag{14-12}$$

where x can be either the increase in oxide thickness or mass during the time t, and k_p is the parabolic rate constant. The parabolic diffusion curve in Figure 14-5 shows how oxidation gradually slows down with time as the protective oxide scale thickens.

The Mechanism of Scale Formation

The parabolic rate law, developed mathematically by Carl Wagner, is the best understood and most generally accepted of all the oxidation equations. In parabolic oxidation the driving force for scale formation is the reduction in Gibbs energy (ΔG), which results from the reaction of metal ions with oxygen ions. The rate-controlling step in the oxidation is solid-state diffusion of cations outward through the scale, or anions inward, or both, with the additional aid of the electric field set up across the oxide scale layer.

For the parabolic equation to hold, the scale must be completely compact and adherent, and must be at thermodynamic equilibrium locally throughout and at its surfaces. The oxide will deviate only slightly from stoichiometry.

14.4 OXIDE SCALES

Multiple Scale Layers

A metal that oxidizes to more than one valence will form a series of oxides, usually in separate layers. For example, iron at high temperatures, with valences of +2 or +3, oxidizes to an inner layer of FeO, a middle layer of Fe_3O_4 (actually $FeO \cdot Fe_2O_3$), and an outer layer of Fe_2O_3. The most metal-rich oxide is always next to the metal, with outer layers successively lower in metal and higher in oxygen. A small concentration gradient exists across each layer, with the higher metal content closer to the metal.

If the oxygen partial pressure in the gas is below the dissociation pressure of some of the oxygen-rich, outer oxides, then only the stable, metal-rich oxides form. (Remember Eq. 14-5?) The innermost oxide is usually p-type because it is the lowest valence metal oxide with a few cations having a higher valence. In contrast, the outermost oxide is usually n-type because it contains mainly the highest valent cations. If the inner scale allows cations to diffuse outward and the outer scale allows anions to diffuse inward, then new oxide forms at the interface between the two oxides.

The relative thicknesses of the different oxide layers remains the same as they grow but usually one layer is much thicker than the others because of differences in diffusion rates of ions through the different oxide crystal structures. Where diffusion controls the growth of each layer, the total scale growth follows the parabolic rate law (Eq. 14-12).

Paralinear Oxidation

Some metals oxidize parabolically but the scale gradually and partially transforms into an outer, unprotective oxide because it sublimes, or transforms into a porous oxide, or cracks, or whatever. The inner scale remains protective and continues to grow until it reaches a thickness where its growth rate exactly equals its transformation rate. The protective layer remains, but further oxidation proceeds linearly as shown in Figure 14-6a.

Stresses in Scales

Growth Stresses.   The Pilling–Bedworth ratio is an indication of growth stresses but it does not predict their magnitude. Polycrystalline scales develop stresses along their grain boundaries when neighboring grains grow at different rates; also, short-circuit diffusion along the grain boundaries may allow more oxide to form there, increasing the compressive stresses. In addition,

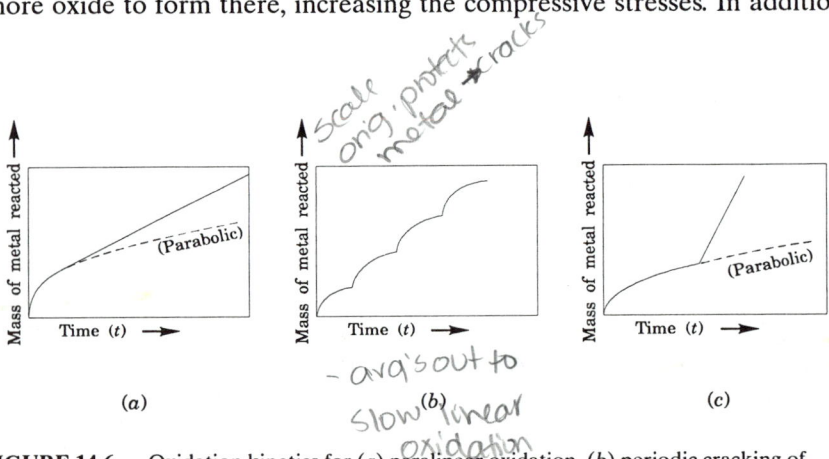

FIGURE 14-6.   Oxidation kinetics for (a) paralinear oxidation, (b) periodic cracking of scale, and (c) breakaway oxidation.

the metal may contain alloy phases or inclusions that oxidize differently from the parent metal so that high stresses are created in the oxide.

Each oxide layer varies from stoichiometry across the layer thickness and this creates stress. For example, FeO (wustite) varies from $Fe_{0.95}O$ next to the metal to as little as $Fe_{0.84}O$ at the FeO/Fe_3O_4 interface at 1370°C (2470°F). Compositional changes will also develop in an alloy as one metal oxidizes more rapidly than another or as oxygen diffuses into the metal from the oxide. These changes generate stress in the metal.

Surface geometry contributes additional stress to a growing, adherent oxide scale. Whether the stress is compressive or tensile depends on the region of growth in the scale and whether the surface is convex or concave.

In Figure 14-7a for a convex surface with new growth at the oxide–gas interface, the metal surface gradually recedes, increasing compressive stresses at the metal–oxide interface, so that the scale may shear or blister depending on how strongly it adheres to the metal.

For convex surfaces with oxide growing at the metal–oxide interface (Fig. 14-7b), stresses from the change in volume during reaction (Pilling–Bedworth ratio) push the oxide away from the metal, relaxing its compressive stresses until the outer part of the oxide surface may even be in tension. Failure can be by decohesion and buckling at the metal–oxide interface or by tensile cracking at the outer surface.

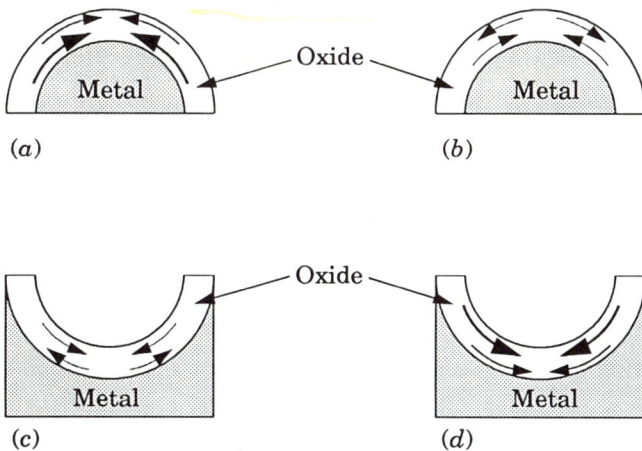

FIGURE 14-7.   Growth stresses developed by oxides growing on curved surfaces, assuming original stresses are compressive. (a) Cation diffusion from a convex surface, (b) anion diffusion to convex surface, (c) cation diffusion from concave surface, and (d) anion diffusion to concave surface.

With concave surfaces, Figure 14-7c illustrates oxide growth at the oxide–gas interface while the metal surface recedes and reduces compressive stresses near the metal, perhaps even putting the oxide there in tension if oxidation continues long enough. Radial cracks in the oxide are sometimes observed near the metal surface, or if adhesive strength is low, the scale detaches itself from the metal.

Figure 14-7d shows oxidation of a concave surface where anion diffusion inward to the metal produces new oxide at the metal–oxide interface. Oxide is pushed away from the metal surface to a smaller circumference, increasing compressive stresses at the oxide–gas interface until they exceed the cohesive strength of the oxide and cause it to shear.

Transformation Stresses. Stresses are created if (a) preferential oxidation alters the composition of an alloy so much that the remaining metal undergoes a crystallographic phase transformation; or (b) a change in temperature causes a crystallographic transformation of either the metal or oxide. The volume change accompanying a crystal transformation creates immense stresses in the oxide.

Thermal Stresses. Stresses generated by cooling or cycling temperatures are a common cause of failure of protective oxide scales. The stresses are directly proportional to the difference in thermal expansion between the oxide and the metal. A few examples of the coefficients of linear thermal expansion are given in Table 14-2. In most cases the thermal expansion of the oxide is less than the metal, so that compressive stresses develop in the oxide during cooling. A scale with different oxide layers also develops stresses at oxide–oxide interfaces.

Stress Relief. Stresses can be relieved by

TABLE 14-2   Coefficients of Linear Thermal Expansion of Some Metals and Oxides

System	Metal (μm/m · K)	Oxide (μm/m · K)	Ratio	Temperature Range(°C)
Al/Al$_2$O$_3$	27.4	7.6	3.6	20–500 (68–900°F)
Co/CoO	14.0	15.0	0.93	25–350 (77–660°F)
Cr/Cr$_2$O$_3$	9.5	7.3	1.30	100–1000 (212–1830°F)
Cu/Cu$_2$O	18.6	4.3	4.3	20–750 (68–1380°F)
Cu/CuO	18.6	9.3	2.00	20–600 (68–1080°F)
Fe/FeO	15.3	12.2	1.25	100–900 (212–1620°F)
Fe/Fe$_2$O$_3$	15.3	14.9	1.03	20–900 (68–1620°F)
Ni/NiO	17.6	17.1	1.03	20–1000 (68–1830°F)
Si/SiO$_2$	2.3	1.8	1.3	500–1000 (900–1830°F)
Ti/TiO$_2$	11.0($\perp$c)	7.2	1.5	25–1000 (77–1830°F)

Sources: Douglass, 1971; Parker, 1967; Brandes, 1983.

a. Development of porosity in the oxide as it grows.
b. Plastic deformation of the oxide, or even deformation of the metal if it is very thin.
c. Fracture of brittle oxide by tensile cracking or shear.
d. Loss of adhesion by the oxide (blistering).

If an oxide scale originally protects the metal but then fails by cracking, spalling, or the like, and re-forms, the oxidation curve looks like the one in Figure 14-6b with a series of parabolic steps that average out approximately to a slow linear oxidation. The successive time periods between failures are remarkably similar.

Pure metals immediately begin to re-form a protective oxide scale that has broken off, but many alloys cannot. The alloying element that reacted to form the original scale may be depleted at the metal surface to such an extent that a second complete protective layer will not form. Breakaway oxidation (Fig. 14-6c) then takes place when a protective scale changes abruptly to a totally unprotective one.

Logarithmic Oxidation of Scales

A logarithmic oxidation law is followed if protective scales gradually decrease their diffusion cross section because cavities or precipitates form in the oxide or because growth stresses crack the oxide in short cracks running parallel to the metal surface. Obstructions in the diffusion path slow down oxidation more than the parabolic equation would predict. (Compare the parabolic curve with the logarithmic curve in Fig. 14-5.)

The logarithmic equation is the same as Equation 14-9, but be aware that logarithmic oxidation of thin films and logarithmic oxidation behavior of thick scales are by completely different mechanisms.

Catastrophic Oxidation

Although many oxidation failures might be described as catastrophes the term "catastrophic oxidation" refers to the special situation where oxidation forms a liquid phase. It is now considered to be a type of hot corrosion, described in Section 14.6.

Catastrophic oxidation results from exposing a metal to a gas containing vapors of a low-melting oxide or from oxidizing an alloy having a component that forms a low-melting oxide. Table 14-3 lists some low-melting oxides that have caused catastrophic oxidation.

The liquid oxide usually forms at the oxide–gas surface, penetrates the scale along grain boundaries, microcracks, or pores to reach the metal, and

TABLE 14-3   Some Low-Melting Oxides

Oxide	Melting point (°C)
V_2O_5	690 (1244°F)
MoO_3	795 (1463°F)
MoO_2-MoO_3 eutectic	778 (1432°F)
Bi_2O_3	825 (1517°F)
PbO	886 (1627°F)
WO_3	1473 (2653°F)

then spreads out by capillary action, detaching the solid scale. The metal with no protective scale then oxidizes linearly in breakaway oxidation fashion (Fig. 14-6c).

• When Venezuela first began producing oil it was priced so cheaply that ships from all over the world stopped there to fill up with bunker fuel. Within months the ships' boilers and stacks were riddled. The fly ash from the oil was almost entirely V_2O_5.

Internal Oxidation

Internal oxidation, sometimes called subscale formation, is the formation of fine oxide precipitates below the surface of an alloy by oxygen that has diffused into the alloy and reacted with one of the alloy metals. Internal oxide only forms if oxygen diffuses inward faster than the reactive metal diffuses outward to the metal surface. Therefore, internal oxides sometimes form only in a certain temperature range and below some critical concentration of the reactive metal in the alloy. Diffusion is rate controlling so the advancement of a subscale layer of internal oxide precipitates is parabolic (Eq. 14-12).

14.5 OTHER GAS–METAL REACTIONS

Metals often oxidize at high temperatures with gases other than oxygen. Reactions with the gases commonly involved in fuel combustion are described in this section.

Water Vapor

Water vapor is an oxidizing atmosphere much like O_2, although water reacts with metal to form hydrogen gas as well as oxide.

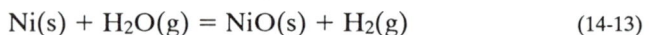

$$Ni(s) + H_2O(g) = NiO(s) + H_2(g) \qquad (14\text{-}13)$$

The reaction follows the parabolic rate law. The reaction above shows that oxide forms if the gas phase is primarily water vapor but the reaction will reverse and decompose the oxide if the gas contains a lot of H_2.

Because water vapor produces H_2 when it reacts, it is not quite as strong an oxidizer as oxygen. Steam oxidizes nickel at only about one-third to one-fifth the rate that oxygen does at 650–1050°C (1200–1925°F). However, steam oxidizes iron faster than O_2 at 700–900°C (1300–1650°F) because only a layer of FeO forms with steam, without a covering layer of the more protective Fe_3O_4, which would form in O_2.

• Iron and steel parts are sometimes blued in steam at around 500°C (1000°F) to improve wear and corrosion resistance as well as for appearance.

Carbon Dioxide and Carbon Monoxide

Carbon dioxide is usually less aggressive than air or water vapor. A typical reaction is

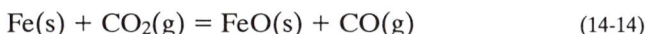

$$Fe(s) + CO_2(g) = FeO(s) + CO(g) \tag{14-14}$$

which produces carbon monoxide (CO) gas. A high CO content in the gas would slow this reaction or even drive it to the left, decomposing the oxide. However, CO itself can oxidize very reactive metals (those that form oxides more stable than CO).

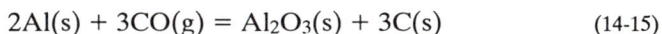

$$2Al(s) + 3CO(g) = Al_2O_3(s) + 3C(s) \tag{14-15}$$

Sulfur-Containing Gases

The gases H_2S, SO_2, S_2 vapor, and the like, usually react with metals to form sulfides, although SO_2 can also form sulfites, sulfates, or even oxides. Sulfides, in general, have more defective structures than oxides and, therefore, grow faster. Furthermore, a metal often forms many different sulfide layers that are likely to spall during cooling as they contract at different rates. Sulfides melt at much lower temperatures than oxides, making them susceptible to catastrophic sulfidation; nickel, silver, and copper are particularly prone to attack. The high-nickel alloys that are so extremely corrosion resistant and oxidation resistant are absolute disasters in sulfur-containing gases. An example of such a failure is shown in Figure 14-8.

• The British Museum, troubled by sulfidation, recently studied the problem and determined that most of their sulfur came from "bioeffluent gases" (flatulence) released by humans.

FIGURE 14-8.   Catastrophic sulfidation of Inconel 601 furnace tube in less than 1 month at 925°C (1700°F) [Reprinted with permission from G. Y. Lai and C. R. Patriarca in *Metals Handbook*, 9th ed., Vol. 13, Joseph R. Davis, Senior Editor. ASM International, Materials Park, OH, 1987, Fig. 2(a), p. 1313.

Nitrogen

Most metals nitride parabolically, with linear or paralinear nitriding being quite rare. Nitrides usually grow slower than oxides, typically one-tenth to one-hundredth as fast. Nitrides are stronger than oxides—in fact, nitrides of most important metals have metallic bonding. Stresses in nitrides are low (Pilling–Bedworth ratios are usually more favorable than for oxides) and nitrides resist thermal cycling better than oxides.

But, unfortunately, nitrides are not as thermodynamically stable as oxides; a nitrided metal exposed to oxygen or air gradually converts to an oxide.

Hydrogen

Steels are decarburized by H_2 at high temperatures (water vapor also decarburizes):

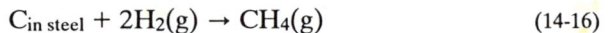

$$C_{in\ steel} + 2H_2(g) \rightarrow CH_4(g) \tag{14-16}$$

Decarburization reduces the tensile strength of the steel and makes high-temperature creep easier. Methane (CH_4) gas may form in voids within the steel and crack it.

14.6 HOT CORROSION

Hot corrosion is an accelerated oxidation–sulfidation attack on a metal when a gas (usually combusion products) forms a molten product (usually a sulfate) on the metal or oxide surface. Formerly called hot salt corrosion, hot corrosion is now broadly defined to include catastrophic oxidation. While rapidly attacking metal, hot corrosion also greatly reduces the metal's creep and fatigue resistance.

- Hot corrosion decreases the stress rupture life of one high-temperature alloy by a factor of 10^5!

Hot corrosion problems occur frequently in jet engines of aircraft that fly over seawater ingesting NaCl and Na_2SO_4, and also in all gas turbines that burn fuels containing sulfur with small amounts of sodium and potassium compounds. Boilers that burn high-ash coal or oil containing sodium vanadates, melting around 570°C (1030°F), suffer "fireside corrosion." Municipal waste incinerators are damaged by chlorides; molten carbonate fuel cells are attacked by alkali metal carbonates; fast breeder reactors produce corrosive fission products of cesium salts. All these are examples of hot corrosion.

*Isothermal Stability Diagrams

Essentially the same as potential–pH (Pourbaix) diagrams, isothermal stability diagrams have the oxidizing power of the environment graphed as a function of the environment acidity expressed in terms of the concentration of one of the acid or basic components.

Figure 14-9 plots the oxygen gas pressure (log scale) versus the log of the

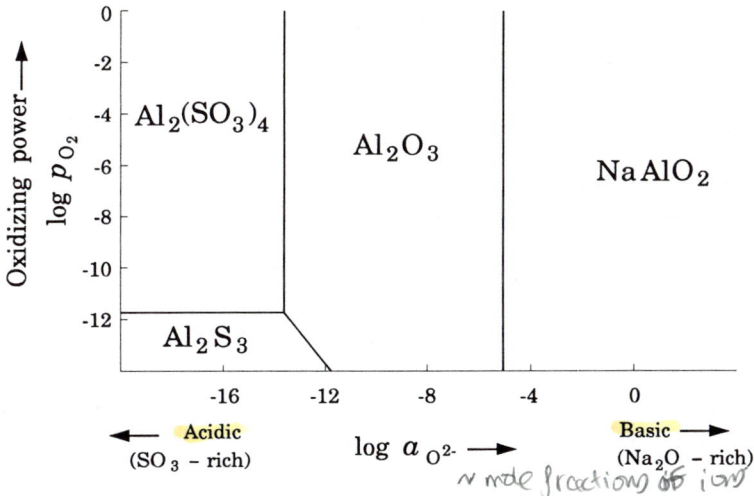

FIGURE 14-9.   Isothermal stability diagram for aluminum in Na_2SO_4 at 900°C (1650°F). Activities of soluble species equal 1.

oxide ion activity (approximately the mole fraction of the anions) in the oxide or salt. Each region is labeled with the compound predominating under those conditions. Boundaries show the conditions of equilibrium between two adjacent phases.

*Acid Fluxing

A salt, which is the reaction product of an acid and a base, can be considered to have an acid component and a basic component. The salt Na_2SO_4, as an example, can be indicated as $Na_2O \cdot SO_3$, which is neutral if the basic and acid components are balanced stoichiometrically. However, if the Na_2SO_4 has some additional oxide (Na_2O or other metal oxide) dissolved in it the salt is basic, or if SO_3 or S^{2-} is in excess the salt is acidic.

In acid fluxing, the predominant hot corrosion mechanism around 700°C (1300°F), the protective oxide scale is dissolved by molten salt that has a low oxide ion activity. The salt has become acidic by dissolving extra SO_3 or some other sulfur-containing species from the gas phase. When the oxide is attacked it goes into solution as its simple ions

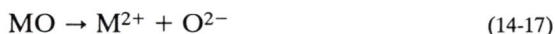

$$MO \rightarrow M^{2+} + O^{2-} \tag{14-17}$$

and the oxide ions (basic) produced will react with the acid in the salt. For aluminum oxide fluxed by acidic Na_2SO_4

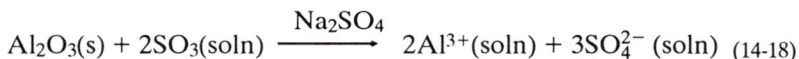

$$Al_2O_3(s) + 2SO_3(soln) \xrightarrow{\text{Na}_2\text{SO}_4} 2Al^{3+}(soln) + 3SO_4^{2-} (soln) \tag{14-18}$$

*Basic Fluxing

Basic fluxing occurs at high temperatures, mainly above 900°C (1650°F), in salt containing dissolved ash or oxides so as to consume O^{2-} ions

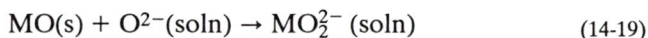

$$MO(s) + O^{2-}(soln) \rightarrow MO_2^{2-} (soln) \tag{14-19}$$

producing complex anions of the metal (aluminate, molybdate, etc.).

The oxide that dissolves at the oxide–salt interface will reprecipitate at the salt–gas interface, where its solubility is lower. (The solubility will be considerably lower if the salt is basic and the gas is acidic or vice versa.) The result is a porous, honeycomb structure (see Fig. 14-10), which gives no protection from gas or salt. The salt will undermine the adjacent scale, lift it, and crack it.

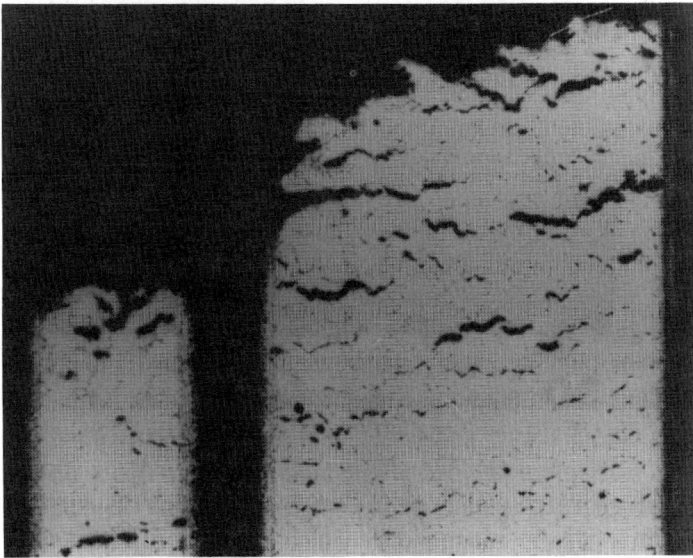

FIGURE 14-10. Aircraft engine turbine blades (*a*) damaged by hot corrosion and (*b*) cross section of damage, 35X. [Reprinted with permission from Michael L. Bauccio in *Metals Handbook*, 9th ed., Vol. 13, Joseph R. Davis, Senior Editor. ASM International, 1987, Fig. 32(*a*) and (*b*), p. 1040.]

Sulfur-Induced Corrosion

An oxide scale covered with molten Na_2SO_4 salt cannot grow by the usual mechanism of $M^+ + e^-$ reacting with O_2 because the solubility of O_2 in molten salt is quite low; the salt shields the oxide from the gas. Instead, the metal ions at the oxide–salt interface steal oxygen from the salt.

$$4M^{2+} + 8e^- + SO_4^{2-} \rightarrow 4MO + S^{2-} \tag{14-20}$$

Sulfide ions form and then migrate rapidly through the oxide via microcracks or other short-circuit paths to get to the metal, where they form metal sulfides. These sulfides may be liquid or a large-volume solid that cracks the scale. The sulfide may form pits, it may penetrate the metal along grain boundaries, and it may form internal sulfides.

Chloride-Induced Effects

Sodium chloride accelerates hot corrosion and in some cases brings it about. Sodium sulfate is not molten below 884°C (1623°F), but the Na_2SO_4-NaCl eutectic melts at around 620°C (1150°F). Chloride penetrates the metal grain boundaries to where the oxygen pressure is low. There it may form volatile chlorides that leave porosity or it can help form internal sulfides in the metal. Chloride weakens the adhesion of the oxide and increases spalling.

Study Problems

14.1   What is the rate-controlling process, if x is the mass of oxide formed in time t, when: (a) x is directly proportional to t, (b) x is directly proportional to $\log t$, and (c) x^2 is directly proportional to t?

*14.2   Suppose molten Na_2SO_4 salt attacks a Cr_2O_3 scale in hot corrosion, dissolving it as Cr^{3+} and O^{2-} ions. (a) Is the salt acidic, basic, or neutral? (b) How will the acidity of the salt change as the oxide dissolves? Explain. (c) How will the maximum solubility of the oxide change as the acidity changes? Explain.

14.3   A sample of scandium metal (density 2.99 g/cm^3 and molar mass 45.0 g/mol) has a weight gain of 14.7 mg when oxidized at 500°C (930°F) for 24 h, forming Sc_2O_3 (density 3.86 g/cm^3 and molar mass 137.9 g/mol). What weight gain would you expect in 1000 h at 500°C (930°F)?

14.4   Would barium be attacked rapidly by N_2 at high temperatures if solid Ba_3N_2 forms? Explain. Densities: Ba = 3.59 Mg/m^3; Ba_3N_2 = 4.78

Mg/m^3; N_2 = 0.44 kg/m^3 at 500°C (930°F). Molar masses: Ba = 137.34 g; Ba_3N_2 = 440.03 g; N_2 = 28.014 g.

14.5 Niobium metal can react with air to form Nb^{2+}, Nb^{4+}, and Nb^{5+} ions. At 300°C (570°F), niobium will react with air to form Nb_2O_5. (a) What oxide growth kinetics would you expect? Explain. (b) Where in the oxide scale does new oxide form? Explain.

14.6 Tin forms a protective oxide scale of SnO_2 at 200°C (390°F) (a high temperature for tin). If a coupon of tin gains 14.3 μg in 4 h, how much would you predict it would gain in 24 h of oxidation?

14.7 Tantalum oxide (Ta_2O_5) is an anion-deficient semiconducting oxide. (a) How is the majority of electrical current conducted through this oxide? Explain. (b) Calculate the Pilling–Bedworth ratio for this oxide. Molar volumes: Ta_2O_5 = 53.9 cm^3; Ta = 10.9 cm^3.

14.8 The molar volume ratio of In_2O_3/In is 2.47. Is a protective oxide likely to form on indium? Explain. Compare oxide thicknesses for 1, 10, and 100 h.

*14.9 A molten Na_2SO_4 layer on top of an oxidized metal gradually becomes acidic. Explain.

14.10 Manganese oxidizes to a protective, semiconducting scale, Mn_2O_3, in which a small fraction of the metal ions are Mn^{4+}. (a) How is current conducted through the oxide? (b) In what range does its Pilling–Bedworth ratio fall? (c) Give the rate equation for the oxidation.

14.11 Tin forms a protective oxide SnO in which a small fraction of cations has a +4 valence. (a) As a semiconductor, is the oxide n- or p-type or amphoteric? Explain. (b) Where should growth of the oxide occur? Explain.

14.12 Manganese forms a protective oxide MnO in which a small fraction of the cations has a +3 valence. (a) As a semiconductor, is the oxide n- or p-type or amphoteric? Explain. (b) Where should growth of the oxide occur? Explain.

14.13 Cobalt oxide (CoO) is a p-type semiconducting oxide with a Pilling–Bedworth ratio of 1.74. (a) Is CoO anion or cation deficient or excess? (b) Diffusion of what species is primarily responsible for growth of CoO? (c) What equation would the oxidation kinetics probably follow?

14.14 Uranium at 200°C (390°F) oxidizes by anion diffusion to form a compact inner U_3O_8 layer and a porous outer layer, also U_3O_8. (a) How would you expect the thickness of the inner layer to change with time? (b) The outer layer?

14.15 Monel (67Ni-33Cu) is heated in a 300°C (575°F) flue gas containing CO and H_2O. How will the CO/H_2O ratio affect the oxidation of the metal to NiO? Explain.

14.16 Lead (molar volume 18.26 cm^3) forms an oxide Pb$_3$O$_4$ (molar volume 75.1 cm^3. (a) Calculate the Pilling–Bedworth ratio for this oxide. (b) What oxidation rate would you expect it to follow? Why?

14.17 In 1-week oxidation tests, a lanthanum sample gained 0.0112 g due to formation of La$_2$O$_3$. What weight gain would you expect in 1 year at this temperature? Explain.

14.18 Chromium oxidizes to Cr$_2$O$_3$. How would you expect the oxidation to change with time? Explain.

14.19 At 1000°C (1830°F) silicon forms oxides SiO and SiO$_2$. (a) Is SiO$_2$ an n- or p-type semiconductor? (b) How and where does growth of SiO$_2$ occur?

14.20 Cuprous oxide (Cu$_2$O) is a semiconducting oxide that contains a small amount of divalent Cu. (a) Would the oxide be an n- or p-type? (b) Would oxide growth occur by anion diffusion, cation diffusion, or both?

14.21 Calcium forms n-type CaO. Cadmium forms n-type CdO. After oxidation for 1 h a Ca sample and a Cd sample each have a weight gain of 0.05000 g. What weight gains would you predict after 100 h? Why?

14.22 At 400°C (750°F) tin oxidizes to SnO$_2$, which contains a small fraction of Sn^{2+} ions. (a) Will the oxide be n-type, p-type, or amphoteric? (b) Oxidation is diffusion controlled. What is diffusing? By what paths?

14.23 Vanadium pentoxide (V$_2$O$_5$) is more correctly written V$_{>2}$O$_5$ since it has a cation excess. (a) Is the oxide an n-type, p-type, or amphoteric semiconductor? Explain. (b) Would oxide growth occur by anion diffusion, cation diffusion, or both?

14.24 Lithium metal (molar volume 13.0 cm^3) reacts in oxygen to form Li$_2$O (molar volume 14.8 cm^3), while in water vapor it forms LiOH (molar volume 16.4 cm^3). (a) What rate equation would you expect lithium oxidation in O$_2$ to follow? Explain. (b) Reaction of Li in water vapor does not follow a parabolic rate. What rate equation would you predict? Explain.

14.25 A common method of removing rusted nuts and bolts is to heat them with a torch. Why should this make removal easier?

14.26 Molybdenum is plentiful, cheap, has a high elastic modulus (50% > iron), and can be alloyed to get high-temperature strength. It has a melting point of 2610°C (4730°F), an atomic weight of 95.9, and a specific gravity of 10.2. Its high-temperature oxide is MoO$_3$, which has a melting point of 795°C (1463°F), molecular weight of 143.9, and specific gravity of 4.5. Estimate the suitability of molybdenum for use in air at 1100°C (1980°F).

15

Oxidation Control

No pure refractory (high-melting) metal performs well in high-temperature air. In fact, not even metals melting as low as 1850°C (3360°F) oxidize satisfactorily, with the sole exception of chromium. Table 15-1 gives the melting points of high-melting pure metals and explains the problems they undergo at high temperatures.

Because all of the pure metals with any high-temperature strength oxidize so badly, they must be protected by alloying or with a coating if they are to be usable. In either case, the object is to form a barrier layer between the metal and the gaseous environment. However, while alloying may improve the oxidation resistance of metals with melting points above iron [1535°C (2795°F)], it always worsens their mechanical properties. Alloying does provide excellent protection for iron-, cobalt-, and nickel-based metals. Unlike alloying, protective coatings can be used on any metal.

15.1 ALLOY THEORY

Alloying to Improve the Oxide

Parabolic oxidation provides the best protection for a metal exposed to a reactive gas. The oxidation rate slows with time and may become negligible if diffusion rates are slow through the protective oxide and if stresses remain low. Slowing the diffusion rate by modifying the defect structure is an obvious approach that has been moderately successful.

The p-type semiconducting oxides, such as NiO, are cation deficient because a few higher valent cations are present in their structures. The more cation vacancies they have, the easier diffusion becomes, and the faster the metal oxidizes. To reduce the number of cation vacancies, lower valence

313

TABLE 15-1   Oxidation of Pure Metals

Metal	Melting Point (°C)	Oxidation Problems
Tungsten	3433 (6211°F)	Catastrophic ox. Paralinear >500°C (950°F)
Rhenium	3106 (5623°F)	Catastrophic oxidation
Osmium	3032 (5490°F)	Embrittled; oxide volatile
Tantalum	3020 (5468°F)	Breakaway ox >500°C (950°F)
Molybdenum	2623 (4753°F)	Catastrophic oxidation
Niobium	2469 (4476°F)	Embrittled; breakaway, linear ox
Iridium	2447 (4437°F)	Embrittled; oxide volatile
Ruthenium	2334 (4233°F)	Embrittled; oxide volatile
Hafnium	2231 (4048°F)	Embrittled. Linear ox >800°C (1500°F)
Rhodium	1963 (3565°F)	Embrittled; oxide volatile
Vanadium	1910 (3470°F)	Catastrophic oxidation
Chromium	1863 (3385°F)	Embrittled by N_2
Zirconium	1855 (3371°F)	Embrittled

cations, such as Li^+, should replace some of the Ni^{2+} in the oxide structure. Two Li^+ ions in the oxide eliminate one cation vacancy to maintain charge balance; the two Li^+ ions counteract two Ni^{3+} ions that would otherwise cause a cation vacancy to form.

Thus, alloying nickel with a small amount of lithium should slow the oxidation, although the amounts of Li used must be small enough that the solubility limit of Li in NiO is not exceeded. Otherwise Li will start forming its own oxide (and Li_2O with its Pilling–Bedworth ratio of 0.57 is not desired). The solubility limits in oxides are generally quite low so that the alloying, or "doping," is severely limited.

- "Doping" is a term borrowed from semiconductor production to describe the deliberate contamination of high-purity semiconductors with extremely small amounts of certain impurities that greatly improve their performance.

Doping a p-type oxide like NiO with higher valent cations, such as Al^{3+}, would be equivalent to adding more Ni^{3+}, thus producing more cation vacancies and making oxidation worse. Alloying with cations having the same valence, such as Co^{2+}, has little effect.

The n-type oxides behave exactly opposite to the p-type. For the n-type oxides that grow by diffusion of anions inward through anion vacancies, typified by ZrO_2, the anion vacancies exist because some cations have a lower valence (Zr^{2+}) than they should have for a perfect oxide structure. Doping the alloy with a metal that will form a higher valence (perhaps V^{5+}) should eliminate some of the anion vacancies and slow oxidation.

With n-type oxides that grow by interstitial diffusion of cations, such as BeO, substituting Al^{3+} for some of the Be^{2+} in the scale leaves the oxide with excess electrons that tend to prevent the metal from ionizing at the metal–oxide interface.

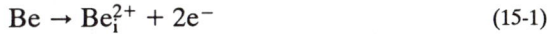

$$Be \rightarrow Be_i^{2+} + 2e^- \qquad (15\text{-}1)$$

The Al^{3+} ion reduces the number of interstitial (subscript i) cations, Be^+, and slows oxidation. On the other hand, doping the metal with an element that forms lower valent cations would just encourage the metal to ionize.

Although this "doping principle" has helped verify the Wagner mechanism of parabolic oxidation, it is very limited in usefulness for developing oxidation-resistant alloys. First, low solubility strictly limits the concentration of foreign cations that can go into solid solution in the oxide. Second, the choice of foreign cations that can be used is restricted by their valence. For p-type oxides, the alloying element should have a lower valence than the metal being oxidized, but most p-type oxides are produced by metals with +1 and +2 valences. The n-type oxides form on metals with high valences but very few alloying elements will have even higher valences, which are needed to slow the oxidation rate.

Alloying for a Different Oxide

Alloying a metal with a large amount of a second metal may allow the added metal to form its own, more protective oxide. The alloying metal may form the only oxide, it may form an oxide in competition with the oxide of the host metal, or it may form a ternary oxide along with the host metal.

To explain the principles simply, the following discussions assume that the alloy is a binary alloy of metals A and B, which each form only a single oxide, AO or BO. Furthermore, the BO oxide is more stable than AO; that is, metal B is more reactive than metal A.

Selective Oxidation.   Selective oxidation is the oxidation of only one component in an alloy. It may be that A is too noble to be oxidized under the given conditions. Since AO is not stable, only BO forms. Alloying a reactive metal, such as iron, with a noble metal, say gold, does not prevent the iron from oxidizing.

However, if both metals theoretically are oxidizable by the gas (B always more so than A), the concentration of one metal may be too low to oxidize. For the reaction

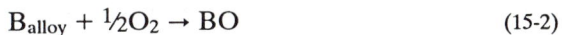

$$B_{alloy} + \tfrac{1}{2}O_2 \rightarrow BO \qquad (15\text{-}2)$$

the driving force is

$$\Delta G = \Delta G^\circ + RT \ln \frac{1}{N'_B \cdot p_{O_2}^{1/2}} \tag{15-3}$$

where ΔG is the Gibbs energy change, ΔG° is the Gibbs energy change at standard conditions, the activity of solid BO is approximately 1 (invarient), the mole fraction of B at the alloy surface is N'_B, and the O_2 pressure is p_{O_2}. Equation 15-3 shows that if N'_B is very small, smaller than some critical concentration N^*_B, ΔG will be positive even though ΔG° is negative, and although B is a reactive metal it will not oxidize. The oxide AO does form because the concentration of A in the binary alloy is high.

As element A oxidizes, its concentration becomes depleted at the metal surface while the concentration of B increases correspondingly. If diffusion in the alloy is so slow that N'_B increases until it reaches N^*_B, BO could eventually form.

Alloying then can prevent the formation of a poorly protective scale and allow the formation of a better oxide, even a less stable one, if the alloying element's concentration is sufficiently high.

- One famous alloy was developed that oxidized only 40% as fast as the competition (Hastelloy X) in dry air at 1100°C (2010°F). But it later turned out to oxidize 500% faster than Hastelloy X in moist air. Therefore it could not handle combustion gases, which always contain water vapor.

Competing Oxides. Both A and B may oxidize if they are above their critical concentrations. Some oxides are completely immiscible; that means that A ions cannot substitute for B ions in the BO oxide (or B ions in AO) but also that A ions cannot even diffuse through a layer of BO. Originally, both AO and BO will grow on the alloy surface but at different rates, so let us assume that BO, the more stable oxide, grows very slowly. Figure 15-1*a* shows this situation. If diffusion in the alloy is rapid enough to prevent A from becoming seriously depleted at the metal surface, AO will continue to grow until it overwhelms the BO so that the situation in Figure 15-1*b* develops: an AO scale with BO islands scattered through it.

However, if the alloy surface becomes depleted of A, the growth of AO slows while BO continues to grow until it forms a complete layer underneath the AO (Fig. 15-1*c*). Pockets of AO at the metal–oxide interface are gradually eliminated by the displacement reaction.

$$AO + B_{alloy} \rightarrow BO + A_{alloy} \tag{15-4}$$

because BO is thermodynamically more stable than AO. Once a barrier layer of slow-growing BO forms, oxidation of the alloy is greatly slowed.

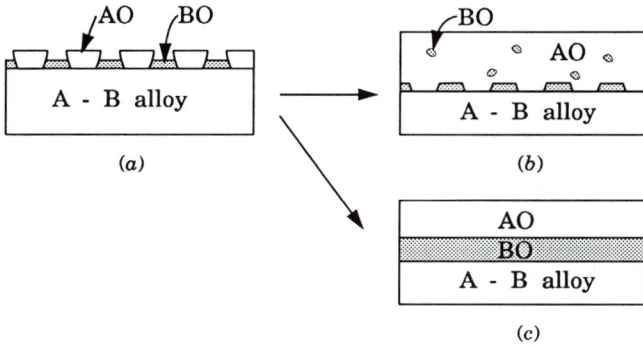

FIGURE 15-1.   Simultaneous growth of competing oxides. (*a*) Both oxides nucleate. (*b*) Later stage if diffusion in alloy is rapid. (*c*) Final stage if diffusion in alloy is slow.

Oxides that are quite miscible in each other will form an (A,B)O scale, although BO (or AO) will precipitate within the scale if its solubility limit is exceeded. If B ions diffuse much more rapidly through the scale than A ions, an outer layer of BO may eventually form.

Protective Double Oxides.   Double oxides are being developed by extensive research because they generally grow very slowly. The most important of these are the spinels, which are double oxides with the general formula MO · Me$_2$O$_3$ having the crystal structure of the mineral called spinel (MgO · Al$_2$O$_3$). The iron oxide Fe$_3$O$_4$ (or FeO · Fe$_2$O$_3$), for example, has an inverse spinel structure that provides good protection. On iron–chromium alloys the spinel can be either stoichiometric FeO · Cr$_2$O$_3$ (also written FeCr$_2$O$_4$) or a Fe$_{3-x}$Cr$_x$O$_4$ solid solution. Although many double oxides tend to be brittle, research now shows that minor alloy additions can improve their high-temperature mechanical properties.

15.2 HIGH-TEMPERATURE ALLOYS

Table 15-2 gives approximate scaling temperatures of metals used in high-temperature service and maximum service temperatures of ceramics. These temperatures are only guidelines. Pollutants in the air (notably sulfur-containing gases), molten salts, stresses on components, and thermal cycling all tend to reduce the practical service temperatures.

• Intermittent service temperatures of ferritic and martensitic stainless steels are actually *higher* than for continuous service!

High chromium contents, which most high-temperature alloys have, pro-

TABLE 15-2   Approximate Scaling Temperatures in Air

Alloy	Scaling Temperature[a]	
	(°C)	(°F)
1010 Steel	480	900
5Cr-0.5Mo Steel	620	1150
9Cr-1.0Mo Steel	620	1150
70Cu-30Zn Brass	700	1300
410 Stainless (12Cr)	730	1350
Hastelloy B	760	1400
Nickel	790	1450
430 Stainless (17Cr)	830	1530
18–8 Stainless (302, 304, 321, 347)	900	1650
316 Stainless	900	1650
Chromium	900	1650
442 Stainless (21Cr)	930	1700
446 Stainless (25Cr)	1040	1900
N-155 (Fe-base superalloy)	1040	1900
309 Stainless	1090	2000
HW (12Cr-60Ni-bal Fe)	1120	2050
310 Stainless	1150	2100
HS-21 (Co-base superalloy)	1150	2100
Hastelloy C	1150	2100
HT (15Cr-35Ni-bal Fe)	1150	2100
HX (17Cr-66Ni-bal Fe)	1150	2100
HU (19Cr-66Ni-bal Fe)	1150	2100
Hastelloy X (Ni-base superalloy)	1200	2200
Ceramics		
Si_3N_4 and SiC ceramics	(1450)[b]	(2650)
Alumina and chromia ceramics	(1750)	(3200)
Yttria and calcia ceramics	(2000)	(3600)
Thoria ceramics	(2500)	(4500)

Source: Adapted from Fontana, 1987; Elliott, 1989.
[a] Scaling temperature at which the material shows a weight gain of 3 g/m² h, generally considered negligible.
[b] Temperatures in parentheses are approximate maximum service temperatures.

vide good oxidation resistance if temperatures do not exceed 950°C (1750°F) for long periods of time. Above this temperature the chromium may vaporize and form CrO_3 instead of a barrier layer of Cr_2O_3 or M_2CrO_4.

Aluminum as an alloying element provides excellent protection by forming Al_2O_3, which is thermodynamically more stable and has a much lower vapor pressure than Cr_2O_3. When both chromium and aluminum are present in an alloy they compete to form the surface scale. For example, an alloy containing 5% Cr and 5% Al will tend to form an alumina scale rather than chromia. Alumina-forming alloys resist halogen vapors and sulfur-containing gases particularly well.

Silicon, like aluminum, works with chromium to improve resiliency of the scale. Silica in the scale forms a glassy surface that is extremely protective and resistant to carburizing and sulfidizing atmospheres.

For alloy steels, which have much higher thermal expansion than plain carbon steels, large amounts of nickel help to lower the thermal expansion and improve compatibility with other metals in an assembly.

Stainless Steels

Both cast and wrought austenitic stainless steels are widely used in the temperature range of 800–950°C (1500–1750°F). The Fe-Cr-Ni alloys have relatively low cost, good mechanical properties, and acceptable oxidation rates. If austenitic stainless steels have their chromium depleted by extensive oxidation, they become magnetic, making this property a useful indicator of the severity of attack.

Sensitization of austenitic stainless steels in the 500–800°C (900–1500°F) range, while greatly increasing susceptibility to intergranular aqueous corrosion, has no effect on high-temperature oxidation. Severe overheating for long periods of time, called "burning," can produce incipient grain boundary melting or intergranular oxidation.

Martensitic stainless steels with 12–14% Cr are resistant to scaling up to about 700°C (1300°F). However, they can become embrittled if tempered incorrectly or cooled slowly through the embrittlement range, which for Type 410 is 370–600°C (700°–1100°F).

Nickel Alloys

Nickel-base alloys receive oxidation resistance from their 15–20% chromium, which forms Cr_2O_3 and $NiCr_2O_4$ rather than NiO. Since high chromium interferes with precipitation hardening, the compositions are a compromise between acceptable oxidation rates and necessary strength. Interstitial carbon gives solid solution strengthening but reduces oxidation resistance so that it is usually kept low, as with austenitic stainless steels.

Superalloys

Superalloys are heat-resistant iron-, nickel-, or cobalt-base alloys that contain chromium for resistance to oxidation and hot corrosion, as well as other alloying elements for high-temperature strength. The maximum service temperature for conventional superalloys in structural applications may be no higher than 950°C (1750°F), although it may exceed 1200°C (2200°F) if the metal is not under load. Overheating will cause melting, the removal of strengthening phases by dissolving them in the solid solution, or extensive oxidation or corrosion.

Superalloys include the Hastelloys, Inconels, Nimonics, and Waspalloy. Some of the alloying elements in superalloys improve strength by solution strengthening or precipitation hardening while resistance to oxidation and corrosion is imparted by chromium, aluminum, and silicon.

Intermetallic Compounds

Intermetallic compounds possess excellent high-temperature strength with superior creep resistance and low density. Those that can selectively oxidize to form the highly protective Al_2O_3, SiO_2, or BeO oxides (see Fig. 15-2) are currently being evaluated for use in the next generation of jet engines and rockets.

The intermetallic $NbAl_3$, for example, selectively oxidizes to form a protective Al_2O_3 scale, but in doing so the Al become so depleted at the metal surface that the next-lower aluminide ($NbAl_2$) forms under the scale. This compound eventually converts to an $NbAlO_4$ inner layer. Intermetallics generally have lower diffusivities than other alloys and also lower oxygen solubilities, so brittle subscale of internal oxide is less likely.

While intermetallic compounds have much tighter stoichiometry than other alloys, the compositions are generally broad enough to allow some significant alloying. Adding Cr to Ni_3Al or $NbAl_3$, for example, promotes Al_2O_3 scale formation at substantially lower Al contents. Boron additions decrease the low-temperature brittleness of Ni_3Al while Ti improves its oxide scale adherence by increasing Al_2O_3 "keying," that is, penetration along grain boundaries. Alloying elements, however, in some cases may lower the melting point of an alloy or its protective oxide and create a problem.

"Pesting" has been reported as a problem for over 50 intermetallic compounds, although some compounds appear to be immune. In a certain temperature range, often about 700–900°C (1300–1650°F) but which is different for each compound, catastrophic oxidation attacks the grain boundaries of the alloy after an incubation period. The intergranular oxidation produces a voluminous oxide powder and causes the individual metal grains to crumble.

15.3 COATING REQUIREMENTS

In designing a coating to protect a metal from hot gases, the coating must satisfy a long list of requirements.

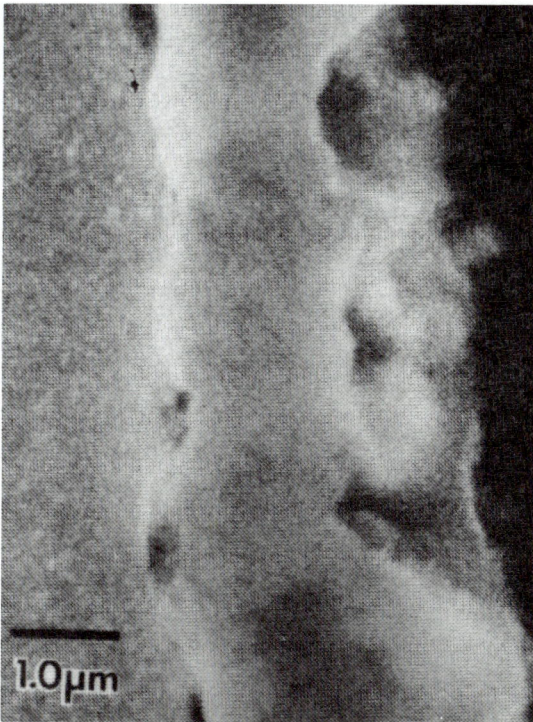

FIGURE 15-2.   Thin, continuous BeO scale formed on NiBe by oxidation in air at 1000°C (1800°F) for 100 h. [Reprinted with permission from J. S. Lee and T. G. Nieh, Figure 3 of *Oxidation Behavior of Nickel Beryllide, in Oxidation of High-Temperature Intermetallics*, p. 274 T. Grobstein and J. Doychak, Eds., The Minerals, Metals and Materials Society, 420 Commonwealth Drive, Warrendale, PA 15086, 1989.]

1. The coating must be stable in the environment.
2. It must adhere well to the metal.
3. Mobilities of reactants through the coating should be low.
4. The coating should not react with the substrate metal to worsen the mechanical properties of either the coating or the metal. Interdiffusion between the two should be slow, although a bit of interdiffusion does improve adherence.
5. The thermal expansion of the coating should be close to that of the metal's.
6. The coating must withstand creep and plastic deformation.
7. The coating should be able to resist impact, erosion in a gas stream, and abrasion.

8. It is desirable, although not a necessity, that the coating should be self-repairing, or at least that defects can be easily repaired.

Coatings may be refractory oxides or they may be metals or compounds that will produce a refractory oxide layer when reacted with the gas.

15.4 OXIDE COATINGS

The metal oxides with high enough melting points and low enough vapor pressures to be generally considered suitable as high-temperature coatings on metals are SiO_2, TiO_2, Al_2O_3, La_2O_3, Y_2O_3, Cr_2O_3, BeO, CaO, ZrO_2, MgO, ThO_2, complex oxides, and spinels. Of these, La_2O_3, BeO, and CaO hydrate rapidly in air. Pure zirconia (ZrO_2) undergoes a crystallographic transformation at 1001°C (1834°F), which would cause loss of adherence, but the structure can be stabilized with CaO, MgO, or Y_2O_3 additions.

For good protection the coatings must restrict diffusion of oxygen and metal ions. Diffusion is lowest in ceramics that bond ionically and deviate only slightly from stoichiometry. Calcia, magnesia, and alumina have particularly low diffusion rates. The spinel oxides, discussed in Section 15.1, often keep diffusion two orders of magnitude slower than simpler oxides, probably because of the complex diffusion paths necessary for ions to get through the structure.

It is difficult to produce compact, pore-free, well-adhering oxide coatings on metals. Because of the innate defects in coatings and danger of damage by erosion or impact, oxide coatings are seldom applied alone. An undercoating of an oxidizable metal provides the self-healing properties so desirable.

Ceramic coatings are used extensively in gas turbines, because while protecting the metals from oxidation they transfer heat so slowly that they also reduce the metal temperature 50–200°C (100–400°F). This slow heat transfer improves performance by reducing cooling air requirements and reduces fabrication costs by eliminating elaborate designs for cooling. Most importantly, the ceramic coatings can improve efficiency by allowing increased turbine inlet temperatures.

Plasma-sprayed zirconia coatings are favored because they are very refractory, with low thermal conductivity and high thermal expansion. Typically, an inner oxidation-resistant and hot-corrosion resistant alloy, such as MCrAlY, provides a bond layer approximately 0.1 mm (4 mils) thick that is coated with an outer layer of stabilized zirconia (ZrO_2-12 wt% Y_2O_3) 0.1–0.5 mm (4–20 mils) thick. The most serious problem with ceramic coatings is their tendency to spall during thermal cycling.

15.5 OXIDIZABLE COATINGS

The most successful high-temperature coatings have been scale-resistant metals that react slowly with the environment to form an additional barrier layer of oxide.

Deposition of coatings (by aqueous electroplating, electroless plating, roll cladding, etc.) can be made at low temperatures that prevent interdiffusion of coating and substrate, or with interdiffusion at high temperatures (by diffusion from powders, hot dipping, flame spraying, etc.). Low-temperature deposition is sometimes followed by a diffusion anneal to improve adherence. If the operating temperature is to be above the melting point of the coating, a diffusion anneal is necessary. For example, spray aluminizing is diffused into the substrate metal before high-temperature service.

With high-temperature deposition the interdiffusion can be reduced by first coating the metal with a thin barrier layer of high-melting metal and then with the coating material.

To prevent evaporation of a coating at high temperatures, the coated metal may be preoxidized at a low temperature to form a continuous, compact, oxide layer.

Diffusion Coatings

A coating metal that diffuses into the substrate to alloy and form intermetallic compounds gives excellent adherence while selective oxidation of the intermetallic phases takes place. For example, silicides that are by far the most-used coatings on refractory metals, are capable of service at over 1700°C (3100°F). Silicon coatings on molybdenum diffuse inward to form successive layers of $MoSi_2$, Mo_5Si_3, and Mo_3Si. The most protective layer is the one that has the largest amount of the element that forms the protective film, which for Mo silicides is the $MoSi_2$ layer that will oxidize to SiO_2 as the main oxide. The inner layers containing less silicon will oxidize progressively faster—a much safer coating than one that fails abruptly.

Silicide coatings have been widely used on molybdenum, niobium, tantalum, and tungsten (see Fig. 15-3) but not without some problems. Silicon dioxide melts at 1713°C (3115°F), considerably lower than many other oxides. Also, at low oxygen pressures a complete protective layer of SiO_2 may not form, leaving the metal susceptible to pinhole attack. Below 1200°C (2200°F) silicides are less protective than at higher temperatures because both the silicon and refractory metal oxidize.

Zinc coatings are frequently used on niobium even though zinc melts at 420°C (787°F). By diffusion of zinc inward, layers of $NbZn_3$, $NbZn_2$, $NbZn_{1.5}$, and $NbZn$ form to give gradually decreasing levels of protection. Oxidation

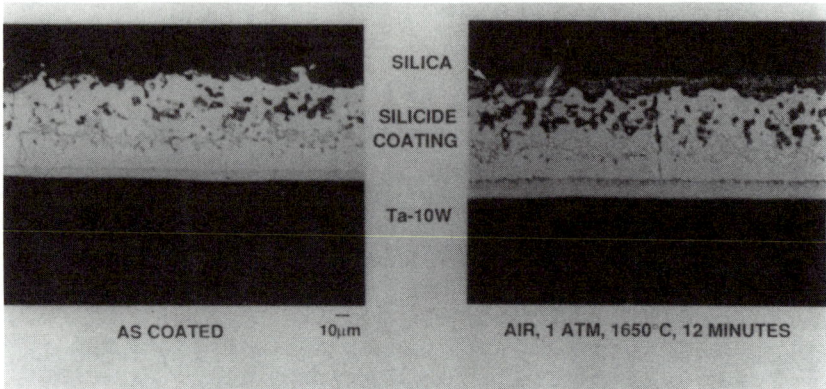

FIGURE 15-3.   Silicide coating on Ta-10W, (*a*) as coated, (*b*) after oxidation in air, 1 atm, 1650°C (3000°F) for 12 min. [Reprinted with permission from C. M. Packer, "Overview of Silicide Coatings for Refractory Metals," in *Oxidation of High-Temperature Intermetallics*, T. Grobstein and J. Doychak, Eds., The Minerals, Metals and Materials Society, 420 Commonwealth Drive, Warrendale, PA 15086, 1989.]

is mainly by parabolic growth of ZnO, limited by the diffusion of Zn through the scale, which is slow even at high temperatures. The self-healing properties of the scale are excellent.

Spray aluminizing has long been used to protect steel at temperatures up to 900°C (1650°F) in air. Figure 15-4 illustrates the various stages that the diffusion coating undergoes in preparation and in service. The original aluminum coating 150–200 μm thick is protected with a layer of water glass or lacquer so that it will not oxidize during the diffusion anneal. In service this layer gradually evaporates and the coating becomes protected by a thin Al_2O_3 layer. Interdiffusion of Fe and Al first allows a series of Al-Fe intermetallics to form and gradually the development of an Al-Fe solid solution (15% Al), which recrystallizes. An example of aluminizing is compared with chromizing and siliconizing in Figure 15-5.

Aluminizing is now being used on Ni- and Co-based superalloys. The diffusion coatings form NiAl and CoAl that oxidize slowly to the very protective Al_2O_3. For sulfurous gases up to 1000°C (1800°F), a Ni-Cr alloy coating, such as Ni-20Cr (Nichrome), 375 μm thick followed by an outer 100 μm of sprayed aluminum is effective. During heat treatment or service the Al diffuses into the Ni-Cr alloy.

Overlay Coatings

Overlay coatings applied to a substrate involve a minimum of interdiffusion in order to prevent problems of dilution of the protective reactants or the

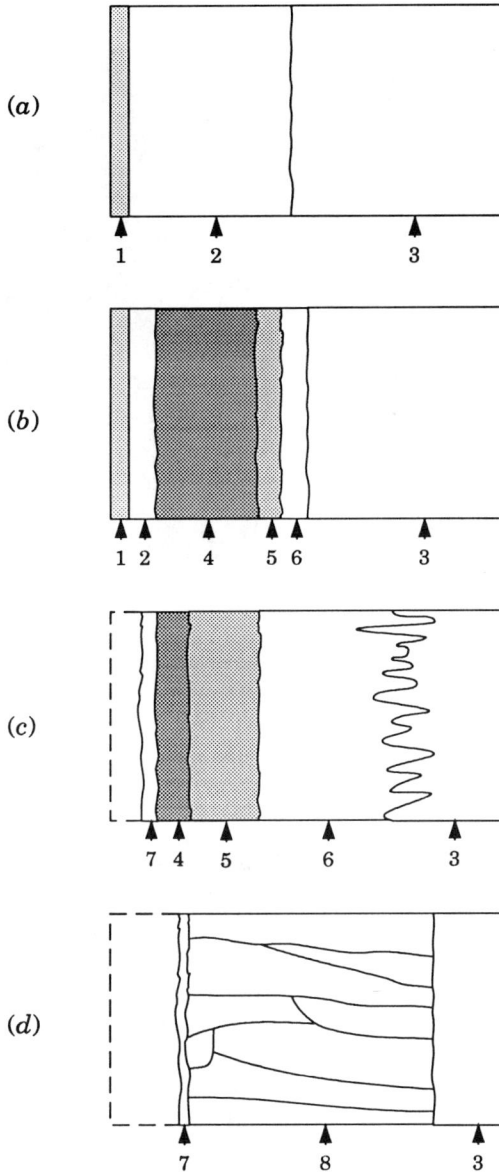

FIGURE 15-4. Spray aluminized coating on steel: (*a*) original coating, (*b*) after initial diffusion anneal, (*c*) after thermal treatment or service, and (*d*) after long service at 950°C (1750°F). (1) protective water glass or lacquer layer; (2) Al-sprayed layer; (3) steel substrate; (4) mixture of Al_3Fe in a solid solution of Fe in Al; (5) layer of brittle Al_3Fe, Al_3Fe_2, and Al_2Fe; (6) solid solution of Al in Fe with Al_2Fe precipitates; (7) Al_2O_3 layer; and (8) solid solution of Al in αFe. [After S. Mrowec and T. Werber, *Gas Corrosion of Metals*, National Center for Scientific, Technical and Economic Information, Warsaw, 1978, Fig. 16.1, p. 514.]

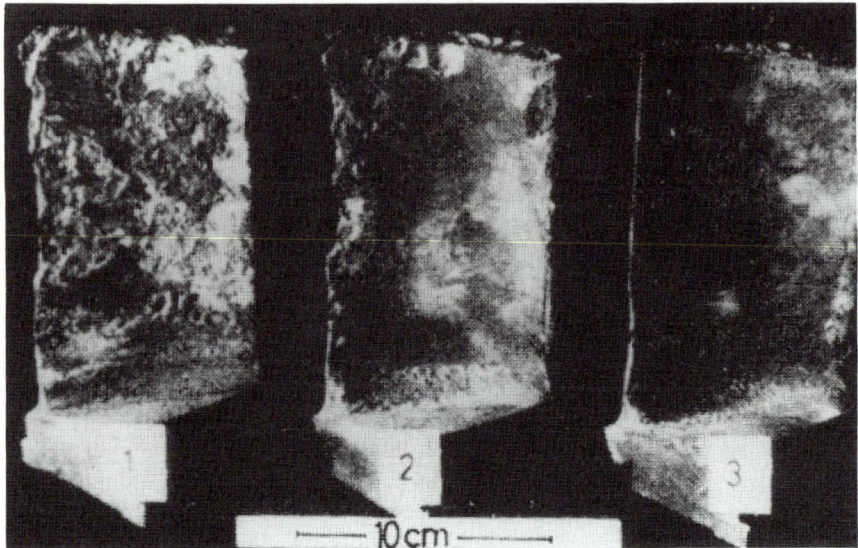

FIGURE 15-5.   Turbine blades after 30,000 h at 710°C (1310°F): (1) chromized, (2) aluminized, and (3) siliconized. [Robert A. Rapp, ed., *High Temperature Corrosion*, NACE-6, National Association of Corrosion Engineers, 1983, Fig. 10, p. 608. Copyright © by NACE.]

worsening of mechanical properties of either the substrate or the coating by
interreaction.

The most important type of overlay coating at present is MCrAlY, with
the M representing Fe, Ni, Co, or their mixtures. Fast-growing Cr_2O_3 first
protects the alloy and is gradually converted to the more stable Al_2O_3, as Al
substitutes into the Cr_2O_3 structure. The small amount of yttrium in the alloy
is included to improve adherence of the Al_2O_3 scale. A 375 μm MCrAlY
coating, such as NiCoCrAlY, is used as an overlay for aircraft turbine parts.
It is excellent in sulfidation environments, such as produced by the burning
of high-sulfur fuels. The MCrAlY coating also holds up well in hot corrosion
situations, especially if Cr is increased and Y decreased. An FeCrAlY coating
is sprayed on tungsten alloy fibers in fiber-reinforced superalloys used for
turbine blades, preventing catastrophic oxidation of the fibers.

Metal carbides also serve as oxidizable barrier coatings. Carbides are
thermodynamically less stable than oxides and gradually convert to a protective oxide at high temperatures. For example, silicon carbide coatings are
used on graphite and on carbon–carbon composites (graphite fibers in a
carbonized resin matrix). The SiC coating forms a silicon oxide surface that
increases the O_2 resistance to 1650°C (3000°F) from 500°C (950°F) uncoated.

Study Problems

15.1 The copper tubes in a boiler developed a Cu_2O scale on the steam side and CuO scale on the fire side. If the tubes are replaced with Alloy 182 (Cu + 1% Cr), any Cr_2O_3 that formed would be completely dissolved in the copper oxides. How would that affect scaling of the tubes? Explain.

15.2 Alloying titanium with niobium improves the oxidation resistance by a factor of up to 30, even though the oxide formed is still the rutile (TiO_2) structure, and not Nb_2O_5. Explain.

15.3 A Mg-Ti alloy oxidizes in a gas where both MgO and TiO_2 can form. Magnesium oxide is more stable and grows faster than TiO_2. The two oxides are immiscible. After extensive oxidation, how would the two oxides be arranged in the scale? Explain.

15.4 A new ferritic stainless steel is proposed, similar to Type 430 (15% Cr) but containing 0.5% Ti to stabilize it. At high temperatures the oxide scale is primarily Cr_2O_3 with some Fe^{3+} and Ti^{4+} dissolved in it. The oxide grows mainly at the gas–scale interface, and simple calculations show that all ions are far larger than the interstitial spaces in the oxide crystal structure. Compare this modified 430 with the regular 430 for resistance to high-temperature oxidation. Explain.

15.5 An A-B alloy oxidizes, forming a thin layer of BO covered by a much thicker layer of AO. The two oxides are completely immiscible. Thermodynamically, BO is more stable than AO. Explain how the oxides formed the way they did.

15.6 Tantalum oxide (Ta_2O_5) is an anion-deficient semiconducting oxide. How will the growth rate be affected if small amounts of Cs^+ are dissolved in the oxide?

15.7 An A-B alloy oxidizes to form a thick layer of BO covered by a much thinner layer of AO. The two oxides are partly miscible. Thermodynamically, BO is more stable than AO. Explain how the oxides formed the way they did.

15.8 How would you expect a small amount of molybdenum impurity to affect the oxidation of cadmium? Explain.

Oxide	Type	Volume Ratio
CdO	n	1.42
MoO_2	p + n	2.10
MoO_3	n	3.27

15.9 An A-B alloy oxidizes. (The BO oxide is more stable than AO.) For the first week only AO forms but then BO begins forming underneath the AO layer. (a) In general, what can you say about the relative concentrations of A and B in the original alloy? (b) Is AO an n- or p-type oxide? Explain. (c) Explain why no BO formed in the first week.

15.10 PdO contains a small amount of Pd^{4+} in its structure. (a) Is the protective oxide n- or p-type? (b) What will be the rate-controlling step in PdO scale formation? (c) How will additions of V^{5+} to the structure affect oxidation? Explain.

15.11 An aluminum–manganese alloy is oxidized. The resulting oxide is largely MnO (molar volume 13.15 cm^3) with pockets of Al_2O_3 (molar volume 25.6 cm^3) within the scale. (a) Explain how this scale formed. (b) Which oxide is more stable? Explain.

15.12 The rate of oxidation of aluminum at 350°C (660°F) depends on the rate that aluminum ions diffuse through the n-type Al_2O_3 oxide layer. Explain how addition of a small amount of silver to the aluminum would affect the oxidation of the metal. (Pure silver forms p-type Ag_2O.)

15.13 An A-B alloy is oxidized. The B metal is just below its critical concentration in the alloy. The oxygen partial pressure in the atmosphere is greater than the equilibrium partial pressure of either AO or BO. The AO oxide is a p-type oxide that is less stable than BO, which is n-type and anion deficient. Diffusion through the oxides is much faster than diffusion through the alloy. (a) What oxide(s) form(s) after a long oxidation period? (b) Where does growth occur?

15.14 Under what conditions would it be possible for an Ag-Cu alloy to oxidize with a layer of Ag_2O only? The Ag_2O layer is thermodynamically less stable than CuO.

15.15 Nickel is alloyed with yttrium. Each metal can form only one stable oxide, with Y_2O_3 being more stable than NiO. How would the following conditions affect which oxide forms? (a) The relative concentrations of Ni and Y in the alloy, (b) the diffusion rates of Ni and Y in the alloy, and (c) the oxygen pressure.

15.16 Molybdenum melts at 2623°C (4753°F) and would be an excellent metal for 1000°C (1800°F) service except that it cannot be used in air above 725°C (1350°F) without a protective coating. Aluminum coatings are now being used. (a) Why can Mo not be used unprotected above 725°C (1350°F)? (b) Since aluminum melts at 660°C (1220°F), how can it provide protection? (c) Would the coating be self-healing? Explain.

15.17 At 350°C (630°F) the oxidation of aluminum is controlled by interstitial diffusion of aluminum ions through the αAl_2O_3 layer. (a) Is the oxide n- or p-type? Why? (b) Will a small amount of Si^{4+} in the oxide layer increase or decrease oxidation? Why? (c) If the oxide layer is x μm thick in 1 day how thick will it be in 2 days? In 3 days?

15.18 How can a silicon coating protect tungsten from high-temperature oxidation? Silicon melts at 1713°C (3115°F), reacts readily with air at all temperatures, and reacts with tungsten at high temperatures to form WSi_2.

15.19 Nickel oxidizes parabolically at high temperatures to form p-type NiO. Addition of 1% Cr_2O_3 to the NiO decreased the oxide conductivity to one-one thousandth of its former value. (a) Explain how the Cr_2O_3 addition would affect the oxidation rate of the nickel. (b) Have you any explanation for the decrease in conductivity?

15.20 The rate of oxidation of zinc at 400°C (750°F) depends on the rate that zinc ions can diffuse through the n-type ZnO oxide layer. Explain how addition of a small amount of lithium to the zinc would affect the oxidation of the metal.

References

Aluminum with Food and Chemicals. 3d ed. 1964. Montreal: Aluminum Company of Canada, Ltd.

Behrens, Dieter, Ed. 1987–1989. *Dechema Corrosion Handbook.* Frankfurt: Dechema.

Bockris, John O'M. 1953. *Electrochemical Constants.* NBS Circular 524. Washington, DC: U.S. Government Printing Office.

Bockris, John O'M. and Amulyn K. N. Reddy. 1970. *Modern Electrochemistry.* New York: Plenum Press.

Brandes, Eric A., Ed. 1983. *Smithells Metals Reference Book.* 6th ed. London: Butterworths.

Britton, S. C. 1952. *The Corrosion Resistance of Tin and Tin Alloys.* Greenford, Middlesex, England: Tin Research Institute.

Carboline Protective Coatings Reference Handbook. (undated) St. Louis, MO: Carboline Co.

Craig, B. D., Ed. 1989. *Handbook of Corrosion Data.* Metals Park, OH: ASM International.

Davis, Joseph R., Ed. 1987. *Metals Handbook.* 9th ed. Vol. 13. Metals Park, OH: ASM International.

De Renzo, D. J., Ed. 1985. *Corrosion Resistant Materials Handbook.* 4th ed. Park Ridge, NJ: Noyes Data Corp.

Dillon, C. P. 1986a. Corrosion control in the process industries with nickel-base and nickel-bearing alloys. In *Proceedings of Materials Engineering Workshop.* Reference Book Series No. 11 001. Toronto, Canada: Nickel Development Institute.

Dillon, C. P. 1986b. *Corrosion Control in the Chemical Process Industries.* New York: McGraw-Hill.

Direct Calculation of Economic Appraisals of Corrosion Control Measures. NACE Standard RP-02-72. 1972. Houston: National Association of Corrosion Engineers.

Douglass, D. L. 1971. Exfoliation and the mechanical behavior of scales. In *Oxidation of Metals and Alloys.* Metals Park, OH: American Society for Metals.

Elliott, Peter. 1989. Practical guide to high-temperature alloys. *Mater. Performance* 28(4):57–66.

Fontana, Mars G. 1987. *Corrosion Engineering*. 3rd ed. New York: McGraw-Hill.

Graver, D. L. 1985. *Corrosion Data Survey*. 6th ed. Houston, TX: National Association of Corrosion Engineers.

Hall, G. R. 1986. Cementitious coatings. In *Encyclopedia of Materials Science and Engineering*, Michael B. Bever, Ed. pp. 1: 571–575. Oxford: Pergamon Press.

Kapusta, Sergio D. 1988. Inhibitors: corrosion. In *Encyclopedia of Chemical Processing and Design*. New York: Marcel Dekker, Inc.

Kubaschewski, O. and B. E. Hopkins. 1962. *Oxidation of Metals and Alloys*. 2d ed. London: Butterworths.

Miller, E. C. 1952. *Liquid Metals Handbook*. 2d ed. Washington, DC: U.S. Government Printing Office.

Moses, A. J. 1978. *The Practicing Scientist's Handbook*. New York: Van Nostrand Reinhold Co.

Parker, Earl R. 1967. *Materials Data Book for Engineers and Scientists*. New York: McGraw-Hill.

Parsons, R. 1959. *Handbook of Electrochemical Constants*. London: Butterworths.

Polar, J. P. 1961. *A Guide to Corrosion Resistance*. New York: Climax Molybdenum Co.

Romanoff, Melvin. 1957. *Underground Corrosion*. NBS Circular 579, Washington, DC: U.S. Department of Commerce.

Schweitzer, Philip A. 1986. *Corrosion Resistance Tables*. 2d ed. New York: Marcel Dekker, Inc.

Schweitzer, Philip A. 1990. *Corrosion Resistance of Elastomers*. New York: Marcel Dekker, Inc.

Seymour, R. B. 1982. *Plastics vs. Corrosives*. New York: John Wiley & Sons, Inc.

Shreir, L. L., Ed. 1976. *Corrosion*. 2d ed. London: Newnes-Butterworths.

Slunder, C. J. and W. K. Boyd. 1983. *Zinc: Its Corrosion Resistance*. 2d ed. New York: International Lead Zinc Research Organization, Inc.

Treseder, R. S., Ed. 1980. *NACE Corrosion Engineer's Reference Book*. Houston, TX: National Association of Corrosion Engineers.

Van Delinder, L. S., Ed. 1984. *Corrosion Basics—An Introduction*. Houston, TX: National Association of Corrosion Engineers.

Weast, R. C., Ed. 1988. *Reference Handbook of Chemistry and Physics*. 69th ed. Boca Raton, FL: CRC Press.

Webster, Harry A. 1992. Economics of cathodic protection. *Mater. Performance* 31(3):25–30.

West, J. M. 1970. *Electrodeposition and Corrosion Processes*. 2d ed. London: Van Nostrand Reinhold.

Wood, William G. coordinator 1982. *Metals Handbook*. 9th ed., Vol. 5. Metals Park, OH: American Society for Metals.

Index

of soils, 137
Response time, of ER probes, 176
Ringworm corrosion, 72–73
Risk, decisions, 4
Rolling:
 effect on grain shape, 52
 of metals, *see* Cold work
Rossini's criterion, 174–175
Rust:
 of automobiles, 1
 converters, 223–224
 corrosion slowed by, 16–17
 definition, 2
 from oxygen reduction, 10
 volume of, 122, 154

Sacrificial protection, 251–252
Safety, 4
Salt, *see* Chloride; Brines; Fused
  salts; Sodium chloride; Sodium
  sulfate
 acid and basic components, 308
 deicing, on concrete, 153–154
 moisture holder, 266–267
 molten eutectics, 158
Sampling, for failure analysis, 182
Sand, particle size, 137
Sawing, samples, 182
Scale, *see* Oxide
 formation, 299
 oxide, 298–304
 porous, from hot corrosion, 308
 protection of steel, 25
 temperatures of formation, 317,
  318
Scanning electron microscope, 183
Scavengers, 245
SCC, *see* Stress-corrosion cracking
Schottky defects, in oxides, 291–292
Screening, *see* Shielding
Season cracking, of brass, 119
Seawater, 135–136
 corrosion rates in, 136
 critical velocities of, 80
 galvanic series in, 55
 inhibitors for, 241
 stainless steel in, 1
 steel in, 1, 11, 53–54

Sediment:
 moisture held by, 266, 268
 pipes containing, 270
Selective leaching, *see* Dealloying
SEM, *see* Scanning electron
  microscope
Semiconductors:
 amphoteric oxides, 295
 intrinsic, 295
 n-type, 294–295
 p-type, 293–294
Sensitization, stainless steels, 65–67
 knife-line attack, 67
 oxidation unaffected, 319
Service:
 simulated, 167, 168
 tests, 165–166
 unalterable, 188
Sewage, 144
Shear, of oxide, 301, 302
SHE, *see* Standard hydrogen
  electrode
Sherardizing, 220
Shielding, of cathodic protection,
  250–251
Ships:
 cathodic protection of, 255
 seawater corrosion of, 1
Shotcrete process, 232
Shot-peening, 90, 114, 283
Siemans, unit of conductance, 250
Silicon:
 in high-temperature alloys, 319
 silicide coatings, 323, 324
 silicon carbide coatings, 326
 silicon dioxide coatings, 322
Silt, particle size, 137
Silver, catastrophic sulfidation, 305
Sketches, of failures, 181
Slag inclusions, in hot working, 52
Slow-strain-rate test, 172
Slurries, causing impingement, 107
Sodium, liquid-metal
  embrittlement by, 127
Sodium chloride, in hot corrosion,
  307, 310
Sodium hydroxide, compatible with
  nickel, 153